COMPLEX VARIABLES

The Wadsworth & Brooks/Cole Mathematics Series

W. Beckner, A Calderón, R. Fefferman, P. Jones, *Conference on Harmonic Analysis in Honor of Antoni Zygmund*
M. Behzad, G. Chartrand, L. Lesniak-Foster, *Graphs and Digraphs*
J. Cochran, *Applied Mathematics: Principles, Techniques, and Applications*
W. Derrick, *Complex Analysis and Applications,* Second Edition
J. Dieudonné, *History of Algebraic Geometry*
R. Durrett, *Brownian Motion and Martingales in Analysis*
S. Fisher, *Complex Variables*
A. Garsia, *Topics in Almost Everywhere Convergence*
R. Salem, *Algebraic Numbers and Fourier Analysis*, and L. Carleson, *Selected Problems on Exceptional Sets*
K. Stromberg, *An Introduction to Classical Real Analysis*

COMPLEX VARIABLES

STEPHEN D. FISHER

NORTHWESTERN UNIVERSITY

Wadsworth & Brooks/Cole Mathematics Series
Belmont   Monterey

Brooks/Cole Publishing Company
A Division of Wadsworth, Inc.

Printed in the United States of America
10 9 8 7 6 5 4 3 2 1

Library of Congress Cataloging-in-Publication Data

Fisher, Stephen D.
  Complex variables.

  (The Wadsworth & Brooks/Cole mathematics series)
  Includes bibliographies and index.
  1. Functions of complex variables. I. Title.
II. Series.
QA331.F589   1986   515.9   86-2580
ISBN 0-534-06168-0

Sponsoring Editor: John Kimmel
Design & Production: Unicorn Production Services, Inc.
Typesetting: Santype International
Printing and Binding: R.R. Donnelley & Sons

For my parents with love and appreciation

Preface

This textbook is intended for undergraduate or graduate students in science and engineering who are taking their first course in complex variables. Its only prerequisite is a three-semester course in calculus; no prior knowledge of Green's theorem or line integrals is needed. A previous course in differential equations would be useful but is not necessary.

The presentation and level of rigor in this book fit this background in several different ways. First, I have not given definitions and theorems in their greatest possible generality, and I have presented only results that are of central importance to elementary complex variables. (However, many important secondary topics are to be found in the exercises that appear at the end of each section.)

Second, although virtually all of the theorems are proved in full there are a few places where I refer the reader to other sources for a fact or a complete proof (for instance, the theorem that a continuous real-valued function on a closed and bounded set attains its maximum and minimum).

Third, the presentation in Chapters 1 and 2 emphasizes areas of complex variables that have much in common with concepts that the student has studied before; for instance, limits of sequences, continuity of functions, convergence of series. The exponential function is developed from its representation in terms of e^x, $\sin y$, and $\cos y$ so that the known continuity and differentiability of these functions can immediately be brought to bear. Furthermore, once analytic functions are defined and their basic properties are developed, I devote a section to complex power series; this includes a derivation of the fact that a convergent power series is an analytic function within its disc of convergence, whose derivative is given by the expected power series. This fact can be obtained later by a different and shorter route but I have found it more sound pedagogically to begin in areas where the student has had some previous experience and only later to move into newer areas.

Fourth, there is a plethora of solved examples (more than 220 altogether) in the text, which illustrate exactly how each concept or theorem is to be applied. In addition to the examples there are numerous exercises at the end of each section, a total of 730 throughout the book. The easiest reinforce basic concepts, while the more challenging extend or expand upon themes in the section or relate ideas from that section to earlier sections. Sections and topics within the exercises marked with an asterisk can be omitted with no loss of generality; they are not needed for future developments.

the exercises marked with an asterisk can be omitted with no loss of generality; they are not needed for future developments.

There is more material in this book than can be included in a course lasting a single quarter or semester. In teaching this material at Northwestern University I have found that in a 10-week quarter I can cover all of Chapters 1, 2, and 3 and many topics from the final two chapters. The variety and extent of the applications and techniques presented in Chapters 4 and 5 allow the instructor to pick his or her favorites and permit ambitious students to read others on their own.

Complex variables is, simultaneously, a practical tool of great utility in the hands of a skilled practitioner and a mathematical structure of enormous beauty and elegance. I hope that readers of this book will come away from it with a residue of each of these facets of the subject.

This preface would not be complete without acknowledging those people who assisted me during the writing of this book. Prof. Robert B. Burckel, Kansas State University, provided me with an infinitude of improvements in the grammer and style of the first draft. Various other helpful stylistic and pedagogical suggestions were made by Profs. M.D. Arthur, Michael Beals, Steven Bell, Carlos Berenstein, Bruce Berndt, James Brennan, David Colton, Michael Cullen, Abel Klein, Steven Krantz, Michael O'Flynn, James Osborn, Kent Pearce, Thomas Porsching, Glenn Schober, Daniel Shea, and Howard Swann. The computer graphics were produced by Benjamin Slivka using the Control Data CYBER 170 model 730 at Northwestern University's Vogelback Computing Center. Randall Kamien provided detailed solutions to the exercises in Chapters 1, 2 and 3; his contribution is very much appreciated. I also want to express my thanks to Prof. Ezra Zeheb of the Department of Electrical Engineering, Technion, Haifa, Israel, for two very informative conversations on the Z-transform and related matters. I also had helpful talks with Prof. Danny Weiss of the Department of Aeronautical Engineering, Technion, and Prof. Nadav Liron of the Department of Mathematics, Technion. They all helped to increase my knowledge of the applications of complex variables.

Special thanks go to my editor at Wadsworth, John Kimmel, who not only got me started on this book but as well stuck with me in what at times seemed to be an endless task. Michael Michaud of Unicorn Productions saw the book through its production; his technical expertise is in evidence throughout.

Typographical errors and mathematical slips are the bane of any author. I would appreciate any such being called to my attention.

A Note to the Student

This textbook presents an introduction to the theory and applications of complex variables. The presentation has been molded by my belief that what you have already studied in calculus can be successfully applied to learning complex variables, which at its basic level is just the calculus of complex-valued functions. Where there are strong, and even obvious, analogies between the new material and calculus I have pointed this out and arranged the presentation to emphasize these analogies. However, there are critical points at which the study of complex variables differs intrinsically from the calculus that you know, and at these points I have provided more details to explain the new material. At all times your comprehension of the subject will be greatly aided by reading with a pencil and paper close at hand. Write things down, fill in computations that may be omitted or only partially worked out, and work through the examples by yourself. And by all means, do as many exercises (assigned or not) as you can. Mathematical knowledge is not gained passively—you must be an active participant in the learning process.

Contents

1

The complex plane

1.1 The complex numbers and the complex plane

The theory and utility of functions of a complex variable ultimately depend in large measure on viewing the usual x- and y-coordinates in the plane as separate components of a single new variable, the complex variable z. This new variable z can then be manipulated in the same way that the conventional numbers are.

The familiar numbers, such as -1, $\sqrt{2}$, $1/3$, and π which are represented by points on a line, will be referred to as **real numbers**. A **complex number** is an expression of the form

$$z = x + iy,$$

where x and y are real numbers and i satisfies the rule

$$(i)^2 = (i)(i) = -1.$$

The number x is called the **real part** of z and is written

$$x = \text{Re } z.$$

The number y, despite the fact that it is also a real number, is called the **imaginary part** of z and is written

$$y = \text{Im } z.$$

Thus, for instance, we have $1 = \text{Re}(1 + 3i)$ and $3 = \text{Im}(1 + 3i)$. The **modulus**, or **absolute value**, of z is defined by

$$|z| = \sqrt{x^2 + y^2}, \qquad z = x + iy.$$

Each complex number $z = x + iy$ corresponds to the point $P(x, y)$ in the xy-plane (Fig. 1.1). The modulus of z then is just the distance from the point $P(x, y)$ to the origin, which in complex-number notation is 0. In this way we see that we have three inequalities relating x, y and $|z|$, namely

$$|x| \leqslant |z|, \qquad |y| \leqslant |z|, \qquad \text{and} \qquad |z| \leqslant |x| + |y|.$$

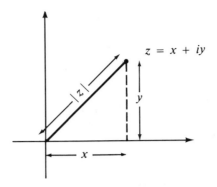

Figure 1.1

The first two of these are obvious; the third is obtained by noting that

$$|z|^2 = x^2 + y^2 \leqslant x^2 + 2|x||y| + y^2 = (|x| + |y|)^2.$$

The **complex conjugate** of $z = x + iy$ is given by

$$\bar{z} = x - iy.$$

Occasionally in engineering books one encounters the notation z^* for $\bar{z}$, as well as the use of j instead of i; we shall not use either of these. For the specific complex number $z = 1 + 3i$, we have

$$\bar{z} = 1 - 3i \qquad \text{and} \qquad |z| = \sqrt{10}$$

(see Fig. 1.2).

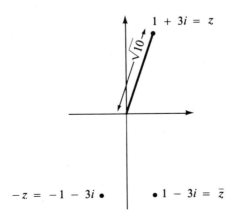

Figure 1.2

Addition, subtraction, multiplication, and division of complex numbers follow the ordinary rules of arithmetic, keeping in mind that $i^2 = -1$ and, as usual, division by zero is not allowed. Specifically, if

$$z = x + iy \quad \text{and} \quad w = s + it,$$

then

$$z + w = (x + s) + i(y + t)$$

$$z - w = (x - s) + i(y - t)$$

$$zw = (xs - yt) + i(xt + ys)$$

$$\frac{z}{w} = \frac{\bar{w}z}{\bar{w}w} = \frac{(xs + yt) + i(ys - xt)}{s^2 + t^2}, \quad w \neq 0.$$

Here, to obtain the formula for the quotient of z and w we used the device of multiplying both numerator and denominator by $\bar{w} = s - it$.

Example 1   To illustrate these rules in a particular case, let us take

$$z = 1 + 3i \quad \text{and} \quad w = -2 + 5i.$$

Then

$$z + w = -1 + 8i$$

$$zw = [(1)(-2) - (5)(3)] + i[(1)(5) + (3)(-2)] = -17 - i$$

$$\frac{z}{w} = \frac{\bar{w}z}{\bar{w}w} = \frac{13 - 11i}{29} = \frac{13}{29} - \frac{11}{29}i.$$

Two facts are of particular importance. The first, which is used repeatedly, is that

$$z\bar{z} = (x + iy)(x - iy) = x^2 + y^2 = |z|^2.$$

The second is that z and $\bar{z}$ have the same absolute value:

$$|z| = \sqrt{x^2 + y^2} = |\bar{z}|.$$

Another useful relation is derived with the following computation.

$$\begin{aligned}
|zw|^2 &= (xs - yt)^2 + (xt + ys)^2 \\
&= x^2s^2 + y^2t^2 + x^2t^2 + y^2s^2 \\
&= (x^2 + y^2)(s^2 + t^2) \\
&= |z|^2|w|^2,
\end{aligned}$$

so that after taking square roots we obtain

$$|zw| = |z||w|.$$

In a similar vein we compute $\overline{zw}$.

$$\overline{zw} = (xs - yt) - i(xt + ys)$$
$$= (x - iy)(s - it)$$
$$= \bar{z}\,\bar{w}.$$

Polar representation

The identification of $z = x + iy$ with the point $P(x, y)$ in the xy-plane has further interest and significance if we make use of the usual polar coordinates in the xy-plane. The polar coordinate system gives

$$x = r \cos \theta \qquad \text{and} \qquad y = r \sin \theta,$$

where $r = \sqrt{x^2 + y^2}$ and θ is the angle measured from the positive x-axis to the line segment from the origin to $P(x, y)$. We immediately see that $r = |z|$ and consequently

$$z = |z| (\cos \theta + i \sin \theta).$$

This is the **polar representation** of z (see Fig. 1.3a).

Example 2   For $z = -1 + i$ we have $|z| = \sqrt{2}$ and $\theta = \frac{3}{4}\pi$. Thus,

$$-1 + i = \sqrt{2} \,(\cos \tfrac{3}{4}\pi + i \sin \tfrac{3}{4}\pi),$$

which the reader can easily verify as correct since $\cos \frac{3}{4}\pi = -\frac{1}{2}\sqrt{2}$ and $\sin \frac{3}{4}\pi = \frac{1}{2}\sqrt{2}$ (see Fig. 1.3b).

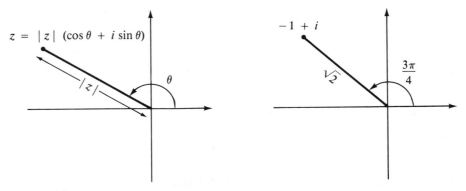

Figure 1.3

By now you have probably noticed that θ could equally well be replaced in the formulas by $\theta + 2\pi$, by $\theta - 4\pi$, or, indeed, by $\theta + 2\pi n$, where n is any integer. This ambiguity about the appropriate angle to use in the polar representation of a complex number is not just a question of semantics. Later we shall see that it causes some fundamental problems. Putting this aside for the moment, however, let's proceed with other properties of the polar representation. Suppose that

$$z = |z|(\cos \theta + i \sin \theta) \quad \text{and} \quad w = |w|(\cos \psi + i \sin \psi)$$

are two complex numbers. Then

$$zw = |z| \, |w|\{(\cos \theta \cos \psi - \sin \theta \sin \psi) + i(\cos \theta \sin \psi + \cos \psi \sin \theta)\}$$

$$|zw|\{\cos(\theta + \psi) + i \sin(\theta + \psi)\}.$$

Moreover,

$$\frac{z}{w} = \frac{|z|(\cos \theta + i \sin \theta)}{|w|(\cos \psi + i \sin \psi)}$$

$$= (|z|/|w|)\{\cos \theta \cos \psi + \sin \theta \sin \psi + i(\cos \psi \sin \theta - \cos \theta \sin \psi)\}$$

$$= (|z|/|w|)\{\cos(\theta - \psi) + i \sin(\theta - \psi)\}.$$

Here we have made use of the trigonometric identities for the sine and cosine of the sum and difference, respectively, of two angles. Hence, the polar representation of the product (or quotient) of two complex numbers is found by multiplying (or dividing) their respective moduli and adding (or subtracting) their respective polar angles (Fig. 1.4). In other words, multiplying w by $z = |z|(\cos \theta + i \sin \theta)$ produces a rotation of w in the counterclockwise direction of θ radians and stretches (or shrinks) $|w|$ by a factor of $|z|$.

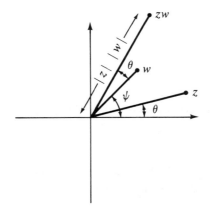

Figure 1.4

Example 3 For $z = -1 + i$ and $w = \sqrt{3} + i$, we have $|z| = \sqrt{2}$, $|w| = 2$, and $\theta = \frac{3}{4}\pi$, $\psi = \pi/6$. Thus,

$$(-\sqrt{3} - 1) + i(\sqrt{3} - 1) = zw = 2\sqrt{2}\left(\cos \frac{11}{12} \pi + i \sin \frac{11}{12} \pi\right)$$

and

$$\frac{(-\sqrt{3} + 1)}{4} + i \frac{(\sqrt{3} + 1)}{4} = z/w = \frac{\sqrt{2}}{2}\left(\cos \frac{7}{12} \pi + i \sin \frac{7}{12} \pi\right).$$

We now use the foregoing to derive **De Moivre's Theorem*** :

$$(\cos \theta + i \sin \theta)^n = \cos n\theta + i \sin n\theta$$

for any positive integer n and any angle θ. The formula is clearly true when $n = 1$; we shall use mathematical induction to prove it true for all n. Suppose there is a positive integer m for which

$$(\cos \theta + i \sin \theta)^m = \cos m\theta + i \sin m\theta.$$

Then

$$
\begin{aligned}
(\cos \theta + i \sin \theta)^{m+1} &= (\cos \theta + i \sin \theta)^m(\cos \theta + i \sin \theta) \\
&= (\cos m\theta + i \sin m\theta)(\cos \theta + i \sin \theta) \\
&= \cos(m + 1)\theta + i \sin(m + 1)\theta
\end{aligned}
$$

by invoking the formula derived above for the polar representation of the product of two complex numbers. Thus, if the equality holds for m, then it holds for $m + 1$. Since we know it is true for $n = 1$, it is true for all positive integers n.

Example 4   Let $\theta = \pi/4$; then $\cos \theta = \sin \theta = \sqrt{2}/2$. Thus,

$$\left(\frac{\sqrt{2}}{2} + i\,\frac{\sqrt{2}}{2}\right)^4 = (\cos \pi/4 + i \sin \pi/4)^4 = \cos \pi + i \sin \pi = -1.$$

Example 5   De Moivre's Theorem can be used to derive trigonometric identities for $\cos n\theta$ and $\sin n\theta$. For instance, by cubing we find that

$$
\begin{aligned}
(\cos \theta + i \sin \theta)^3 &= \cos^3 \theta + 3i \cos^2 \theta \sin \theta - 3 \cos \theta \sin^2 \theta - i \sin^3 \theta \\
&= (\cos^3 \theta - 3 \cos \theta \sin^2 \theta) + i(3 \cos^2 \theta \sin \theta - \sin^3 \theta).
\end{aligned}
$$

However, De Moivre's Theorem gives $(\cos \theta + i \sin \theta)^3 = \cos 3\theta + i \sin 3\theta$. After we equate the real and imaginary parts of these expressions we find that

$$\cos 3\theta = \cos^3 \theta - 3 \cos \theta \sin^2 \theta = 4 \cos^3 \theta - 3 \cos \theta$$
$$\sin 3\theta = 3 \cos^2 \theta \sin \theta - \sin^3 \theta = 3 \sin \theta - 4 \sin^3 \theta.$$

Similar formulas can be derived for $\cos 4\theta$, $\sin 4\theta$, $\cos 5\theta$, etc. (See Exercise 19.)

We define an **argument** of the nonzero complex number z to be any angle θ for which

$$z = |z|(\cos \theta + i \sin \theta)$$

whether or not it lies in the range $[0, 2\pi)$; we write $\theta = \arg z$. To repeat

$$\arg z = \theta \qquad \text{is equivalent to} \qquad z = |z|(\cos \theta + i \sin \theta).$$

---

* Abraham De Moivre, 1667–1754.

A concrete choice of arg z is made by defining **Arg** z to be that number θ_0 in the range $[-\pi, \pi)$ such that

$$z = |z|(\cos\theta_0 + i\sin\theta_0).$$

We may then write

$$\text{Arg}(zw) = \text{Arg } z + \text{Arg } w \qquad (\text{mod } 2\pi),$$

where the expression (mod 2π) means that the two sides of this last formula differ by some integer multiple of 2π. For example, if $z = -1 + i$ and $w = i$, then $zw = -1 - i$, so $\text{Arg}(zw) = -3\pi/4$. However, $\text{Arg } z = 3\pi/4$ and $\text{Arg } w = \pi/2$, so that $\text{Arg } z + \text{Arg } w = (5/4)\pi = -3\pi/4 + 2\pi$.

Complex numbers as vectors

If $z = x + iy$ and $w = s + it$ are two nonzero complex numbers, then

$$|z - w| = \sqrt{(x-s)^2 + (y-t)^2}$$

is nothing but the distance in the xy-plane from the point $P(x, y)$ to the point $Q(s, t)$. Moreover, the sum $z + w = (x + s) + i(y + t)$ and the difference $z - w = (x - s) + i(y - t)$ correspond exactly to the addition and subtraction of the vectors **OP** and **OQ** (Fig. 1.5).

We note, too, that the angle α between the vector **OP** and the vector **OQ** is found by using the usual dot product of two vectors:

$$\begin{aligned}
\cos\alpha &= (\mathbf{OP} \cdot \mathbf{OQ})/|\mathbf{OP}||\mathbf{OQ}| \\
&= (xs + yt)/\sqrt{x^2 + y^2}\sqrt{s^2 + t^2} \\
&= \text{Re}(z\bar{w})/|z||w|.
\end{aligned}$$

In particular, **OP** and **OQ** are perpendicular if and only if $\text{Re}(z\bar{w}) = 0$. The relations among the lengths of the sides of the triangle formed by z, w, and

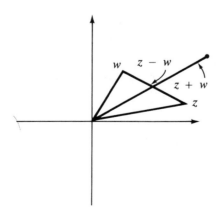

Figure 1.5

$z - w$, which is just the law of cosines, is formulated here as

$$|z - w|^2 = (x - s)^2 + (y - t)^2$$
$$= x^2 + s^2 + y^2 + t^2 - 2(xs + yt)$$
$$= |z|^2 + |w|^2 - 2 \operatorname{Re}(z\bar{w}).$$

In summary, we see that the usual xy-plane has a natural interpretation as the location of the complex variable $z = x + iy$ and all the rules for the geometry of the vectors $P(x, y)$ can be recast in terms of z. Henceforth, then, we refer to the xy-plane as the **complex plane** or simply, the plane. The x-axis will be called the **real axis** and the y-axis will be called the **imaginary axis**.

Exercises for Section 1.1

1. Let $z = 1 + 2i$, $w = 2 - i$, and $\zeta = 4 + 3i$. Compute (a) $z + 3w$; (b) $-2w + \zeta$; (c) z^2; (d) $w^3 + w$; (e) $\operatorname{Re}(\zeta^{-1})$; (f) w/z; (g) $\zeta^2 + 2\zeta + 3$.

2. Use the quadratic formula to solve these equations; express the answers as complex numbers. (a) $z^2 + 36 = 0$; (b) $2z^2 + 2z + 5 = 0$; (c) $5z^2 + 4z + 1 = 0$; (d) $z^2 - z = 1$; (e) $z^2 = 2z$.

3. Sketch the locus of those points w with (a) $|w| = 3$; (b) $|w - 2| = 1$; (c) $|w + 2|^2 = 4$; (d) $|w + 2| = |w - 2|$; (e) $|w^2 - 2w - 1| = 0$; (f) $\operatorname{Re}[(1 - i)\bar{z}] = 0$; (g) $\operatorname{Re}[z/(1 + i)] = 0$.

4. Find $\operatorname{Re}(1/z)$ and $\operatorname{Im}(1/z)$ if $z = x + iy$, $z \neq 0$. Show that $\operatorname{Re}(iz) = -\operatorname{Im} z$ and $\operatorname{Im}(iz) = \operatorname{Re} z$.

5. Give the polar representation for (a) $-1 + i$; (b) $1 + i\sqrt{3}$; (c) $-i$; (d) $(2 - i)^2$; (e) $|4 + 3i|$; (f) $\sqrt{5} - i$; (g) $-2 - 2i$; (h) $\sqrt{2}/(1 + i)$; (i) $[(1 + i)/\sqrt{2}]^4$.

6. Give the complex number whose polar coordinates (r, θ) are (a) $(\sqrt{3}, \pi/4)$; (b) $(1/\sqrt{2}, \pi)$; (c) $(4, -\pi/2)$; (d) $(2, -\pi/4)$; (e) $(1, 4\pi)$; (f) $(\sqrt{2}, 9\pi/4)$.

7. Let a, b, and c be real numbers with $a \neq 0$ and $b^2 < 4ac$. Show that the two roots of $ax^2 + bx + c = 0$ are complex conjugates of each other.

8. Show that one value of $\arg(1/z)$ is $-\theta$, where θ is any value of $\arg z$, $z \neq 0$.

9. Show that $|z| = 1$ if and only if $1/z = \bar{z}$.

10. Let z and w be complex numbers with $zw = 0$. Show that either z or w is zero.

11. Show that $|z + w|^2 - |z - w|^2 = 4 \operatorname{Re}(z\bar{w})$ for any complex numbers z, w.

12. Let $z_1, z_2, \ldots, z_n$ be complex numbers. Establish the following formu-

las by mathematical induction
(a) $|z_1 z_2 \cdots z_n| = |z_1||z_2| \cdots |z_n|$
(b) $\mathrm{Re}(z_1 + z_2 + \cdots + z_n) = \mathrm{Re}(z_1) + \mathrm{Re}(z_2) + \cdots + \mathrm{Re}(z_n)$
(c) $\mathrm{Im}(z_1 + z_2 + \cdots + z_n) = \mathrm{Im}(z_1) + \mathrm{Im}(z_2) + \cdots + \mathrm{Im}(z_n)$
(d) $\overline{z_1 z_2 \cdots z_n} = \bar{z}_1 \bar{z}_2 \cdots \bar{z}_n$.

13. Determine which of the following sets of three points constitute the vertices of a right triangle (a) $3 + 5i$, $2 + 2i$, $5 + i$; (b) $2 + i$, $3 + 5i$, $4 + i$; (c) $6 + 4i$, $8 + 4i$, $7 + 5i$.

14. Show that $\cos \theta = \cos \psi$ and $\sin \theta = \sin \psi$ if and only if $\theta - \psi$ is an integer multiple of 2π.

15. Show that the triangle with vertices at 0, z, and w is equilateral if and only if $|z|^2 = |w|^2 = 2\,\mathrm{Re}(z\bar{w})$.

16. Let z_0 be a nonzero complex number. Show that the locus of points tz_0, $-\infty < t < \infty$, is the straight line through z_0 and 0.

17. Show that if $w \neq 0$, then $|z/w| = |z|/|w|$.

18. Prove the identity $1 + z + z^2 + \cdots + z^n = (1 - z^{n+1})/(1 - z)$ valid for all z, $z \neq 1$.

19. Show that $\cos n\theta$ can be expressed as a combination of powers of $\cos \theta$ with integer coefficients. (**Hint:** Use De Moivre's Theorem and the fact that $\sin^2 \theta = 1 - \cos^2 \theta$.)

The Schwarz* inequality

20. Let B and C be nonnegative real numbers and A a complex number. Suppose that $0 \leqslant B - 2\,\mathrm{Re}(\bar{\lambda}A) + |\lambda|^2 C$ for all complex numbers λ. Conclude that $|A|^2 \leqslant BC$. (**Hint:** If $C = 0$, show that $A = 0$. If $C \neq 0$, then choose $\lambda = A/C$.)

21. Let $a_1, \ldots, a_n$ and $b_1, \ldots, b_n$ be complex numbers. Establish the **Schwarz inequality**:

$$\left| \sum_{j=1}^{n} a_j \bar{b}_j \right|^2 \leqslant \left\{ \sum_{j=1}^{n} |a_j|^2 \right\} \left\{ \sum_{j=1}^{n} |b_j|^2 \right\}.$$

(**Hint:** For all complex numbers λ, we have $0 \leqslant \sum_{j=1}^{n} |a_j - \lambda b_j|^2$. Expand this and apply Exercise 20 with $A = \sum_{j=1}^{n} a_j \bar{b}_j$, $B = \sum_{j=1}^{n} |a_j|^2$, $C = \sum_{j=1}^{n} |b_j|^2$.)

22. Verify the Schwarz inequality directly for the case $n = 2$.

23. When does equality hold in the Schwarz inequality?

---

* Hermann Amandus Schwarz, 1843–1921.

24.   Use the Schwarz inequality to establish that

$$\left\{\sum_{j=1}^{n}|a_j+b_j|^2\right\}^{1/2} \leqslant \left\{\sum_{j=1}^{n}|a_j|^2\right\}^{1/2} + \left\{\sum_{j=1}^{n}|b_j|^2\right\}^{1/2}.$$

(**Hint:** Expand $\sum_{j=1}^{n}|a_j+b_j|^2$ and apply the Schwarz inequality.)

1.1.1   A formal view of the complex numbers★

The complex numbers can be developed in a formal way from the real numbers. A complex number z is defined to be an ordered pair (x, y) of real numbers; we write $z = (x, y)$. Two complex numbers $z_1 = (x_1, y_1)$ and $z_2 = (x_2, y_2)$ are *equal* when $x_1 = x_2$ and $y_1 = y_2$. The basic arithmetic operations of addition and multiplication are defined respectively, by,

addition: $z_1 + z_2 = (x_1 + x_2, y_1 + y_2)$ (1)

multiplication: $z_1 z_2 = (x_1 y_1 - x_2 y_2, x_1 y_2 + x_2 y_1)$.   (2)

The **additive identity** is $0 = (0, 0)$ since $(0, 0) + (x, y) = (x, y) + (0, 0) = (x, y)$ for all (x, y). The **multiplicative identity** is $1 = (1, 0)$ since $(1, 0)(x, y) = (x, y)(1, 0) = (x, y)$ for all (x, y). Further, it is elementary but somewhat tedious to show that the arithmetic operations of addition and multiplication are **commutative**:

$$z_1 + z_2 = z_2 + z_1; \qquad z_1 z_2 = z_2 z_1$$

and **associative**:

$$(z_1 + z_2) + z_3 = z_1 + (z_2 + z_3); \qquad (z_1 z_2)z_3 = z_1(z_2 z_3).$$

Each $z = (x, y)$ has a unique **additive inverse** $-z = (-x, -y)$ since $z + (-z) = 0$. A nonzero $z = (x, y)$ necessarily satisfies the condition $x^2 + y^2 > 0$ and its unique **multiplicative inverse** is

$$z^{-1} = \left(\frac{x}{x^2 + y^2}, \frac{-y}{x^2 + y^2}\right),$$

since $(z)(z^{-1}) = (1, 0) = 1$.

The mathematical system of the complex numbers so constructed is one example of a **field**. There are many other examples of fields besides the complex numbers; for instance, the real numbers themselves form a field, as do the rational numbers.

The complex number $(0, 1)$ has the interesting property that its square is -1: $(0, 1)(0, 1) = (-1, 0)$. Further, $(0, 1)$ and $(0, -1)$ are the only two complex numbers with this property (see the exercises below). We denote $(0, 1)$ by the symbol *i*. Each complex number then can be written

$$\begin{aligned} z = (x, y) &= (x, 0) + (0, y) \\ &= (x, 0) + (y, 0)(0, 1) \\ &= (x, 0) + (y, 0)i. \end{aligned}$$

Complex numbers of the form $(a, 0)$ are just the real numbers with their usual rules of arithmetic:

$$(a, 0) + (b, 0) = (a + b, 0)$$
$$(a, 0)(b, 0) = (ab, 0)$$

and it is entirely natural to identify $(a, 0)$ with a. In this way we have

$$z = (x, y) = (x, 0) + i(y, 0) = x + iy.$$

This brings us back to the point where we began Section 1.

Exercises for Section 1.1.1

Throughout, we assume the usual rules of arithmetic for the real numbers; in particular, $a^2 > 0$ for any nonzero real number a.

1. Show directly from rule (2) for multiplication that $z^2 = (-z)^2$.

2. Suppose that $z = (x, y)$ and $z^2 = (-1, 0)$. Show that $z = i$ or $z = -i$.

3. Solve the equation $z^2 = (0, 1)$.

4. Suppose that z^2 is real and negative, that is, $z^2 = (a, 0)$, $a < 0$. Show that $z = (0, b)$ and find b in terms of a.

5. Show by computation that addition of complex numbers is associative: $(z_1 + z_2) + z_3 = z_1 + (z_2 + z_3)$; and commutative: $z_1 + z_2 = z_2 + z_1$.

6. Show by computation that multiplication of complex numbers is associative: $(z_1 z_2)z_3 = z_1(z_2 z_3)$; and commutative: $z_1 z_2 = z_2 z_1$.

7. Define the absolute value, $|z|$, of $z = (x, y)$ by $|z| = \sqrt{x^2 + y^2}$. Show directly that $|z_1 z_2| = |z_1||z_2|$.

8. Define the complex conjugate, $\bar{z}$, of $z = (x, y)$ by $\bar{z} = (x, -y)$. Show that $z\bar{z} = (|z|^2, 0)$.

9. Show that $z_1 z_2 = 0$ implies that either z_1 or z_2 is zero.

10. Let $z = (x, y)$. Show that (a) $|x| \leqslant |z|$, (b) $|y| \leqslant |z|$, (c) $|z| \leqslant |x| + |y|$.

1.2   Some geometry

The triangle inequality

We begin with an important inequality that has a simple geometric interpretation. Suppose $z = x + iy$ and $w = s + it$ are two complex numbers. Then

$$\begin{aligned}
|z + w|^2 &= (x + s)^2 + (y + t)^2 \\
&= x^2 + s^2 + y^2 + t^2 + 2(xs + yt) \\
&= |z|^2 + |w|^2 + 2\,\mathrm{Re}(z\bar{w}) \\
&\leqslant |z|^2 + |w|^2 + 2|z\bar{w}| \\
&= |z|^2 + |w|^2 + 2|z|\,|w| \\
&= (|z| + |w|)^2.
\end{aligned}$$

Taking the square root of both sides yields the inequality

$$|z + w| \leqslant |z| + |w|.$$

This is the **triangle inequality**, since it simply expresses the fact that any one side of a triangle is not longer than the sum of the lengths of the other two sides (see Fig. 1.6).

If ζ and ξ are two (other) complex numbers, then by putting $z = \zeta - \xi$ and $w = \xi$ we get $|\zeta| \leqslant |\zeta - \xi| + |\xi|$ or

$$|\zeta| - |\xi| \leqslant |\zeta - \xi|.$$

Likewise

$$|\xi| - |\zeta| \leqslant |\zeta - \xi|,$$

which together yield a variation of the triangle inequality,

$$\big||\zeta| - |\xi|\big| \leqslant |\zeta - \xi|.$$

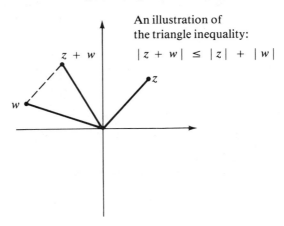

An illustration of the triangle inequality:

$$|z + w| \leq |z| + |w|$$

Figure 1.6

Straight lines

The equation of a (nonvertical) straight line, $y = mx + b$, m and b real, can be formulated as

$$0 = \mathrm{Re}((m + i)z + b).$$

More generally, if $a = A + iB$ is a nonzero complex number and b is any complex number (not just a real number), then

$$0 = \mathrm{Re}(az + b)$$

is exactly the straight line $Ax - By + \mathrm{Re}(b) = 0$; this formulation also includes the vertical lines, $x = \mathrm{Re}\ z = $ constant. (See Fig. 1.7.)

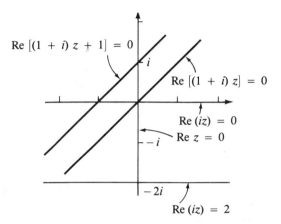

Re $[(1 + i) z + 1] = 0$

Re $[(1 + i) z] = 0$

Re $(iz) = 0$

Re $z = 0$

Re $(iz) = 2$

Figure 1.7

Roots and powers

The computation of the fractional powers of a nonzero complex number is possible with the techniques developed in Section 1 of this chapter. The attempt to find the roots of such equations as $x^2 + 1 = 0$ is where the whole subject of complex numbers first arose; here we will see a certain "completeness" evidenced by the complex numbers but not by the real numbers. Suppose w is a nonzero complex number and n is a positive integer. A complex number z satisfying the equation $z^n = w$ is called an **nth root of w**. We shall determine all the distinct nth roots of w.

Let $w = |w|(\cos \psi + i \sin \psi)$ be the polar representation of w, where we specify that ψ lies in the range $[-\pi, \pi)$. Let $z = |z|(\cos \theta + i \sin \theta)$; the relation $z^n = w$ and De Moivre's Theorem from Section 1 then yield three equations:

$$|z|^n = |w|, \qquad \cos(n\theta) = \cos \psi \qquad \text{and} \qquad \sin(n\theta) = \sin \psi.$$

Thus, we must have $|z| = |w|^{1/n}$; θ is not so well determined. Of course, one possibility for θ is $\theta = \psi/n$; however, there are others. We define

$$\theta_k = \psi/n + k(2\pi/n), \qquad k = 0, 1, \ldots, n - 1.$$

Then $n\theta_k = \psi + 2\pi k$ and so $\cos n\theta_k = \cos \psi$ and $\sin n\theta_k = \sin \psi$. We define complex numbers $z_0, \ldots, z_{n-1}$ by the rule

$$z_k = |w|^{1/n}(\cos \theta_k + i \sin \theta_k), \qquad k = 0, 1, \ldots, n - 1.$$

Then each of $z_0, \ldots, z_{n-1}$ is distinct (see Exercise 14, Section 1) and each satisfies

$$z_k^n = w, \qquad k = 0, 1, \ldots, n - 1.$$

Moreover, these complex numbers $z_0, \ldots, z_{n-1}$ are the only possible roots of the equation $z^n = w$. For if

$$\cos n\theta = \cos \psi, \qquad \sin n\theta = \sin \psi,$$

then again by Exercise 14, Section 1, we have $n\theta = \psi + 2\pi j$ for some integer j. The values $j = 0, \ldots, n - 1$ yield distinct numbers $\cos \theta_j + i \sin \theta_j$, whereas other values of j just give a repetition of numbers already obtained. The geometric picture of the nth roots of w is very simple: the n roots lie on the circle centered at the origin of radius $\rho = |w|^{1/n}$; the roots are equally spaced on this circle with one of the roots having polar angle $\theta_0 = (\text{Arg } w)/n$.

Example 1   Find the 12th roots of 1.

Solution   We have $w = 1 = \cos 0 + i \sin 0$; thus, the modulus of all the 12th roots is 1. The roots are equally spaced on the circle of radius 1 centered at the origin. One root is $z_0 = 1$; the others have polar angles of $2\pi/12, 4\pi/12, 6\pi/12, \ldots, 22\pi/12$, respectively. □

Example 2   Find the 5th roots of $i + 1$.

Solution   We have

$$w = 1 + i = \sqrt{2}(\cos \pi/4 + i \sin \pi/4)$$

so that the modulus of all the 5th roots of $i + 1$ is $2^{1/10} \doteq 1.0717$, the real, positive 10th root of 2. One of the roots is located with polar angle $\pi/20$ and the others have polar angles of $\pi/20 + 2\pi/5$, $\pi/20 + 4\pi/5$, $\pi/20 + 6\pi/5$, and $\pi/20 + 8\pi/5$, respectively. (See Fig. 1.8.) □

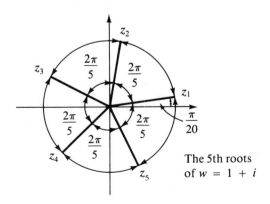

The 5th roots of $w = 1 + i$

Figure 1.8

Another equation involving roots can be solved in a similar fashion. Suppose n and m are nonzero integers, m positive, and w is a nonzero complex number. The equation

$$z^{n/m} = w$$

is taken to mean

$$z^n = w^m, \quad \text{if } n \text{ is positive,}$$

$$z^{-n} = w^{-m}, \quad \text{if } n \text{ is negative.}$$

Accordingly, there are exactly $|n|$ distinct roots and these roots are equally spaced on the circle of radius $|w|^{m/n}$ centered at the origin, with one of the roots having polar angle $(m \text{ Arg } w)/n$.

Example 3   Find the solutions z of the equation

$$z^{-3/4} = 2 - i.$$

Solution   This equation is equivalent to

$$z^3 = (2 - i)^{-4} = \frac{-7 + 24i}{625}.$$

The roots of this equation have modulus

$$\rho = (625)^{-1/3} = |2 - i|^{-4/3} \doteq 0.3420.$$

One root is located with polar angle

$$\theta_0 = \frac{1}{3} \text{Arg}\left(\frac{-7 + 24i}{625}\right) = \frac{4 \text{ Arg}(2 - i)}{-3} = \frac{2\pi}{9}$$

and the other two roots have polar angle $\theta_0 + 2\pi/3$ and $\theta_0 + 4\pi/3$, respectively. $\qquad\square$

Circles

A circle is the set of all points equidistant from a given point, the center. If z_0 is the center and r the radius, then the circle of radius r and center z_0 is described by the equation $|z - z_0| = r$. There are, however, other ways to use complex numbers to describe circles.

If p and q are distinct complex numbers, then those complex numbers z with

$$|z - p| = |z - q|$$

are equidistant from p and q. The locus of these points is precisely the straight line that is the perpendicular bisector of the line segment joining p to q. However, if ρ is a positive real number not equal to 1, then those z with

$$|z - p| = \rho|z - q|$$

form a circle. To see this, we may suppose that $0 < \rho < 1$ (otherwise, divide both sides of the equation by ρ). Let $z = w + q$ and $c = p - q$; then the equation becomes

$$|w - c| = \rho|w|.$$

Upon squaring and transposing terms, this can be written as

$$|w|^2(1 - \rho^2) - 2 \operatorname{Re} w\bar{c} + |c|^2 = 0.$$

We complete the square of the left side and find that

$$(1 - \rho^2)|w|^2 - 2 \operatorname{Re} w\bar{c} + |c|^2/(1 - \rho^2) = |c|^2\rho^2/(1 - \rho^2).$$

Equivalently,

$$|w - c/(1 - \rho^2)| = |c|\rho/(1 - \rho^2).$$

Thus, w lies on the circle of radius $R = |c|\rho/(1 - \rho^2)$ centered at the point $c/(1 - \rho^2)$ and so z lies on the circle of the same radius R centered at the point $z_0 = (p - \rho^2 q)/(1 - \rho^2)$.

Example 4   To confirm in one special case what we did above, let us look at the locus of points z with

$$|z - i| = (1/2)|z - 1|.$$

After multiplying both sides by 2 and squaring, we find that

$$4\{|z|^2 - 2 \operatorname{Re}(z\bar{i}) + |i|^2\} = |z|^2 - 2\operatorname{Re} z + |1|^2,$$

or after simplifying,

$$3|z|^2 - 8y + 2x = -3.$$

More algebra yields

$$3x^2 + 2x + 1/3 + 3y^2 - 8y + 16/3 = -3 + 17/3 = 8/3.$$

Thus, the locus is

$$(x + 1/3)^2 + (y - 4/3)^2 = 8/9.$$

This is a circle of radius $2\sqrt{2}/3$ centered at $-1/3 + 4i/3$. Now from the computation that preceded the example, we have $p = i$, $q = 1$, and $\rho = 1/2$. The radius should be

$$R = \frac{|p - q|\rho}{1 - \rho^2} = \frac{\sqrt{2}(1/2)}{3/4} = \frac{2\sqrt{2}}{3}$$

and the center at

$$z_0 = \frac{p - \rho^2 q}{1 - \rho^2} = \frac{-(1/4) + i}{3/4} = -\frac{1}{3} + \frac{4}{3}i,$$

which, of course, is just what we found. Note that the center of the circle is on the line through 1 and i (Fig. 1.9).

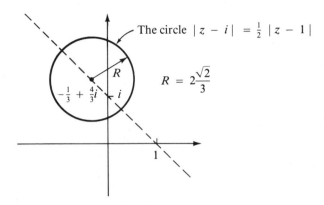

The circle $|z - i| = \frac{1}{2}|z - 1|$

$R = 2\dfrac{\sqrt{2}}{3}$

$-\frac{1}{3} + \frac{4}{3}i$

Figure 1.9

We now apply the information derived above to produce a beautiful geometric pattern: two families of mutually perpendicular circles.

Let C_1 be the family of circles of the form

$$|z - p| = \rho|z - q|, \qquad 0 < \rho < \infty,$$

where we include the case $\rho = 1$ (which yields a straight line) for completeness. Let L be the perpendicular bisector of the line segment from p to q. We take C_2 to be the family of circles through p and q and centered on the line L (see Fig. 1.10). We shall show that each circle in the family C_1 is perpendicular to each circle in the family C_2 at their two points of intersection. The computation is considerably simplified if we locate the origin at the point of intersection of the line L and the line L', which passes through p and q. We can then take L' to be the real axis and L to be the imaginary axis and, in this way, we may assume that $0 < p = -q$. A circle from the family C_1 is then centered at a point on the real axis and because there is no loss in assuming $0 < \rho < 1$, the center of that circle is at the point $s = p(1 + \rho^2)/(1 - \rho^2)$. The center of the circle from the family C_2 is at the point $t = i\alpha$ (α real) and this circle must pass through p and $-p$. Let $z = x + iy$ be on both circles. Then

$$|z - p| = \rho|z + p|$$

so that

$$x^2(1 - \rho^2) - 2px(1 + \rho^2) + p^2(1 - \rho^2) + y^2(1 - \rho^2) = 0.$$

Set $v = (1 + \rho^2)/(1 - \rho^2)$; then the equation above becomes

$$x^2 - 2pvx + p^2 + y^2 = 0.$$

On the other hand, since $z = x + iy$ is also on the second circle, we have

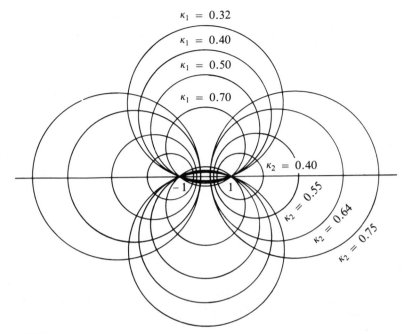

$\kappa_1 = 0.32$

$\kappa_1 = 0.40$

$\kappa_1 = 0.50$

$\kappa_1 = 0.70$

$\kappa_2 = 0.40$

$\kappa_2 = 0.55$

$\kappa_2 = 0.64$

$\kappa_2 = 0.75$

Figure 1.10

$$|z - i\alpha| = |p - i\alpha| = |-p - i\alpha| = \sqrt{p^2 + \alpha^2}$$

so that

$$x^2 + y^2 - 2\alpha y = p^2.$$

We must show that the segment from $i\alpha$ to z is perpendicular to the segment from s to z. However,

$$2 \operatorname{Re}[\overline{(z - i\alpha)}(z - s)]$$
$$= 2x(x - s) + 2y(y - \alpha); \quad s = pv$$
$$= 2(x^2 + y^2 - pvx - \alpha y) = x^2 + y^2 - 2pvx + x^2 + y^2 - 2\alpha y$$
$$= -p^2 + p^2 = 0,$$

which is the desired statement.

These last computations have an important physical interpretation. Imagine a mass m placed at a point p in the plane. This mass produces a gravitational field (in the plane) whose effect on a mass m_1 is directed radially toward p with a strength inversely proportional to the square of the distance to p and directly proportional to the product, mm_1, of the masses. Thus, the force on a particle of mass one located at the point z is

$$F(z) = m(p - z)(|p - z|)^{-3}.$$

In this case, the straight lines arg $F(z) = $ constant are the lines of force and the circles $|F(z)| = $ constant are the curves where the force has constant modulus (see Fig. 1.11).

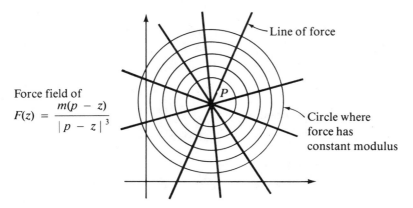

Figure 1.11

Suppose now that we place a mass m_p at p and a mass m_q at q. The corresponding force fields are

$$F_p(z) = m_p(p - z)(|p - z|)^{-3}$$

and

$$F_q(z) = m_q(q - z)(|q - z|)^{-3}.$$

Thus, the curve on which $|F_p(z)/F_q(z)|$ is a constant is nothing but the set of those z with

$$(m_p/m_q)^{1/2}|z - q|/|z - p| = \text{constant}.$$

We showed above that this is a circle centered on the line through p and q. Further, the set of those z for which $\arg(F_p(z)/F_q(z))$ is a constant is the curve $\arg[(p - z)/(q - z)] = \text{constant}$. By Exercise 31 of this section this curve is a circle through p and q with its center on the perpendicular bisector of the segment from p to q. Thus, the circles of the family C_1 are exactly the curves along which the ratio of the strengths of the fields remains constant, whereas the circles of the family C_2 are exactly the curves along which the fields make a constant angle to each other.

Exercises for Section 1.2

In Exercises 1 to 10 describe the locus of points z satisfying the given equation

1.	$|z + 1| = |z - 1|$	2.	$|z - 4| = 4|z|$
3.	$\text{Re}[(4 + i)z + 6] = 0$	4.	$\text{Im}(2iz) = 7$
5.	$|z + 2| + |z - 2| = 5$	6.	$|z - i| = \text{Re } z$
7.	$\text{Re}(z^2) = 4$	8.	$|z - 1|^2 = |z + 1|^2 + 6$
9.	$|z^2 - 1| = 0$	10.	$|z + 1|^2 + 2|z|^2 = |z - 1|^2$

In Exercises 11 to 17 write the equation of the given circle or straight line in complex number notation. For example, the circle of radius 4 centered at the point $3 - 2i$ is given by the equation $|z - (3 - 2i)| = 4$.

11.   The circle of radius 2 centered at $4 + i$.

12.   The straight line through 1 and $-1 - i$.

13.   The vertical line containing $-3 - i$.

14.   The circle through 0, $2 + 2i$, and $2 - 2i$.

15.   The circle through 1, i, and 0.

16.   The perpendicular bisector of the line segment joining $-1 + 2i$ and $1 - 2i$.

17.   The straight line of slope -2 through $1 - i$.

18.   Show that the two lines $\text{Re}(az + b) = 0$ and $\text{Re}(cz + d) = 0$ are perpendicular if and only if $\text{Re}(a\bar{c}) = 0$.

19.   Let p be a positive real number and let Γ be the locus of points z satisfying $|z - p| = cx$, $z = x + iy$. Show that Γ is (a) an ellipse if $0 < c < 1$; (b) a parabola if $c = 1$; (c) a hyperbola if $1 < c < \infty$.

20.   Let z_1 and z_2 be distinct complex numbers. Show that the locus of points $tz_1 + (1 - t)z_2$, $-\infty < t < \infty$, describes the line through z_1 and z_2. The values $0 \leqslant t \leqslant 1$ give the line segment joining z_1 and z_2.

21.   Let α be a complex number with $0 < |\alpha| < 1$. Show that the set of all z with
         (a) $|z - \alpha| < |1 - \bar{\alpha}z|$ is the disc $\{z : |z| < 1\}$
         (b) $|z - \alpha| = |1 - \bar{\alpha}z|$ is the circle $\{z : |z| = 1\}$
         (c) $|z - \alpha| > |1 - \bar{\alpha}z|$ is the set $\{z : |z| > 1\}$.
         (**Hint:** Square both sides and simplify.)

22.   Let z and w be nonzero complex numbers. Show that $|z + w| = |z| + |w|$ if and only if $z = sw$ for some positive real number s.

In Exercises 23 to 29 follow the technique outlined in the text to find all solutions of the given equation.

23. $z^5 = i$ 24. $z^{3/2} = 8$ 25. $(z + 1)^4 = 1 - i$

26. $z^{10/3} = 2$ 27. $z^8 = -1$ 28. $z^{-5/4} = (1 + i)/\sqrt{2}$

29. $z^3 = 8$

30.   Let b and c be complex numbers. Show that the roots of the quadratic equation $z^2 + bz + c = 0$ are complex conjugates of each other if and only if the quantity $b^2 - 4c$ is real and negative, b is real, and c is positive.

31.   Let C be a circle and let A and B be any two distinct points on C.
        Show that if P is selected on the smaller arc of C joining A to B, then
        the angle from the segment AP to the segment BP is independent of
        P. This angle is $\pi/2$ if A and B are on opposite ends of a diameter. The
        result remains true if "smaller" is replaced by "larger."

32.   Let $z_1, \ldots, z_n$ be complex numbers. Show by mathematical induction
        that $|z_1 + \cdots + z_n| \leqslant |z_1| + \cdots + |z_n|$.

Translation and scaling★

33.   Let C be a circle or a straight line. Show that the same is true of the
        locus of points $z + \beta$, $z \in C$, and β a fixed complex number.

34.   Let C be a circle or a straight line. Show that the same is true of the
        locus of points αz, $z \in C$, and α a fixed nonzero complex number.

Inversion★

35.   Let L be the line $y = a$, $a > 0$. Show that the locus of points $1/z$, $z \in L$,
        is the circle of radius $1/2a$ centered at $-i/2a$.

36.   Let L be a line through the origin. Show that the locus of points $1/z$,
        $z \in L$, is a line through the origin. What is the relationship of the
        slopes of the two lines?

37.   Let C be the circle $|z - c| = r$, $0 < r < c$. Show that the locus of
        points $1/z$, $z \in C$, is the circle centered at $c/(c^2 - r^2)$, of radius
        $r/(c^2 - r^2)$.

38.   Let C be the circle $|z - r| = r$, $r > 0$. Show that the locus of points
        $1/z$, $z \in C$, is the vertical line through $1/2r$.

1.3   Subsets of the plane

To understand the fundamentals of complex variables it is necessary to single
out several special types of subsets of the complex plane which will be used in
our discussion of analytic and harmonic functions in subsequent chapters.
This section gives the definitions and basic properties of these sets.

Open sets

The set consisting of all points z satisfying $|z - z_0| < R$ is called the **open disc**
of radius R centered at z_0. A point w_0 in a set D in the complex plane is called
an **interior point** of D if there is some open disc centered at w_0 that lies entirely
within D (see Fig. 1.12). A set D is called **open** if all of its points are interior
points.

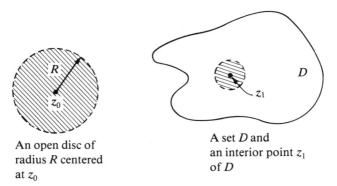

An open disc of
radius R centered
at z_0

A set D and
an interior point z_1
of D

Figure 1.12

Example 1   Each open disc $D = \{z: |z - z_0| < R\}$ is an open set. For if $r = |w_0 - z_0| < R$, choose $\varepsilon = (R - r)/3$. Then for any z with $|z - w_0| < \varepsilon$, we have

$$|z - z_0| \leqslant |z - w_0| + |w_0 - z_0| < \varepsilon + r = (R - r)/3 + r < R$$

by the triangle inequality. Thus, the open disc of radius ε centered at w_0 lies within the set D. Hence, each point of D is an interior point and so D is open.

Example 2   The set $R = \{z: \operatorname{Re} z > 0\}$ is an open set. To see this let $w_0 \in R$; then $\sigma_0 = \operatorname{Re} w_0 > 0$. Let $\varepsilon = (1/2)\sigma_0$ and suppose that $|z - w_0| < \varepsilon$. Then $-\varepsilon < \operatorname{Re}(z - w_0) < \varepsilon$ and so

$$\operatorname{Re} z = \operatorname{Re}(z - w_0) + \operatorname{Re} w_0 > -\varepsilon + \sigma_0 = (1/2)\sigma_0 > 0.$$

Consequently, z also lies in R. Hence each point w_0 of R is an interior point and so R is open.

Example 3   The set $\{z: |\operatorname{Im} z| > 1\}$ is an open set. This can be shown in the manner of Example 2.

Example 4   The set of all points $z = x + iy$ with $x^2 < y$ is also an open set. Let $z_0 = x_0 + iy_0$ be in this set; then there is a positive δ with $x_0^2 + \delta < y_0$. We may also assume that δ is so small that $2x_0 \delta + \delta^3 < 1 + \delta$. (We shall need this latter inequality only at one rather technical point.) Suppose now that $z = x + iy$ satisfies $|z - z_0| < \delta^2$. Then $\delta^2 > |x - x_0|$ and $\delta^2 > |y - y_0|$. Therefore,

$$\begin{aligned} x^2 &< (x_0 + \delta^2)^2 = x_0^2 + 2x_0 \delta^2 + \delta^4 \\ &< y_0 - \delta + 2x_0 \delta^2 + \delta^4 \\ &< y - \delta^2 - \delta + 2x_0 \delta^2 + \delta^4. \end{aligned}$$

However, the choice of δ gives $-\delta^2 - \delta + 2x_0 \delta^2 + \delta^4 < 0$ so that $x^2 < y$ whenever $z = x + iy$ satisfies $|z - z_0| < \delta^2$.

Example 5   The set of all z with $\operatorname{Re} z \leqslant 6$ is not an open set.

The boundary of a set

The set just described in Example 5 is not open because, for instance, the point $w_0 = 6$ is in it, but every open disc centered at w_0, no matter how small the radius, must contain a point z with $\operatorname{Re} z > 6$. The point $w_0 = 6$ is called a boundary point because it is located on the edge of the set, at the place where the set "almost" meets its complement. The precise definition follows.

A point p is a **boundary point** of a set S if every open disc centered at p contains both points of S and points not in S. The set of all boundary points of a set S is called the **boundary** of S (Fig. 1.13).

Let us find the boundary of each of the sets in the examples above (see Fig. 1.13).

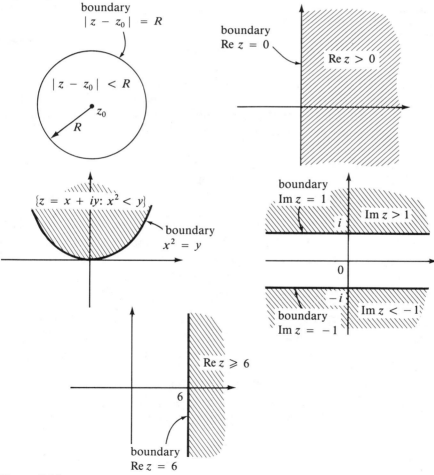

Figure 1.13

Example 6   The boundary of the open disc of radius R centered at z_0 is the circle of radius R centered at z_0.

Example 7   The boundary of the set of those z with Re $z > 0$ is the imaginary axis.

Example 8   The boundary of the set of those z with $|\operatorname{Im} z| > 1$ is the set of those z with $|\operatorname{Im} z| = 1$.

Example 9   The boundary of the set of those $z = x + iy$ with $x^2 < y$ is the parabola $y = x^2$.

Example 10   The boundary of the set of those z with Re $z \leqslant 6$ is the vertical line Re $z = 6$.

Closed sets

You have observed by now that the open sets discussed above contain none of their boundary. Indeed this is an elementary theorem, which we set out below. However, what about exactly the reverse situation? A set C is called **closed** if it contains its boundary. These definitions now prepare us for this result.

> **Theorem:** A set D is open if and only if it contains no point of its boundary. A set C is closed if and only if its complement $D = \{z : z \notin C\}$ is open.

Proof. To establish the validity of this theorem, we start by supposing that D is an open set. Let p be a boundary point of D. If it happens that p is in D, then because D is open there is an open disc centered at p that lies within D. Thus, p is not in the boundary of D. This contradiction shows that p is not in D. Conversely, suppose D is a set that contains none of its boundary points; we must show that D is open. If $z_0 \in D$, then z_0 cannot be a boundary point of D, so that there is some disc centered at z_0 that is either a subset of D or a subset of the complement of D. The latter is impossible since z_0 itself is in D. Hence, the disc lies in D and we have shown that each point of D is an interior point; therefore, D is open.

The second assertion of the theorem follows immediately from the fact that the boundary of a set coincides exactly with the boundary of the complement of that set; this in turn is a direct consequence of the definition of boundary point.   □

The reader should be wary here—there are many sets that are neither open nor closed since they contain part, but not all, of their boundary. For example, the set D of those z with Re $z \leqslant 6$ *and* Im $z > 2$ is neither open nor closed since its boundary consists of those w with either Re $w = 6$ and Im $w \geqslant 2$ or Im $w = 2$ and Re $w \leqslant 6$ (Fig. 1.14).

There is also one technical point here: The complex plane itself has no boundary and so by definition is both open and closed. It happens to be the only nonempty subset of the plane that is both open and closed. (The truth of this statement depends on a fundamental property of the real numbers.) Since in the following we will have no need for this fact about the plane we will not attempt to justify it.

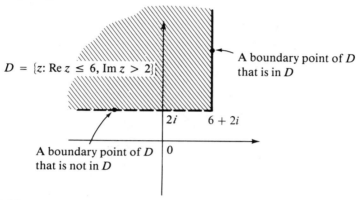

$D = \{z : \text{Re } z \leq 6, \text{Im } z > 2\}$

A boundary point of D that is in D

$2i$ $6 + 2i$

A boundary point of D that is not in D 0

Figure 1.14

Connected sets

A **polygonal curve** is the union of a finite number of directed line segments $\mathbf{P_1 P_2}, \mathbf{P_2 P_3}, \ldots, \mathbf{P_{n-1} P_n}$, where the terminal point of one is the initial point of the next (except for the last) (see Fig. 1.15). An open set D is **connected** if each pair p, q of points in D may be joined by a polygonal curve lying entirely within D. That is, there are points $P_2, \ldots, P_{n-1}$ in D such that all the line segments $\mathbf{pP_2}, \mathbf{P_2 P_3}, \ldots, \mathbf{P_{n-1} q}$ lie in D.

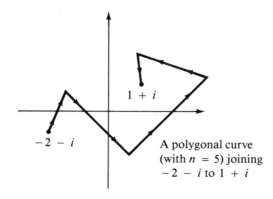

$1 + i$

$-2 - i$

A polygonal curve (with $n = 5$) joining $-2 - i$ to $1 + i$

Figure 1.15

Example 11   An open disc is connected.

Example 12   The set of those z with Re $z > 0$ is connected.

Example 13   The set of those $z = x + iy$ with $x^2 < y$ is connected.

Example 14   The set of those z with $|\operatorname{Im} z| > 1$ is not connected.

Example 15   The set of those z with Re $z < 6$ is connected.

Example 16   The set of those z with Re $z \neq 0$ is not connected.

For instance, in Example 16 we see that the points $p = 1$ and $q = -1$ lie in the given set but any polygonal curve joining p to q must necessarily cross the imaginary axis (where Re $z = 0$) and hence this curve cannot lie entirely within the set Re $z \neq 0$.

An open connected set is called a **domain**. Domains are the natural setting for the study of analytic and harmonic functions.

A set S is **convex** if the line segment **pq** joining each pair of points p, q in S also lies in S. In particular, we see immediately that any convex open set is connected (Fig. 1.16).

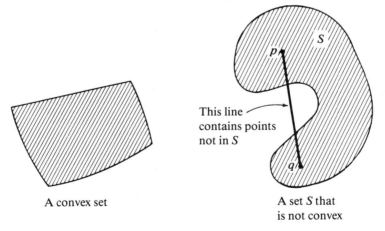

This line contains points not in S

A convex set   A set S that is not convex

Figure 1.16

Example 17   Each open disc is convex.

Example 18   Those $z = x + iy$ with $x^2 > y$ is not convex.

Example 19   Both of the sets Re $z > 0$ and Re $z \leqslant 6$ are convex.

An **open half-plane** is defined to be those points strictly to one side of a straight line; that is, those points z for which $\mathrm{Re}(az + b) > 0$, say. Each open half-plane is a convex set and an open set as well. A **closed half-plane** is the open half-plane plus the defining line; that is, those z with $\mathrm{Re}(az + b) \geqslant 0$. Each closed half-plane is convex as well as closed (Fig. 1.17).

It is instructive to see why each half-plane (open or closed) is convex. If p and q are two points, then

$$tq + (1 - t)p, \quad 0 \leqslant t \leqslant 1$$

describes the line segment from p to q. If p and q are in the open half-plane given by $\mathrm{Re}(az + b) > 0$, then

$$\mathrm{Re}(a(tq + (1 - t)p) + b) = t\,\mathrm{Re}(aq + b) + (1 - t)\mathrm{Re}(ap + b) > 0$$

so that the line segment from p to q lies in the same open half-plane. The case of a closed half-plane is almost identical.

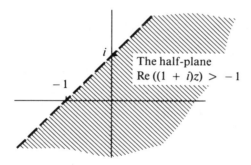

The half-plane
Re $((1 + i)z) > -1$

Figure 1.17

The point at infinity

A helpful and frequently used convention is to add the "**point at infinity**" to the complex plane. This is understood in the following way: A set D contains the point at infinity if there is a large number M such that D contains all the points z with $|z| > M$. For instance, the open half-plane Re $z > 0$ does not contain the point at infinity but the open set $D = \{z : |z + 1| + |z - 1| > 1\}$ does. One "reaches" the point at infinity by letting $|z|$ increase without bound, with no restriction at all on arg z. One way to visualize all this is to let $w = 1/z$ and think about $|w|$ being very small: an open set containing the point at infinity will become an open set containing $w_0 = 0$. Further, the statement "z approaches infinity" is identical with the statement "w converges to zero." The point at infinity is denoted by the usual symbol for infinity: ∞.

Exercises for Section 1.3

For each of the sets in Exercises 1 to 8 (a) describe the interior and the boundary, (b) state whether the set is open or closed or neither open nor closed, (c) state whether the interior of the set is connected (if it has an interior).

1. $A = \{z = x + iy : x \geqslant 2 \text{ and } y \leqslant 4\}$

2. $B = \{z : |z| < 1 \text{ or } |z - 3| \leqslant 1\}$

3. $C = \{z = x + iy : x^2 < y\}$

4. $D = \{z : \text{Re}(z^2) = 4\}$

5. $E = \{z : z\bar{z} - 2 \geqslant 0\}$

6. $F = \{z : z^3 - 2z^2 + 5z - 4 = 0\}$

7. $G = \{z = x + iy : |z + 1| \geqslant 1 \text{ and } x < 0\}$

8. $H = \{z = x + iy : -\pi \leqslant y < \pi\}$

9.   Let α and β be complex numbers with $\alpha \neq 0$. Describe the set of points $\alpha z + \beta$ as z varies over (a) the first quadrant, $\{z = x + iy : x > 0 \text{ and } y > 0\}$; (b) the upper half-plane, $\{z = x + iy : y > 0\}$; (c) the disc $\{z : |z| < R\}$. Show that in each case that the resulting set is open and connected. (**Hint:** first investigate the set αz.)

10.   Describe the set of points z^2 as z varies over the second quadrant: $\{z = x + iy : x < 0 \text{ and } y > 0\}$. Show that this is an open, connected set. (**Hint:** use the polar representation of z.)

11.   A set S in the plane is **bounded** if there is a positive number M such that $|z| \leqslant M$ for all z in S; otherwise, S is **unbounded**. In Exercises 1 to 8, six of the given sets are unbounded. Find them.

12.   Which, if any, of the sets given in Exercises 1 to 8 contains ∞?

13.   (a) Show that the union of two nonempty open sets is open. Do the same, replacing "open" with "closed." Do the same, replacing "union" with "intersection."
      (b) Repeat part (a) replacing "two" with "finitely many."

14.   Let D_1 and D_2 be domains with a nonempty intersection. Show that $D_1 \cup D_2$ is a domain.

15.   Let $\Omega_1 = \{z : 1 < |z| < 2 \text{ and } \text{Re } z > -\frac{1}{2}\}$ and $\Omega_2 = \{z : 1 < |z| < 2 \text{ and } \text{Re } z < \frac{1}{2}\}$. Show that both Ω_1 and Ω_2 are domains but $\Omega_1 \cap \Omega_2$ is not.

16.   Let D be a domain and let p and q be points of D. Show that there is a polygonal curve joining p to q whose line segments are either horizon-

tal or vertical (both types can be used). (**Hint:** Replace a "slanting" segment by (perhaps many) horizontal and vertical segments; see Fig. 1.18.)

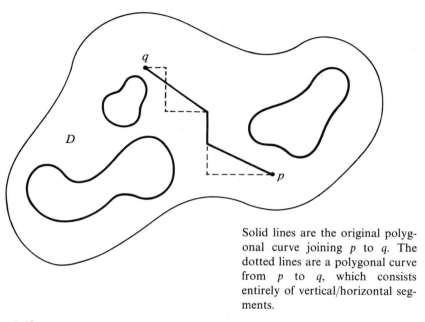

Solid lines are the original polygonal curve joining p to q. The dotted lines are a polygonal curve from p to q, which consists entirely of vertical/horizontal segments.

Figure 1.18

17.   Fix a nonzero complex number z_0. Show that the set D obtained from the plane by deleting the ray $\{tz_0 : 0 \leqslant t < \infty\}$ is a domain.

18.   An open set D is **star-shaped** if there is a point p in D with the property that the line segment from p to z lies in D for each z in D. (a) Show that the disc $\{z : |z - z_0| < r\}$ is star-shaped. (b) Show that any convex set is star-shaped.

19.   Determine which of the following sets is star-shaped
       (a) $D = \{z : |z - 1| < 2 \text{ or } |z + 1| < 2\}$
       (b) $D = \{z = x + iy : x > 0 \text{ and } |z| > 1\}$
       (c) $D = \{z : |z| > 1\}$
       (d) $D = \{z = x + iy : x > 0 \text{ and } [x > y + 1 \text{ or } x > 1 - y]\}$.

20.   Show that each star-shaped set is connected.

Separation of a point and convex set⋆

Let C be a closed convex set and z_0 a point *not* in C. It is a fact that there is a point p in C with $r = |z_0 - p| \leqslant |z_0 - q|$ for all q in C. (This last statement requires a bit of proof; we shall assume its validity.)

21.   Show that the only point of C in the disc $|z_0 - z| \leqslant r$ is the point p.

22.   Let L be the perpendicular bisector of the line segment from z_0 to p. Show that no point of C lies on L or in the half-plane, determined by L, which contains z_0.

23.   Conclude from 22 that L **separates** z_0 from C: z_0 and C lie in the two open half-planes determined by L, but not in the same open half-plane.

24.   Show that each closed convex set is the intersection of all the closed half-planes that contain it.

Topological properties⋆

25.   Show that the boundary of any set D is itself a closed set.

26.   Show that if $p \in D$, then p is either an interior point of D or a boundary point of D.

27.   Show that a set D coincides with its boundary if and only if D is closed and D has no interior points.

28.   Show that if D is a set and E is a *closed* set containing D, then E must contain the boundary of D.

29.   Show that if D is a set and S is an *open* set that is a subset of D, then S must be composed entirely of interior points of D.

30.   Let C be a bounded closed convex set and let D be the complement of C. Show that D is a domain.

1.4   Functions and limits

A **function** of the complex variable z is a rule that assigns a complex number to each z within some specified set D; D is called the **domain of definition** of the function. The collection of all possible values of the function is called the **range** of the function. Thus, a function f has as its domain of definition some subset of the complex plane and as its range some other, usually entirely different, subset of the complex plane. We frequently write $w = f(z)$ to distinguish the independent complex variable z from the dependent complex variable w.

Example 1 $f(z) = 4z^2 + 2z + 1$ has as its domain of definition the entire complex plane. Its range is also the entire complex plane, for if w is any complex number then the equation $f(z) = w$ is nothing but the quadratic equation

$$4z^2 + 2z + 1 - w = 0.$$

This equation is solved by use of the quadratic formula. Its solutions are

$$z_j = \frac{-2 + s_j}{8} \qquad j = 1, 2,$$

where s_1, s_2 are the two square roots of $4 - 16(1 - w) = -12 + 16w$. (It is possible that $s_1 = s_2$; this occurs only when $w = \frac{3}{4}$.)

Example 2 $f(z) = 1/(z - 1)$ has as its domain of definition all complex numbers except $z = 1$, where it is not defined. Its range consists of all complex numbers w except $w = 0$, since $f(z) = w = 1/(z - 1)$ is solved by $z = 1 + (1/w)$; this is a complex number as long as $w \neq 0$.

Example 3 $g(z) = |z|^2$ has the complex plane as its domain of definition; its range consists of all nonnegative real numbers.

Example 4 $h(z) = i(2 - (\text{Im } z)^{-1})$ is defined for all z except those on the real axis. For a pure imaginary number w a solution of $w = h(z) = i(2 - (\text{Im } z)^{-1})$ is any complex number z with $\text{Im } z = (2 + iw)^{-1}$. This defines many complex numbers as long as $w \neq 2i$. Hence, the range of h is all purely imaginary numbers except $w = 2i$.

Example 5   Show that the range of the function $w = T(z) = (1 + z)/(1 - z)$ on the disc $|z| < 1$ is the set of those w whose real part is positive.

Solution   We have

$$\text{Re } w = \text{Re}\left(\frac{1 + z}{1 - z}\right) = \text{Re}\,\frac{(1 + z)(1 - \bar{z})}{|1 - z|^2} = \frac{1 - |z|^2}{|1 - z|^2}.$$

This last quantity is positive when $|z| < 1$. This shows that the range of T is a subset of those w with $\text{Re } w > 0$. Now let w' be any point with $\text{Re } w' > 0$; we shall show that $z' = (w' - 1)/(w' + 1)$ satisfies $|z'| < 1$. Indeed, $1 > |(w' - 1)/(w' + 1)|$ exactly when $|w' + 1|^2 > |w' - 1|^2$. We expand both $|w' + 1|^2$ and $|w' - 1|^2$ and obtain

$$|w'|^2 + 2 \text{ Re } w' + 1 > |w'|^2 - 2 \text{ Re } w' + 1.$$

This is a correct inequality because $\text{Re } w' > 0$. Thus, $z' = (w' - 1)/(w' + 1)$ lies in the disc $|z| < 1$ and

$$Tz' = \frac{1 + z'}{1 - z'} = \frac{1 + \dfrac{w' - 1}{w' + 1}}{1 - \dfrac{w' - 1}{w' + 1}} = \frac{2w'}{2} = w'. \qquad \square$$

(**Note:** $T(z) = (1 + z)/(1 - z)$ is an example of a **linear fractional transformation**. Linear fractional transformations will be studied in some detail in Section 3 of Chapter 3.)

As with functions of a real variable, the domain of definition of a function is usually easier to determine than the range and we shall often be satisfied with a general description of certain properties of the range rather than an explicit rule for each ·point in it. For instance, we may be able to say that the range is open or connected or convex.

Graphs

For real-valued functions of a real variable, like those studied extensively in calculus, the device of displaying the graph of the function is an extremely useful tool in visualizing the behavior of the function. Such a technique is not as readily available for functions of a complex variable. *If* the function f has only real values, then its graph in 3-space (x, y, t), $t = f(x + iy)$, can be sketched (see Fig. 1.19). If, however, the function has complex values then this type of picture is not possible (at least in our world!). Moreover, almost all the functions we will deal with have complex values. One way to proceed is to graph $|f|$. A more useful way, however, is to use two complex planes, one for the domain variable z and a second for the range variable $w = f(z)$. For instance,

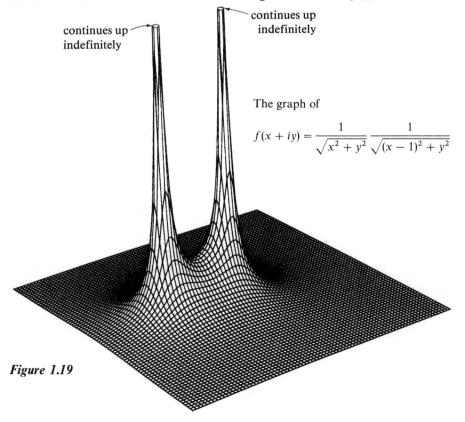

continues up indefinitely

continues up indefinitely

The graph of

$$f(x + iy) = \frac{1}{\sqrt{x^2 + y^2}} \frac{1}{\sqrt{(x - 1)^2 + y^2}}$$

Figure 1.19

we saw in Example 5 that the function $T(z) = (1 + z)/(1 - z)$ maps the disc $\{z : |z| < 1\}$ onto those w with Re $w > 0$; this can be "graphed" as shown in Figure 1.20. This type of picture is very helpful in understanding the behavior of the complex-valued function of a complex variable and we will employ it frequently.

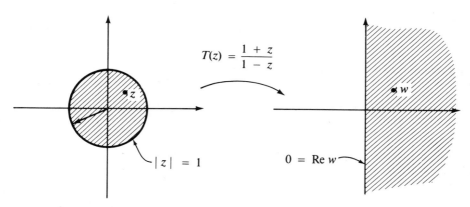

$$T(z) = \frac{1 + z}{1 - z}$$

$|z| = 1$ $0 = $ Re w

Figure 1.20

Limits

The concepts of the limit of a sequence of complex numbers, or the limit of a function of a complex variable, or even of continuity of a function of a complex variable, are almost identical to those for a real variable. We begin with the limit of a sequence.

Let $\{z_n\}_{n=1}^{\infty}$ be a sequence of complex numbers. We say that $\{z_n\}$ has the complex number A as a **limit**, or that $\{z_n\}$ **converges** to A, and we write,

$$\lim_{n \to \infty} z_n = A \qquad \text{or} \qquad z_n \to A$$

if given any positive number ε, there is an integer N such that

$$|z_n - A| < \varepsilon \quad \text{for all } n \geqslant N.$$

A sequence that does not converge, for any reason whatever, is called **divergent**. If $z_n = x_n + iy_n$ and $A = s + it$, then $z_n \to A$ if and only if $x_n \to s$ and $y_n \to t$. This is due to three inequalities noted in Section 1; namely, we have

$$|x_n - s| \leqslant |z_n - A|, \qquad |y_n - t| \leqslant |z_n - A|$$

and

$$|z_n - A| \leqslant |x_n - s| + |y_n - t|.$$

We can equivalently state that the sequence $\{z_n\}$ of complex numbers converges to the complex number A if and only if, whenever D is any open disc centered at A, then all but a finite number of the points $\{z_n\}$ lie in D.

Example 6   The sequence $z_n = 1 + i/n$ converges to 1.

Example 7   The sequence $z_n = (-1/2)^n + i(1 - 1/2n)$ converges to i.

Example 8   The sequence $z_n = (1/n)(\cos(n\pi/4) + i \sin(n\pi/4))$ converges to zero.

Example 9   The sequence $z_n = n - 1/n$ diverges.

Example 10   The sequence $z_n = i^n$ diverges because its terms are $i, -1, -i, 1$, in that order, repeated infinitely often.

One simple consequence of the definition of convergence is this.

$$\text{If } z_n \to A, \text{ then } |z_n| \to |A|.$$

The converse of this assertion is generally false; for example, $|(-1)^n + i/n| \to 1$ but the sequence $\{(-1)^n + i/n\}$ itself has no limit.

You will no doubt recall from calculus results about the sum, product, and quotient of convergent sequences of real numbers. Similar statements hold for sequences of complex numbers; we collect these results in the next theorem.

Theorem 1: Let $\{z_n\}$ and $\{w_n\}$ be convergent sequences of complex numbers with limits A and B, respectively. Let λ be a complex number. The sequences $\{z_n + \lambda w_n\}$ and $\{z_n w_n\}$ then also both converge with limits $A + \lambda B$ and AB, respectively. Furthermore, if $B \neq 0$, then the sequence $\{z_n/w_n\}$ converges to A/B.

Proof. The proof of each of these results is quite direct. For example, to show that $z_n w_n \to AB$, we write $z_n w_n - AB = (z_n - A)B + (w_n - B)z_n$. Let N_1 be chosen so big that $|z_n - A| < \varepsilon'$ if $n \geq N_1$ and N_2 chosen so big that $|w_n - B| < \varepsilon'$ if $n \geq N_2$. Then $|z_n| \leq \varepsilon' + |A| < 1 + |A|$ if $\varepsilon' < 1$ and for $n \geq N = N_1 + N_2$ we obtain

$$\begin{aligned}
|z_n w_n - AB| &\leq |z_n - A||B| + |w_n - B||z_n| \\
&< \varepsilon'|B| + \varepsilon'(1 + |A|) \\
&= \varepsilon'(1 + |A| + |B|) \\
&< \varepsilon
\end{aligned}$$

if ε' is chosen initially to satisfy $\varepsilon'(1 + |A| + |B|) < \varepsilon$. $\square$

Example 11   The sequence $z_n = 1 + i[1 - (2/n)]$ converges to $1 + i$ as $n \to \infty$.

Hence, the sequence

$$z_n^2 = \frac{4}{n} - \frac{4}{n^2} + i\left(2 - \frac{4}{n}\right)$$

converges to $2i = (1 + i)^2$ as $n \to \infty$.

Example 12   Suppose that a_0, a_1, a_2, and a_3 are complex numbers and $\{z_n\}$ is a sequence with $z_n \to A$. Then the sequence $\{a_0 + a_1 z_n + a_2 z_n^2 + a_3 z_n^3\}_{n=1}^\infty$ converges to $a_0 + a_1 A + a_2 A^2 + a_3 A^3$ as $n \to \infty$ as can be seen by several applications of results in Theorem 1.

Suppose next that f is a function defined on a subset S of the plane. Let z_0 be a point either in S or in the boundary of S. We say that f has **limit** L at the point z_0, and we write

$$\lim_{z \to z_0} f(z) = L \text{ or } f(z) \to L \text{ as } z \to z_0$$

if, given $\varepsilon > 0$, there is a $\delta > 0$ such that

$$|f(z) - L| < \varepsilon \text{ whenever } z \in S \text{ and } |z - z_0| < \delta.$$

It is worth stressing here that f has the limit L at the point z_0 exactly when the numbers $f(z)$ approach L as z approaches z_0 *from any direction*. This is substantially different from the case of a function of a real variable where the real variable can approach only from the left or right. The complex variable z may approach z_0 from infinitely many directions. Once again, it will be informative to look at some examples.

Example 13   The function $f(z) = |z|^2$ has limit 4 at the point $z_0 = 2i$.

Example 14   The function $g(z) = 1/(1 - z)$ has limit $(1/2)(1 + i)$ at $z_0 = i$.

Example 15   The function $h(z) = \text{Re}(z^4 + 4)$ has limit 0 at $z_0 = 1 + i$.

Example 16   The function $f(z) = (z^4 - 1)/(z - i)$ has limit $-4i$ at $z_0 = i$, since $z^4 - 1 = (z - i)(z + i)(z - 1)(z + 1)$ and so $f(z)$ simplifies to $(z + i)(z + 1)(z - 1)$ so long as $z \neq i$.

Example 17   The function $f(z) = z/\bar{z}$, $z \neq 0$, has no limit at $z_0 = 0$. For if z is real then $f(z) = 1$, while if z is purely imaginary, $z = iy$, then $f(iy) = -1$. Such a function cannot have a limit at $z_0 = 0$.

We say that the function f has a **limit** L at ∞, and we write

$$\lim_{z \to \infty} f(z) = L$$

if, given $\varepsilon > 0$, there is a large number M such that

$$|f(z) - L| < \varepsilon \quad \text{whenever} \quad |z| \geqslant M.$$

Note that we only require that $|z|$ is large; there is no restriction at all on arg z.

Example 18 $\lim\limits_{z \to \infty} 1/z^m = 0$ if $m = 1, 2, \ldots$.

Example 19 $\lim\limits_{z \to \infty} [(z^3 + 1)/(z^4 + 5z^2 + 3)] = 0$ since the ratio can be written as

$$(1/z + 1/z^4)/(1 + 5/z^2 + 3/z^4)$$

all of whose terms except the 1 go to zero as $|z| \to \infty$.

Example 20 $\lim\limits_{z \to \infty} [(x + y^3)/(x^2 + y^3)]$ does not exist, for if we take $z = x$, then the expression goes to zero, but if we take $z = iy$, then the expression is identically 1.

The following theorem on the sum, product, and quotient of functions that have limits at z_0 is a counterpart to Theorem 1.

> **Theorem 2:** Suppose that f and g are functions with limits L and M, respectively, at z_0. Let λ be a complex number. Then the functions $f + \lambda g$ and fg have the limits $L + \lambda M$ and LM, respectively, at the point z_0 and, if $M \neq 0$, then the function f/g has the limit L/M at the point z_0.

Continuity

Suppose again that f is a function defined on a subset S of the complex plane. If $z_0 \in S$, then f is **continuous** at z_0 if

$$\lim_{z \to z_0} f(z) = f(z_0).$$

That is, f is continuous at z_0 if the values of $f(z)$ get arbitrarily close to the value $f(z_0)$, so long as z is in S and z is sufficiently close to z_0. If it happens that f is continuous at all points of S then we say f is **continuous on** S. The function f is continuous at ∞ if $f(\infty)$ is defined and $\lim\limits_{z \to \infty} f(z) = f(\infty)$.

Let us review Examples 13 to 17 in the context of "continuity":

Example 21 $f(z) = |z|^2$ is continuous at every point of the complex plane.

Example 22 $g(z) = 1/(1 - z)$ is continuous at all points of the plane except $z = 1$.

Example 23 $h(z) = \mathrm{Re}(z^4 + 4)$ is continuous at all points of the plane.

Example 24 $f(z) = (z^4 - 1)/(z - i)$ is continuous on the whole plane if we define $f(i) = -4i$.

Example 25 $h(z) = z/\bar{z}$ is continuous everywhere except $z = 0$. Further, there is no way to define $h(0)$ to make h continuous at $z_0 = 0$ since $h(z) = 1$ for all z that are real and $h(z) = -1$ for z that are purely imaginary. Such a function cannot be continuous at $z = 0$, the point where the real axis meets the imaginary axis.

Example 26 The function $g(z) = 1/(|z| + 1)$ is continuous at all points of the plane and at ∞ as well, if we set $g(\infty) = 0$.

You will no doubt recall that the sum and product of continuous functions is continuous. This follows, in fact, directly from Theorem 2. We state this formally in the next theorem.

Theorem 3: Suppose that f and g are functions both of which are continuous at the point z_0. Let λ be a complex number. Then the functions $f + \lambda g$ and fg are also continuous at the point z_0. Further, if $g(z_0) \neq 0$, then the function f/g is continuous at z_0. Finally, if h is a function continuous at each point of some disc centered at the point $w_0 = f(z_0)$, then the composition $h(f(z))$ is continuous at z_0.

The proofs of these facts are left to the exercises at the end of this section. They allow us to assert that each **polynomial**

$$p(z) = a_0 + a_1 z + \cdots + a_n z^n$$

is continuous on the complex plane; here $a_0, \ldots, a_n$ are complex numbers. Further, if p and q are two polynomials, then their quotient $r = p/q$ is also continuous at all points at which $q(z) \neq 0$. The ratio of two polynomials is called a **rational function**.

Each complex-valued function f can be written as $f = u + iv$, where u and v are real-valued functions: $u(z) = \mathrm{Re}\, f(z)$, $v(z) = \mathrm{Im}\, f(z)$. In this way the statements about limits, continuity, etc., of f can be recast as statements about u and v. For example, f is continuous at a point z_0 if and only if both u and v are continuous at z_0. Some exercises on these topics are included at the end of this section.

Infinite series

Just as the notion of the convergence of a sequence of complex numbers is almost the same as the convergence of a sequence of real numbers, so the notion of the sum of an **infinite series of complex numbers** is virtually the same as that of the sum of an infinite series of real numbers. Specifically, if the numbers $z_1, z_2, \ldots$, are complex numbers we define their **nth partial sum** by

$$s_n = \sum_{j=1}^{n} z_j = z_1 + \cdots + z_n, \quad n = 1, 2, \ldots.$$

We then examine the behavior of the sequence $\{s_n\}$. If the sequence $\{s_n\}$ has a limit s, then we say that the infinite series $\sum_{j=1}^{\infty} z_j$ **converges** and has sum s; this is written

$$\sum_{j=1}^{\infty} z_j = s.$$

If for any reason the sequence $\{s_n\}$ does not have a limit then we say that the series $\sum_{j=1}^{\infty} z_j$ **diverges**. The convergence (or divergence) of the infinite series $\sum_{j=1}^{\infty} z_j$ of complex numbers can be formulated in terms of the convergence (or divergence) of two infinite series of real numbers. This follows directly by writing $z_j = x_j + iy_j$ so that

$$s_n = \sum_{j=1}^{n} z_j = \sum_{j=1}^{n} x_j + i \sum_{j=1}^{n} y_j = \sigma_n + i\tau_n.$$

As we noted earlier in this section, the sequence $\{s_n\}$ converges if and only if *both* of the sequences $\{\sigma_n\}$ and $\{\tau_n\}$ of real numbers converge, and, this being the case, we have

$$s = \lim s_n = \lim \sigma_n + i \lim \tau_n.$$

However, the convergence of the two sequences $\{\sigma_n\}$ and $\{\tau_n\}$ is exactly the statement that the two infinite series $\sum_{j=1}^{\infty} x_j$ and $\sum_{j=1}^{\infty} y_j$ both converge. We thus arrive at this result: Let $z_j = x_j + iy_j; j = 1, 2, 3, \ldots$. The infinite series $\sum_{j=1}^{\infty} z_j$ converges, $z_j = x_j + iy_j$, if and only if both

$$\sum_{j=1}^{\infty} x_j \quad \text{and} \quad \sum_{j=1}^{\infty} y_j$$

converge. Furthermore, if $\sum z_j$ converges, then

$$\sum_{j=1}^{\infty} z_j = \sum_{j=1}^{\infty} x_j + i \sum_{j=1}^{\infty} y_j.$$

There is more that can be said here. Since

$$|x_j| \leqslant |z_j|, \quad |y_j| \leqslant |z_j|$$

we see that the convergence of the series $\sum_{j=1}^{\infty} |z_j|$ of nonnegative numbers implies, by the use of the usual comparison test for infinite series of real numbers, the convergence of both the series

$$\sum_{j=1}^{\infty} |x_j| \quad \text{and} \quad \sum_{j=1}^{\infty} |y_j|.$$

Hence, both the series

$$\sum_{j=1}^{\infty} x_j \quad \text{and} \quad \sum_{j=1}^{\infty} y_j$$

converge and so

$$\sum_{j=1}^{\infty} (x_j + iy_j)$$

also converges. Furthermore, we have

$$|s_n| = \left| \sum_{j=1}^{n} z_j \right| \leqslant \sum_{j=1}^{n} |z_j|$$

for all n. We conclude that if $\sum_{j=1}^{\infty} |z_j|$ converges, then $\sum_{j=1}^{\infty} z_j$ converges and

$$\left| \sum_{j=1}^{\infty} z_j \right| \leqslant \sum_{j=1}^{\infty} |z_j|.$$

This is a useful criterion for convergence, since there are several tests available from calculus (comparison, ratio, root, integral; see the exercises) for the convergence of an infinite series of nonnegative numbers.

Example 27 If α is not equal to 1, then the equality

$$1 + \alpha + \cdots + \alpha^n = \frac{1 - \alpha^{n+1}}{1 - \alpha}$$

is easily verified by multiplying both sides of the equation by $1 - \alpha$. If $|\alpha| < 1$, then we have $\alpha^{n+1} \to 0$ as $n \to \infty$ so that

$$1 + \alpha + \alpha^2 + \cdots = \sum_{j=0}^{\infty} \alpha^j = \frac{1}{1 - \alpha}, \quad |\alpha| < 1.$$

This is the **geometric series**, with ratio α.

Example 28 The series

$$\sum_{j=1}^{\infty} j \left(\frac{1 + 2i}{3} \right)^j$$

converges since

$$\left| j \left(\frac{1 + 2i}{3} \right)^j \right| = j \left(\frac{\sqrt{5}}{3} \right)^j$$

and the series

$$\sum_{j=1}^{\infty} j \left(\frac{\sqrt{5}}{3} \right)^j$$

converges, as can be seen by use of, say, the ratio test.

Example 29   The series

$$\sum_{k=1}^{\infty} \frac{1}{k} (i)^k$$

converges even though the series

$$\sum_{k=1}^{\infty} 1/k$$

diverges. Note that

$$i^k = \begin{cases} i & \text{if} \quad k = 1, 5, 9, \ldots \\ -1 & \text{if} \quad k = 2, 6, 10, \ldots \\ -i & \text{if} \quad k = 3, 7, 11, \ldots \\ 1 & \text{if} \quad k = 4, 8, 12, \ldots \end{cases}$$

Hence

$$\sum_{k=1}^{\infty} i^k = \left(-\frac{1}{2} + \frac{1}{4} - \frac{1}{6} + \cdots \right) + i\left(1 - \frac{1}{3} + \frac{1}{5} - \frac{1}{7} + \cdots \right)$$

and both series in the parentheses converge by the alternating-series test.

Example 30   The series

$$\sum_{k=1}^{\infty} \frac{1}{2^k} (1 + i)^{2k}$$

diverges since

$$\frac{1}{2^k} (1 + i)^{2k} = \left\{ \frac{(1 + i)^2}{2} \right\}^k = i^k$$

and so the terms of the series do not go to zero. Thus, the series must diverge (see Exercise 30).

Some sums⋆

The identity at the beginning of Example 27 has significance beyond the geometric series. Let x be any real number not equal to one. Then

$$1 + x + x^2 + x^3 + \cdots + x^n = \frac{1 - x^{n+1}}{1 - x}.$$

We can manipulate this identity to obtain other useful identities. For instance, differentiate both sides with respect to x, then multiply both sides by x, and add 1. The result is the identity

$$1 + x + 2x^2 + 3x^3 + 4x^4 + \cdots + nx^n = 1 - (n + 1)\frac{x^{n+1}}{1 - x} + x\frac{1 - x^{n+1}}{(1 - x)^2}.$$

Repeating this three-step process yields another identity, this time for $1 + x + 4x^2 + 9x^3 + \cdots + n^2x^n$.

Exercises for Section 1.4

Limits

In Exercises 1 to 8 find the limit of each sequence that converges; if the sequence diverges, explain why.

1. $z_n = \left(\dfrac{1 + i}{\sqrt{3}}\right)^n$

2. $z_n = \left(\dfrac{1 + i}{\sqrt{2}}\right)^n$

3. $z_n = n\left(\dfrac{i}{2}\right)^n$

4. $z_n = \text{Log}\left(1 + \dfrac{1}{n}\right)$

5. $z_n = n + \dfrac{i}{n}$

6. $z_n = \dfrac{\cos n\theta + i \sin n\theta}{n}$, θ fixed

7. $z_n = \text{Arg}\left(1 + \dfrac{\alpha}{n}\right)$, α fixed

8. $z_n = n\left\{1 - \cos\left(\dfrac{\theta}{n}\right) - i \sin\left(\dfrac{\theta}{n}\right)\right\}$, θ fixed

In Exercises 9 to 14 find the limit of each function at the given point or explain why it does not exist.

9. $f(z) = |1 - z|^2$ at $z_0 = i$

10. $f(z) = \text{Arg } z$ at $z_0 = -1$

11. $f(z) = (1 - \text{Im } z)^{-1}$ at $z_0 = 8$ and then at $z_0 = 8 + i$

12. $f(z) = (z - 2)\log|z - 2|$ at $z_0 = 2$

13. $f(z) = \dfrac{|z|^2}{z}$, $z \neq 0$, at $z_0 = 0$

14. $f(z) = \dfrac{z^3 - 8i}{z + 2i}$, $z \neq -2i$, at $z_0 = -2i$

Continuity

In Exercises 15 to 20 find all points of continuity of the given function

15. $f(z) = \begin{cases} \dfrac{z^3 + i}{z - i}, & z \neq i \\ -3, & z = i \end{cases}$

16. $f(z) = \begin{cases} \dfrac{z^4 - 1}{z - i}, & z \neq i \\ 4i, & z = i \end{cases}$

17. $f(z) = (\text{Im } z - \text{Re } z)^{-1}$

18. $g(z) = (1 - |z|^2)^{-3}$

19. $h(z) = \begin{cases} z & \text{if } |z| \leqslant 1 \\ |z|^2 & \text{if } |z| > 1 \end{cases}$

20. $h(z) = \bar{z}^3$

In Exercises 21 to 24 find the limit at ∞ of the given function, or explain why it does not exist.

21. $f(z) = \dfrac{1}{|z| - 1}$

22. $h(z) = \dfrac{|z|}{z}, z \neq 0$

23. $g(z) = \dfrac{4z^6 - 7z^3}{(z^2 - 4)^3}$

24. $h(z) = \text{Arg } z, z \neq 0$

25. Let f and g be continuous at z_0. Show that $f + g$ and fg are also continuous at z_0. If $g(z_0) \neq 0$, show that $1/g$ is continuous at z_0.

26. Establish the following result. A function f is continuous at a point z_0 of its domain of definition S if and only if, given $\varepsilon > 0$, there is a $\delta > 0$ such that $|f(z) - f(z_0)| < \varepsilon$ for all $z \in S$ with $|z - z_0| < \delta$.

27. Let g be defined on a set containing the range of a function f. If f is continuous at z_0 and g is continuous at $f(z_0)$, then $g(f(z))$ is continuous at z_0.

28. Suppose that $f = u + iv$ is continuous at z_0. Show that each of the functions u, v, $u - iv$, and $(u^2 + v^2)^{1/2}$ are also continuous at z_0. Conversely, show that if u and v are continuous at z_0, then so is f.

29. Show that the function f has the limit L at ∞ if and only if the function $\tilde{f}(z) = f(1/z)$ has the limit L at 0.

Infinite series

30. Suppose that $\sum_{n=1}^{\infty} a_n$ converges. Show that $\lim_{n \to \infty} a_n = 0$.

In Exercises 31 to 36 determine whether the given infinite series converges or diverges.

31. $\displaystyle\sum_{n=1}^{\infty} \left(\dfrac{1 + 2i}{\sqrt{6}}\right)^n$

32. $\displaystyle\sum_{n=1}^{\infty} n\left(\dfrac{1}{2i}\right)^n$

33. $\displaystyle\sum_{n=1}^{\infty} \left(\dfrac{2 + i}{\sqrt{5}}\right)^n$

34. $\displaystyle\sum_{n=1}^{\infty} \dfrac{1}{2 + i^n}$

35. $\displaystyle\sum_{n=2}^{\infty} n(n - 1)\beta^{n-2}, |\beta| < 1$

36. $$\sum_{n=1}^{\infty} \frac{1}{n^2 + i^n}$$

37.   Show that each of the following series converges for all z.

(a) $\displaystyle\sum_{n=0}^{\infty} \frac{z^n}{n!}$ (b) $\displaystyle\sum_{n=0}^{\infty} (-1)^n \frac{z^{2n}}{(2n)!}$ (c) $\displaystyle\sum_{n=1}^{\infty} \frac{z^{2n+1}}{(2n+1)!}$

38.   Suppose that the series $\sum_{n=1}^{\infty} a_n$ converges. Let $|z| < 1$. Show that the series $\sum_{n=1}^{\infty} a_n z^n$ is convergent; indeed, that $\sum_{n=1}^{\infty} |a_n z^n|$ converges. (**Hint:** By 30, $|a_n| \leqslant M$ for some M and all n.)

The root and ratio tests$\star$

In Exercises 39 and 40, $\{c_n\}$ is a sequence of positive numbers.

39.   (**Root test**)   Suppose that $\lim_{n \to \infty} (c_n)^{1/n} = A$ exists. Show that the series $\sum c_n$ converges if $0 \leqslant A < 1$ and diverges if $A > 1$. (**Hint:** If $A < 1$, then there is a B with $A < B < 1$. Hence, $c_n < B^n$ if $n \geqslant N$. (Why?) If $1 < A$, then $c_n \geqslant 1$ for all $n \geqslant N'$. (Why?))

40.   (**Ratio test**)   Suppose that $\lim_{n \to \infty} (c_{n+1})/(c_n) = C$ exists. Show that the series $\sum c_n$ converges if $0 \leqslant C < 1$ and diverges if $C > 1$. (**Hint:** If $C < 1$, then for D, $C < D < 1$, we have $c_{n+1} \leqslant Dc_n$ for all $n \geqslant N$. (Why?) Therefore, $c_{N+k} \leqslant D^k c_N$, $k = 0, 1, 2, \ldots$ and $\sum c_j$ converges. If $C > 1$, then $c_{n+1} \geqslant c_n$ for all $n \geqslant N'$ (Why?) and so $\{c_k\}$ does not converge to zero.)

1.5   The exponential, logarithm, and trigonometric functions

The exponential function

The exponential function is one of the most important functions in complex analysis. Its definition is simple:

$$e^z = e^x(\cos y + i \sin y), \qquad z = x + iy.$$

The form $\exp(z)$ is used sometimes, especially if z itself has some complicated form. The definition of e^z allows us immediately to derive a most significant property of the exponential function; namely, for any two complex numbers z and w we have

$$e^{z+w} = e^z e^w.$$

To see this, we write, as usual, $z = x + iy$ and $w = s + it$. Then, making use of two basic trigonometric identities for the sine and cosine of the sum of two numbers, we find that

$$e^{z+w} = e^{x+s}[\cos(y+t) + i\,\sin(y+t)]$$
$$= e^x e^s[(\cos y \cos t - \sin y \sin t) + i(\sin y \cos t + \sin t \cos y)]$$
$$= e^x(\cos y + i\,\sin y)e^s(\cos t + i\,\sin t)$$
$$= e^z e^w.$$

The basic definition allows us to draw several more conclusions about the exponential function. First, e^z is a continuous function of z. The functions e^x, $\cos y$, and $\sin y$ are continuous functions of x and y. Consequently, $\mathrm{Re}(e^z) = e^x \cos y$ and $\mathrm{Im}(e^z) = e^x \sin y$ are both continuous, and hence e^z is continuous at all points of the plane. Second,

$$|e^z| = e^{\mathrm{Re}\,z}$$

because

$$|e^z| = ((e^x \cos y)^2 + (e^x \sin y)^2)^{1/2}$$
$$= e^x = e^{\mathrm{Re}\,z}.$$

In particular

$$|e^{it}| = 1, \quad t \text{ real}$$

Since $e^{it} = \cos t + i\,\sin t$, t real, we see that as t increases, e^{it} moves on the circle of radius 1 centered at the origin in a counterclockwise direction making one complete circuit when t increases by 2π (Fig. 1.21). In particular, of course, we have

$$e^{2\pi i m t} = 1 \quad \text{for} \quad m = 0, \pm 1, \ldots$$

and

$$e^{\pi i} = -1.$$

Second, the function $f(z) = e^z$ never has the value zero since neither e^x nor e^{iy} are ever zero. On the other hand, if w is any nonzero complex number, then the equation

$$e^z = w$$

has infinitely many solutions. This can be seen, for example, by writing w in polar form

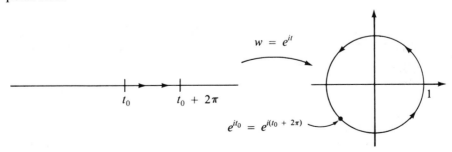

Figure 1.21

$$w = r(\cos \psi + \sin \psi)$$

and then setting $x = \ln r$ and $y = \psi + 2\pi m$, where m is any integer, positive or negative, and $\ln r$ is the natural logarithm of r; ln is log to the base e, studied in calculus. Thus,

$$e^{x+iy} = e^{\ln r} e^{i(\psi + 2\pi m)}$$
$$= r[\cos \psi + i \sin \psi]$$
$$= w.$$

Furthermore, *every* solution z of the equation $e^z = w$ has the form given above. For if $e^z = w$, then

$$r = |w| = |e^z| = e^x$$

so $x = \ln r$. Consequently

$$\cos y + i \sin y = e^{iy} = \cos \psi + i \sin \psi.$$

Thus, $y = \psi + 2\pi m$ for some integer m (see Exercise 14 in Section 1).

The mapping of $f(z) = e^z$ thus carries the complex z-plane onto the complex w-plane with the origin deleted; each point w_0 has infinitely many pre-images z, each of the form $z_0 + 2\pi i m$, $m = 0, \pm 1, \pm 2, \ldots$, where z_0 is *any* solution of $e^{z_0} = w_0$. In particular, $f(z) = e^z$ carries each strip $y_0 \leqslant \operatorname{Im} z < y_0 + 2\pi$, $-\infty < \operatorname{Re} z < \infty$, onto the w-plane with the origin removed. The function $f(z) = e^z$ is **one-to-one** on that strip, that is, distinct points have distinct images. For $e^{z_1} = e^{z_2}$ if and only if $e^{z_1 - z_2} = 1$; this occurs exactly when $z_1 - z_2 = 2\pi i k$ for some integer k (Fig. 1.22).

The function $f(z) = e^z$ maps each horizontal line $y = c$ onto a ray from the origin to infinity, specifically, the ray $\{r \cos c + ir \sin c : 0 < r < \infty\}$ since $\exp(x + ic) = e^x \cos c + ie^x \sin c$ and e^x increases from 0 to ∞ as x increases from $-\infty$ to ∞; of course, e^x is always positive. Furthermore, e^z maps each vertical line $x = c$ onto the circle centered at the origin of radius e^c. This is because $\exp(c + iy) = e^c (\cos y + i \sin y)$ and $\cos y + i \sin y$ travels along the circle of radius 1 centered at the origin in a counterclockwise direc-

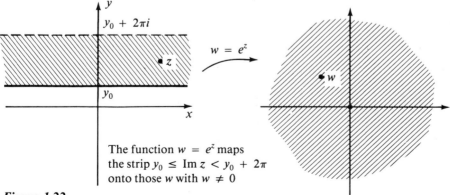

The function $w = e^z$ maps
the strip $y_0 \leq \operatorname{Im} z < y_0 + 2\pi$
onto those w with $w \neq 0$

Figure 1.22

tion as y increases. Each point on this circle is the image of infinitely many points on the vertical line since both $\cos y$ and $\sin y$ are periodic, with period 2π.

The logarithm function

The inverse of the exponential function is the logarithm function. For a nonzero complex number z we *define* **log** z to be any complex number w with $e^w = z$. The preceding discussion of the exponential function leads us immediately to the relationship

$$\log z = \ln|z| + i \arg z, \quad z \neq 0.$$

However, it is obvious that this is not a single complex number, but rather a set of complex numbers, each two elements of which differ by an integer multiple of $2\pi i$. If we want to be definite, we use

$$\text{Log } z = \ln|z| + i \text{ Arg } z.$$

You should be wary here; it may happen that $\text{Log}(z_1 z_2) \neq \text{Log } z_1 + \text{Log } z_2$ exactly because $\text{Arg}(z_1 z_2)$ need not equal $\text{Arg } z_1 + \text{Arg } z_2$. The function $\text{Log } z$ is called the **principal branch** of the logarithm of z. Other choices of $\arg z$ yield other values of $\log z$ (more on this below). To investigate the continuity of $\log z$ we delete from the plane any ray beginning at the origin (for example, delete all the nonpositive real numbers) and let D be the domain remaining. Let z_0 be any point of D and choose and then fix any value for $\arg z_0$, say $\arg z_0 = \theta_0$. We then define in D a **branch** of $\log z$ by the rule $\log z = \ln|z| + i \arg z$ with the additional specification that $\log z_0 = \ln|z_0| + i\theta_0$. Then within D the values of $\arg z$ lie in a uniquely determined open interval of length 2π, which contains θ_0; furthermore, $\arg z$ is continuous in D with this specification. The function $\ln|z|$ is a continuous function on D as well, so that the branch of $\log z$ determined by the specific choice $\arg z_0 = \theta_0$ is continuous on D. For example, if we delete the nonpositive real axis (see Fig. 1.23), and we pick $z_0 = 1 + i$ and $\theta_0 = (9/4)\pi$, then the branch of $\log z$ that is

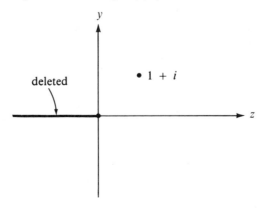

Figure 1.23

determined has arg z in the range $(\pi, 3\pi)$. Clearly, there are infinitely many possible branches of log z in D, each continuous.

The definition of log z allows us to complete the discussion of roots begun in Section 2. For a nonzero complex number a, we define a^z by the rule

$$a^z = e^{z \log a}.$$

This results in a many-valued "function" but it does agree with the usual definition in the special case when a and z are both positive and real.

Example 1   Find all the values of $(-1)^i$.

Solution   Using the relation $\log(-1) = (2n + 1)\pi i$, $n = 0, \pm 1, \ldots$, we have

$$(-1)^i = e^{i \log(-1)} = e^{-(2n+1)\pi}, \quad n = 0, \pm 1, \pm 2, \ldots. \qquad \square$$

Example 2   Find the solutions of $z^{1+i} = 4$.

Solution   We write this equation as

$$e^{(1+i)\log z} = 4$$

so that $(1 + i) \log z = \ln 4 + 2\pi n i$, $n = 0, \pm 1, \ldots$. Hence,

$$\begin{aligned}
\log z &= (1 - i)[\ln 2 + \pi n i] \\
&= (\ln 2 + \pi n) + i(\pi n - \ln 2).
\end{aligned}$$

Thus,

$$\begin{aligned}
z &= 2e^{\pi n}(\cos(\pi n - \ln 2) + i \sin(\pi n - \ln 2)) \\
&= 2e^{\pi n}\{(-1)^n \cos(\ln 2) + i(-1)^{n+1} \sin(\ln 2)\} \\
&= (-1)^n 2 e^{\pi n}\{\cos(\ln 2) - i \sin(\ln 2)\}, \ n = 0, \pm 1, \ldots. \qquad \square
\end{aligned}$$

Example 3   Prove the formula

$$\lim_{n \to \infty} (1 + z/n)^n = e^z, \quad z \text{ a complex number}$$

Solution   We look at $n \operatorname{Log}(1 + z/n)$ for large n. We have

$$n \operatorname{Log}(1 + z/n) = n \ln|1 + z/n| + n \operatorname{Arg}(1 + z/n).$$

The real part satisfies

$$n \ln|1 + z/n| = \tfrac{1}{2}n \ln[1 + (2x/n) + (x^2 + y^2)/n^2] \to x$$

as $n \to \infty$ by l'Hôpital's Rule, for example. Next, if $z = r(\cos \theta + i \sin \theta)$ and $\psi_n = \operatorname{Arg}(1 + z/n)$, then from Figure 1.24 we find that

$$\tan \psi_n = \frac{r/n \sin \theta}{1 + r/n \cos \theta}$$

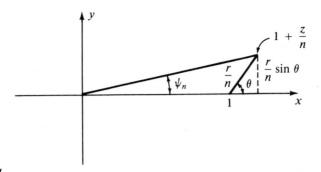

Figure 1.24

and hence

$$n \tan \psi_n = \frac{r \sin \theta}{1 + r/n \cos \theta}.$$

Because $\psi_n \to 0$ as $n \to \infty$ we know that $\psi_n/\tan \psi_n \to 1$ (say, l'Hôpital's Rule again). Hence,

$$n\psi_n = (n \tan \psi_n)(\psi_n/\tan \psi_n) \to r \sin \theta = y \quad \text{as } n \to \infty.$$

Consequently,

$$n \operatorname{Log}(1 + z/n) \to x + iy = z, \quad \text{as } n \to \infty.$$

and so $(1 + z/n)^n \to e^z$, as we wished to show. $\qquad\qquad\square$

Trigonometric functions

The trigonometric functions of z are defined in terms of the exponential function. We begin with the cosine and the sine of z:

$$\cos z = \frac{1}{2}(e^{iz} + e^{-iz})$$

$$\sin z = \frac{1}{2i}(e^{iz} - e^{-iz}).$$

If z is real, $z = x$, then $\cos z$ and $\sin z$ agree with the usual definitions of cosine and sine. We note that

$$\begin{aligned}
\cos(z + 2\pi) &= \tfrac{1}{2}(e^{i(z+2\pi)} + e^{-i(z+2\pi)}) \\
&= \tfrac{1}{2}(e^{iz} + e^{-iz}) \\
&= \cos z.
\end{aligned}$$

Likewise, $\sin(z + 2\pi) = \sin z$ for all z. Thus,

$$\cos(z + 2\pi k) = \cos z$$
$$\sin(z + 2\pi k) = \sin z$$

for all z and any integer k. Furthermore, $2\pi k$ is the only number α with $\cos(z + \alpha) = \cos z$ for all z. For if

$$\cos(z + \alpha) = \cos z \quad \text{for all } z$$

then

$$e^{iz}e^{i\alpha} + e^{-iz}e^{-i\alpha} = e^{iz} + e^{-iz}$$

and so

$$e^{iz}(e^{i\alpha} - 1) = e^{-iz}(1 - e^{-i\alpha})$$
$$= e^{-i\alpha}e^{-iz}(e^{i\alpha} - 1).$$

If $e^{i\alpha} - 1$ is not zero, then it can be cancelled from both sides leaving

$$e^{iz} = e^{-i\alpha}e^{-iz}, \quad \text{all } z.$$

But then setting $z = 0$ gives $1 = e^{-i\alpha}$, contradicting the fact that $1 \neq e^{i\alpha}$. Hence, it must be the case that $e^{i\alpha} = 1$ and so $\alpha = 2\pi k$ for some integer k. Consequently, 2π is the basic period of $\cos z$ and, likewise, of $\sin z$.

In the exercises at the end of this section, the reader is asked to establish the formulas

$$\cos(x + iy) = \cos x \cosh y - i \sin x \sinh y,$$
$$\sin(x + iy) = \sin x \cosh y + i \cos x \sinh y,$$

where

$$\cosh u = \tfrac{1}{2}(e^u + e^{-u}), \quad u \text{ real}$$
$$\sinh u = \tfrac{1}{2}(e^u - e^{-u}), \quad u \text{ real}$$

Let us use the formula given above for $\sin(x + iy)$ to establish some basic properties of the function $f(z) = \sin z$. Note that

$$\sin(\bar{z}) = \sin(x - iy)$$
$$= \sin x \cosh(-y) + i \cos x \sinh(-y)$$
$$= \underline{\sin x \cosh y - i \cos x \sinh y}$$
$$= \overline{\sin z}.$$

Let us restrict $z = x + iy$ so that $0 \leqslant x < \pi$ and $y \geqslant 0$. In this restricted region, we shall show that the function $\sin z$ is one-to-one. That is, we now shall show that if

$$\sin(x_1 + iy_1) = \sin(x_2 + iy_2),$$

where x_1 and x_2 lie in $[0, \pi)$ and y_1 and y_2 are nonnegative, then $x_1 = x_2$ and $y_1 = y_2$. To see this observe that

$$2i \sin(x + iy) = e^{ix}e^{-y} - e^{-ix}e^y.$$

Hence, if $\sin(x_1 + iy_1) = \sin(x_2 + iy_2)$, then

$$e^{ix_1}e^{-y_1} - e^{-ix_1}e^{y_1} = e^{ix_2}e^{-y_2} - e^{ix_2}e^{y_2}.$$

Thus,

$$e^{ix_1}e^{-y_1} - e^{ix_2}e^{-y_2} = e^{-ix_1}e^{y_1} - e^{-ix_2}e^{y_2}$$
$$= e^{-ix_1}e^{-ix_2}e^{y_1}e^{y_2}[e^{ix_2}e^{-y_2} - e^{ix_1}e^{-y_1}].$$

If $e^{ix_1}e^{-y_1} - e^{ix_2}e^{-y_2} \neq 0$, then we obtain

$$1 = -e^{-ix_1}e^{-ix_2}e^{y_1}e^{y_2}.$$

The absolute value of the left side is 1 and that of the right side is $\exp(y_1 + y_2)$; this implies that $y_1 + y_2 = 0$. This in turn implies that $-1 = \exp[-ix_1 - ix_2]$, which can be true only when $x_1 + x_2 = \pi(2m + 1)$ for some integer m. However, we have restricted x_1, x_2 to lie in $[0, \pi)$ and y_1, y_2 to lie in $[0, \infty)$. The only conclusion, then, is that $x_1 = x_2 = 0$ and $y_1 = y_2 = 0$. Otherwise,

$$e^{ix_1}e^{-y_1} = e^{ix_2}e^{-y_2}$$

so that $y_1 = y_2$ and $x_1 - x_2$ is an integer multiple of 2π. This again implies that $x_1 = x_2$.

Let us now examine $f(z) = \sin z$ on the strip $0 \leqslant x \leqslant \pi/2$ and $0 \leqslant y < \infty$. We see first that

$$\text{Re}(\sin z) = \sin x \cosh y \geqslant 0$$

and

$$\text{Im}(\sin z) = \cos x \sinh y \geqslant 0.$$

Hence, the values of $\sin z$ lie in the first quadrant (Fig. 1.25). Next, on the ray $0 \leqslant y < \infty$, $x = 0$, we have

$$\sin(iy) = i \sinh y = \frac{i}{2}(e^y - e^{-y}).$$

The function $\sinh(y)$ is zero when $y = 0$ and is increasing (its derivative is positive) so that $\sin(iy)$ assumes all the values $i\beta$, $0 \leqslant \beta < \infty$, as y increases from 0 to ∞. Next, $\sin x$ increases from 0 to 1 as x increases from 0 to $\pi/2$. Finally,

$$\sin(\pi/2 + iy) = \cosh y$$

and this increases from 1 to ∞ as y increases from 0 to ∞. Thus, the boundary of the strip $0 \leqslant x \leqslant \pi/2$, $0 \leqslant y < \infty$ is carried by $\sin z$ onto the boundary of

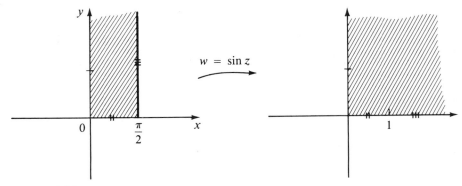

Figure 1.25

the first quadrant (see Fig. 1.25). We now show that the interior is mapped onto the interior; we already know it is mapped into. Let us write $\sin z = w = \sigma + i\tau$. The vertical segment $z = x_0 + iy$, $y \geqslant 0$, is mapped onto that portion of the hyperbola

$$\frac{\sigma^2}{(\sin x_0)^2} - \frac{\tau^2}{(\cos x_0)^2} = 1,$$

which lies in the first quadrant. This shows us how to solve the equation

$$\sin z = a + ib, \quad a, b \text{ positive.}$$

First choose x so that the point $a + ib$ lies on the hyperbola consisting of those $w = \sigma + i\tau$ with

$$\frac{\sigma^2}{(\sin x)^2} - \frac{\tau^2}{(\cos x)^2} = 1;$$

then choose y so that $a = \sin x \cosh y$ and $b = \cos x \sinh y$.

Consequently, we see that the function $f(z) = \sin z$ maps the semi-infinite strip $0 \leqslant x \leqslant \pi/2$ and $0 \leqslant y < \infty$ in a one-to-one fashion *onto* the whole first quadrant, mapping the boundary of the strip onto the boundary of the first quadrant. This type of information, and this mapping in particular, will be very important in the solution of boundary-value problems. Finally, because $\sin(-\bar{z}) = -\overline{\sin z}$, we may conclude that $f(z) = \sin z$ maps the semi-infinite strip $\{-\pi/2 < x < \pi/2, \ y > 0\}$ both one-to-one and onto the upper half-plane, $\{w = s + it : -\infty < s < \infty, t > 0\}$.

Inverse trigonometric functions

We have just seen that the function $w = \sin z$ maps the strip $\{x + iy : 0 < x < \pi/2, \ 0 < y < \infty\}$ onto the first quadrant $\{w = \sigma + i\tau : \sigma, \tau > 0\}$ and distinct z_1 and z_2 in the strip have distinct images w_1 and w_2 in the first quadrant. Thus, given a $w = \sigma + i\tau$ with $\sigma, \tau > 0$ there is one and only one z in the strip $\{x + iy : 0 < x < \pi/2, \ 0 < y < \infty\}$ with $w = \sin z$. Consequently, the function $\sin z$ has an inverse function, naturally called arcsin w, defined at least for w in the first quadrant. Let us pursue this a little further. We have the formula for $\sin z$

$$w = \sin z = \frac{1}{2i}(e^{iz} - e^{-iz}),$$

and so

$$e^{2iz} - 2iwe^{iz} - 1 = 0,$$

which is a quadratic equation in the variable e^{iz}. Solving by the quadratic formula we obtain

$$e^{iz} = iw + \sqrt{1 - w^2},$$

and so

$$z = -i \log(iw + \sqrt{1 - w^2}),$$

provided an appropriate branch of the logarithm is chosen. A careful examination of the mapping $w = \sin z$ (see the exercises) shows that $\sin z$ maps the
strip

$$\left\{ x + iy : -\frac{\pi}{2} < x < \frac{\pi}{2}, \ -\infty < y < \infty \right\}$$

both one-to-one and onto the region D obtained from the plane by deleting
the two intervals $(-\infty, -1]$ and $[1, \infty)$ (see Fig. 1.26). Thus, on D we can
solve uniquely for z in terms of w and the formula is valid. Interchanging the
roles of z and w we write

$$\text{Arcsin } z = -i \ \text{Log}(iz + \sqrt{1 - z^2}), \quad z \in D$$

Similarly, we find that

$$\text{Arccos } z = -i \ \text{Log}(z + \sqrt{z^2 - 1})$$

and

$$\text{Arctan } z = \frac{i}{2} \ \text{Log}\left(\frac{1 - iz}{1 + iz}\right), \quad z \neq \pm i$$

with appropriate interpretations of the resulting logarithms and roots.

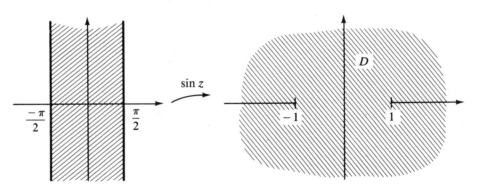

$\sin z$

D

Figure 1.26

Exercises for Section 1.5

Find the value(s) of the given expression in Exercises 1 to 14

1. $e^{i\pi/4}$	2. $e^{5\pi i/4}$	3. $\log(1 + i\sqrt{3})$
4. $\log(-i)$	5. $(1 + i)^i$	6. 2^{-1-i}
7. $e^{-7\pi i/3}$	8. $\exp(\text{Log}(3 + 2i))$ 9. $\text{Log}(4 - 4i)$

10. $\text{Log}(-1)$ 11. $i^{\sqrt{3}}$ 12. $\log(\sqrt{3} - i)$

13. $\log((1 - i)^4)$ 14. $\exp\left[\pi\left(\dfrac{i+1}{\sqrt{2}}\right)^4\right]$

15.   Show that z is one of the values of $\log(e^z)$. Are there other values? If so, what are they?

16.   Establish the formulas

$$\cos(x + iy) = \cos x \cosh y - i \sin x \sinh y,$$
$$\sin(x + iy) = \sin x \cosh y + i \cos x \sinh y,$$

where

$$\cosh u = \tfrac{1}{2}(e^u + e^{-u}), \quad u \text{ real}$$
$$\sinh u = \tfrac{1}{2}(e^u - e^{-u}), \quad u \text{ real.}$$

17.   Show that $\cos z = 0$ if and only if $z = \pi/2 + n\pi$, $n = 0, \pm 1, \pm 2, \ldots$; show that $\sin z = 0$ if and only if $z = n\pi$, $n = 0, \pm 1, \pm 2, \ldots$. That is, extending $\sin z$ and $\cos z$ from the real axis to the whole plane does not introduce any new zeros.

18.   Verify that

$$\cos(z + w) = \cos z \cos w - \sin z \sin w$$

and

$$\sin(z + w) = \sin z \cos w + \cos z \sin w$$

for all complex numbers z and w.

19.   Show that both $\cos z$ and $\sin z$ are unbounded if $z = iy$ and $y \to \infty$. Show as well that

$$|\cos(x + iy)| \leqslant e^y \quad \text{if} \quad y \geqslant 0, \; -\infty < x < \infty;$$

and

$$|\sin(x + iy)| \leqslant e^y \quad \text{if} \quad y \geqslant 0, \; -\infty < x < \infty.$$

20.   Prove that $\cos^2 z + \sin^2 z = 1$ for all z.

21.   Define $\cosh z$ and $\sinh z$ by

$$\cosh z = \tfrac{1}{2}(e^z + e^{-z})$$
$$\sinh z = \tfrac{1}{2}(e^z - e^{-z}).$$

Show that the following identities hold
(i) $\cosh^2(z) - \sinh^2(z) = 1$
(ii) $\cosh z = \cos(iz)$
(iii) $\sinh z = -i \sin(iz)$
(iv) $|\cosh z|^2 = \sinh^2 x + \cos^2 y$
(v) $|\sinh z|^2 = \sinh^2 x + \sin^2 y$

22.   Show that $\sin(-z) = -\sin z$ and $\cos z = \cos(-z)$ for all z.

Mappings with the exponential, logarithm, and trigonometric functions

23.   Show that $F(z) = e^z$ maps the strip $S = \{x + iy : -\infty < x < \infty, -\pi/2 \leqslant y \leqslant \pi/2\}$ onto the region $\Omega = \{w = s + it : s \geqslant 0, w \neq 0\}$ and that F is one-to-one on S (see Fig. 1.27). Furthermore, show that F maps the boundary of S onto all the boundary of Ω except $w = 0$. Explain what happens to each of the horizontal lines $\{\text{Im } z = \pi/2\}$ and $\{\text{Im } z = -\pi/2\}$.

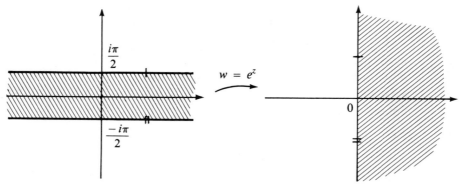

Figure 1.27

24.   Let D be the domain obtained by deleting the ray $\{x : x \leqslant 0\}$ from the plane and let $G(z)$ be a branch of $\log z$ on D. Show that G maps D onto a horizontal strip of width 2π,

$$\{x + iy : -\infty < x < \infty, c_0 < y < c_0 + 2\pi\},$$

and that the mapping is one-to-one on D.

25.   Show that $w = \sin z$ maps the strip $-\pi/2 < x < \pi/2$ both one-to-one and onto the region obtained by deleting from the plane the two rays $(-\infty, -1]$ and $[1, \infty)$ (see Fig. 1.26). (**Hint:** Use Exercise 22 and the fact that $\sin(\bar{z}) = \overline{\sin(z)}$.)

26.   Show that the function $w = \cos z$ maps the strip $\{0 < x < \pi\}$ one-to-one and onto the region D shown in Figure 1.26. Use this to define the inverse function to $\cos z$. Derive the formula for arccos z given in the text.

27.   Let $0 < \alpha < 2$. Show that an appropriate choice of $\log z$ for $f(z) = z^\alpha = \exp[\alpha \log z]$ maps the domain $\{x + iy : y > 0\}$ both one-to-one and onto the domain $\{w : 0 < \arg w < \alpha\pi\}$. Show that f also carries the boundary to the boundary (see Fig. 1.28).

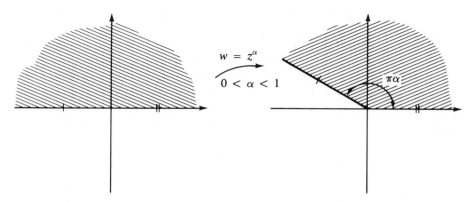

Figure 1.28

28. Show that the function $w = g(z) = e^{z^2}$ maps the lines $x = y$ and $x = -y$ onto the circle $|w| = 1$. Show further that g maps each of the two pieces of the region $\{x + iy : x^2 > y^2\}$ onto the set $\{w : |w| > 1\}$ and each of the two pieces of the region $\{x + iy : x^2 < y^2\}$ onto the set $\{w : |w| < 1\}$.

Inverse trigonometric functions

29. Show directly that if ζ is any value of

$$-i \log(iz + \sqrt{1 - z^2})$$

then $\sin \zeta = z$. Likewise, show if ξ is any value of

$$\frac{i}{2} \log\left(\frac{1 - iw}{1 + iw}\right)$$

then $\tan \xi = w$.

30. Use the result in Exercise 29 and your knowledge of the branches of the logarithmic function to explain the branches of arcsin z.

1.6 Line integrals and Green's Theorem

The fundamental theorems of complex variables depend on line integrals and so this section is devoted to that topic and to formulating Green's Theorem, the basic theorem about line integrals. Several consequences of Green's Theorem are also covered.

Curves

A **curve** γ is a continuous complex-valued function $\gamma(t)$ defined for t in some interval $[a, b]$ in the real axis. The curve γ is **simple** if $\gamma(t_1) \neq \gamma(t_2)$ whenever

$a \leqslant t_1 < t_2 < b$ and it is **closed** if $\gamma(a) = \gamma(b)$. The famous **Jordan* Curve Theorem** asserts that the complement of the range of a curve that is both simple and closed consists of two disjoint open connected sets, one bounded and the other unbounded. The bounded piece is the **inside** of the curve and the unbounded piece the **outside**. Despite the almost painful obviousness of this statement, the theorem is hard to prove.† We shall accept it as true.

Suppose γ is a curve; we separate the complex number $\gamma(t)$ into its real and imaginary parts and write $\gamma(t) = x(t) + iy(t)$, $a \leqslant t \leqslant b$. The functions $x(t)$ and $y(t)$ are real-valued functions of the real variable t and thus they may (or may not) be differentiable. If both $x(t)$ and $y(t)$ are differentiable at t_0, then we say $\gamma(t)$ is differentiable at t_0 and we set $\gamma'(t_0) = x'(t_0) + iy'(t_0)$; this is consistent with the usual rules in calculus for differentiating a vector-valued function of a real variable. A curve γ is **smooth** if $\gamma(t)$ has the added property that $\gamma'(t)$ not only exists but is also continuous on $[a, b]$, the derivatives at a and b being taken from the right and left, respectively. A curve is **piecewise smooth** if it is composed of a finite number of smooth curves, the end of one coinciding with the beginning of the next. That is, the curve γ is piecewise smooth if in the interval $[a, b]$ there are points $t_0, t_1, \ldots, t_n$ with $a = t_0 < t_1 < \cdots < t_{n-1} < t_n = b$ such that $\gamma'(t)$ is continuous on each closed interval $[t_k, t_{k+1}]$, $k = 0, 1, \ldots, n - 1$. It is not required that $\gamma'(t)$ be continuous on all of $[a, b]$.

Note: It is very common and convenient to refer to the range of $\gamma(t)$ as the curve γ and to $\gamma(t)$ itself as the **parametrization** of the curve. With this use of the word curve, a curve becomes a concrete geometric object such as a circle or a straight line segment and hence is easily visualized. The difficulty with this view is that a particular curve has many different parametrizations. The curves we will use are generally composed of straight line segments and/or arcs of circles. These have standard parametrizations as illustrated in the examples below.

Example 1   Fix z_0 and z_1 in the plane and let $\gamma(t) = tz_1 + (1 - t)z_0$, $0 \leqslant t \leqslant 1$. This is a smooth simple curve and its range is the straight line segment joining z_0 to z_1 (in that order).

Example 2   Fix a point p in the plane and a positive number R; the curve $\gamma(t) = p + Re^{it}$, $0 \leqslant t \leqslant 2\pi$, is a smooth simple closed curve. The range is precisely the circle of radius R centered at p; the circle is traversed in the counterclockwise direction.

Example 3   The square with vertices at $z_0, iz_0, -z_0$, and $-iz_0$ is the range of the piecewise smooth, simple closed curve $\gamma(t)$ given by the rule

---

* Camille Jordan, 1838–1922.
† For a proof, see Newman, M. H. A. *Elements of the topology of plane sets of points.* Cambridge: Cambridge University Press, 1961.

$$\gamma(t) = \begin{cases} tiz_0 + (1-t)z_0, & 0 \leqslant t \leqslant 1 \\ (t-1)(-z_0) + (2-t)(iz_0), & 1 \leqslant t \leqslant 2 \\ (t-2)(-iz_0) + (3-t)(-z_0), & 2 \leqslant t \leqslant 3 \\ (t-3)(z_0) + (4-t)(-iz_0), & 3 \leqslant t \leqslant 4. \end{cases}$$

Example 4   Look at Figure 1.29. There is a piecewise smooth closed curve γ whose range is traced out by traveling counterclockwise on the circle $|z| = 1$ from $z = 1$ all the way around to $z = 1$ and then along the real axis to $z = \varepsilon$, $\varepsilon > 0$, and then traveling clockwise all around the circle $|z| = \varepsilon$ back to $z = \varepsilon$ and finally returning to $z = 1$ along the x-axis. You might wish to write down the rule for this γ.

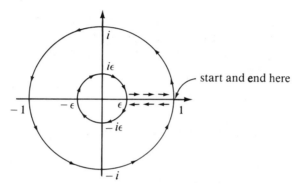

Figure 1.29

Each curve γ is **oriented** by increasing t. The curve γ **begins** at $\gamma(a)$, is traversed as t increases from a to b, γ **ends** at $\gamma(b)$. The **reverse orientation** is given to γ by beginning at $\gamma(b)$ and ending at $\gamma(a)$; this curve is denoted by $-\gamma$ and is given by $-\gamma(t) = \gamma(a + b - t)$, $a \leqslant t \leqslant b$. A simple closed curve γ is **positively oriented** if for each point p on the inside of γ the argument of $\gamma(t) - p$ increases by 2π as t increases from a to b. Equivalently, γ is positively oriented if, as you walk along γ in the direction of the orientation of γ, the inside of γ is on your left. For example, a circle is positively oriented when it is traversed counterclockwise; the same is true for a triangle or a rectangle.

Suppose $g(t) = \sigma(t) + i\tau(t)$ is a continuous complex-valued function on the interval $[a, b]$. We define the **integral of g over $[a, b]$** by

$$\int_a^b g(t)dt = \int_a^b \sigma(t)dt + i \int_a^b \tau(t)dt.$$

This definition, like the definition of the derivative of a complex-valued function of the real variable t, is consistent with the definition of the integral of a vector-valued function of the real variable t studied in calculus. Note, for instance, that

$$\mathrm{Re}\left\{ \int_a^b g(t)dt \right\} = \int_a^b \{\mathrm{Re}\ g(t)\}dt,$$

since both these expressions equal $\int_a^b \sigma(t)dt$.

Suppose now that γ is a smooth curve and u is a continuous function on the range of γ. We define the **line integral of u along γ** by

$$\int_\gamma u(z)dz = \int_a^b u(\gamma(t))\gamma'(t)dt,$$

where the right side is the integral of the complex-valued function $u(\gamma(t))\gamma'(t)$ from a to b and is computed as discussed above. For a piecewise smooth curve γ we define the line integral of u along γ by

$$\int_\gamma u(z)dz = \sum_{j=0}^{n-1} \int_{t_j}^{t_{j+1}} u(\gamma(s))\gamma'(s)ds.$$

The points $t_0, t_1, \ldots, t_n$ come from the definition of "piecewise smooth"; by assumption $\gamma'(s)$ is continuous on each segment $[t_j, t_{j+1}], j = 0, 1, \ldots, n-1$.

Line integrals have the familiar properties of definite integrals studied in calculus. For example,

$$\int_\gamma \{Au(z) + Bv(z)\}dz = A \int_\gamma u(z)dz + B \int_\gamma v(z)dz$$

if A and B are complex numbers and u, v are continuous functions on the range of γ. For the curve $-\gamma$, we have

$$\int_{-\gamma} u(z)dz = \int_a^b u(\gamma(a + b - t))\{\gamma(a + b - t)\}'dt$$

$$= -\int_a^b u(\gamma(a + b - t))\gamma'(a + b - t)dt; \quad \text{put } s = a + b - t$$

$$= -\int_a^b u(\gamma(s))\gamma'(s)ds = -\int_\gamma u(z)dz.$$

Suppose γ_1 and γ_2 are two curves with parameter intervals $[a_1, b_1]$ and $[a_2, b_2]$, respectively. If $\gamma_1(b_1) = \gamma_2(a_2)$, then the **sum** of γ_1 and γ_2 is the curve

$$(\gamma_1 + \gamma_2)(t) = \begin{cases} \gamma_1(t), & a_1 \leqslant t \leqslant b_1 \\ \gamma_2(t + a_2 - b_1), & b_1 \leqslant t \leqslant b_1 + b_2 - a_2 \end{cases}$$

Further, a simple computation shows that

$$\int_{\gamma_1 + \gamma_2} u(z)dz = \int_{\gamma_1} u(z)dz + \int_{\gamma_2} u(z)dz$$

provided, of course, that u is continuous on the range of both γ_1 and γ_2.

Let g be a complex-valued continuous function on $[a, b]$; we shall show the inequality

$$\left| \int_a^b g(t)dt \right| \leqslant \int_a^b |g(t)|dt.$$

The inequality is obviously true if $\int_a^b g(t)dt = 0$ so we may assume that

$\int_a^b g(t)dt \neq 0$. Let

$$\theta = \text{Arg}\left(\int_a^b g(t)dt\right)$$

and define $h(t) = e^{-i\theta}g(t)$, $a \leqslant t \leqslant b$. We then have

$$0 < \left|\int_a^b g(t)dt\right| = e^{-i\theta}\int_a^b g(t)dt$$

$$= \int_a^b e^{-i\theta}g(t)dt = \int_a^b h(t)dt, \quad h(t) = e^{-i\theta}g(t).$$

Hence, $\int_a^b h(t)dt$ is positive and so

$$\left|\int_a^b g(t)dt\right| = \text{Re}\int_a^b h(t)dt$$

$$= \int_a^b (\text{Re } h(t))dt$$

$$\leqslant \int_a^b |h(t)|dt = \int_a^b |g(t)|dt.$$

Therefore, we find that

$$\left|\int_\gamma u(z)dz\right| = \left|\int_a^b u(\gamma(t))\gamma'(t)dt\right|$$

$$\leqslant \int_a^b |u(\gamma(t))| \, |\gamma'(t)|dt. \tag{1}$$

Recall now from calculus that if $\gamma(t) = x(t) + iy(t)$, then the length of the curve that is the range of $\gamma(t)$ is given by

$$\text{length }(\gamma) = \int_a^b \sqrt{(x'(t))^2 + (y'(t))^2} \, dt$$

$$= \int_a^b |\gamma'(t)|dt. \tag{2}$$

This allows us to make the very important estimate

$$\left|\int_\gamma u(z)dz\right| \leqslant \left(\max_{z \in \gamma} |u(z)|\right)\text{length }(\gamma). \tag{3}$$

The inequality in (3) follows directly from that in (1) by replacing $|u(\gamma(t))|$ by its maximum value on the interval, $a \leqslant t \leqslant b$; this is the first factor on the right in (3). The integral that remains is that in (2) and it yields the length of the curve.

Example 5   Compute $\int_\gamma (z^2 - 3|z| + \text{Im } z)dz$, where γ is the quarter-circle centered at the origin and extending from 2 to $2i$.

Solution   We have $\gamma(t) = 2e^{it}$, $0 \leqslant t \leqslant \pi/2$, and so $\gamma'(t) = 2ie^{it}$ and

$$\int_\gamma (z^2 - 3|z| + \text{Im } z)dz = \int_0^{\pi/2} (4e^{2it} - 6 + 2 \sin t)2ie^{it} \, dt$$

$$= \left(\frac{8}{3} e^{3it} - 12e^{it} + \frac{1}{i} e^{2it} - 2t \right)\Big|_0^{|\pi/2}$$

$$= \frac{28}{3} - \pi - \frac{38}{3} i;$$

here we used the formula

$$\sin t = \frac{1}{2i} (e^{it} - e^{-it}).$$ □

Example 6   Compute $\int_\gamma \cos z \, dz$, where γ is the line segment from $-(\pi/2) + i$ to $\pi + i$.

Solution   The curve is given by

$$\gamma(t) = t(\pi + i) + (1 - t)\left(-\frac{\pi}{2} + i \right) = \frac{3\pi}{2} t - \frac{\pi}{2} + i, \quad 0 \leqslant t \leqslant 1.$$

Furthermore,

$$\cos(x + iy) = \cos x \cosh y - i \sin x \sinh y.$$

(See the formula in Exercise 16 of Section 5.) Hence,

$$\int_\gamma \cos z \, dz = \int_0^1 \left\{ \cos\left(\frac{3\pi}{2} t - \pi/2 \right)\cosh(1) - i \sin\left(\frac{3\pi}{2} t - \frac{\pi}{2} \right)\sinh(1) \right\} \frac{3\pi}{2} \, dt$$

$$= \left\{ \cosh(1)\sin\left(\frac{3\pi}{2} t - \frac{\pi}{2} \right) + i \sinh(1)\cos\left(\frac{3\pi}{2} t - \frac{\pi}{2} \right) \right\}\Big|_0^1$$

$$= \cosh(1) - i \sinh(1).$$ □

Example 7   Estimate

$$\left| \int_\gamma \frac{1}{z^2 + 4} \, dz \right|,$$

where γ is the semicircle $Re^{i\theta}$, $-\pi \leqslant \theta \leqslant 0$, $R > 2$.

Solution   On the semicircle, we have

$$\left| \frac{1}{z^2 + 4} \right| \leqslant \frac{1}{R^2 - 4},$$

since $|z^2 + 4| \geq |z|^2 - 4$ by the triangle inequality (see Section 2). The length of the semicircle is πR so that the absolute value of the integral cannot exceed $\pi R/(R^2 - 4)$.   □

Example 8   Estimate $|\int_\gamma e^{-z} \, dz|$, where γ is the vertical line segment from $-i + 1$ to $i + 1$.

Solution   On γ we have $|e^{-z}| = e^{-x} = e^{-1}$. The length of γ is 2 so the integral cannot exceed $2/e$ in absolute value.   □

Example 9   Let u be a continuous function in the disc $|z - z_0| < r$ and let γ_ε be the circle $|z - z_0| = \varepsilon$. Show that

$$\lim_{\varepsilon \to 0} \frac{1}{2\pi i} \int_{\gamma_\varepsilon} \frac{u(z)}{z - z_0} \, dz = u(z_0).$$

Solution   We parametrize γ_ε by $\gamma(t) = z_0 + \varepsilon e^{it}, 0 \leq t \leq 2\pi$; then

$$\frac{1}{2\pi i} \int_{\gamma_\varepsilon} \frac{u(z)}{z - z_0} dz = \frac{1}{2\pi i} \int_0^{2\pi} \frac{u(z_0 + \varepsilon e^{it})}{\varepsilon e^{it}} i\varepsilon e^{it} \, dt$$

$$= \frac{1}{2\pi} \int_0^{2\pi} u(z_0 + \varepsilon e^{it}) dt.$$

Thus,

$$\left| \frac{1}{2\pi i} \int_{\gamma_\varepsilon} \frac{u(z)}{z - z_0} dz - u(z_0) \right| = \left| \frac{1}{2\pi} \int_0^{2\pi} \{u(z_0 + \varepsilon e^{it}) - u(z_0)\} dt \right|$$

$$\leq \max_{0 \leq t \leq 2\pi} \{|u(z_0 + \varepsilon e^{it}) - u(z_0)|\}.$$

This last quantity goes to zero as $\varepsilon \to 0$ exactly because u is continuous at z_0 (see Exercise 26 in Section 4).   □

We digress briefly at this point to state a differentiation formula that will be needed. Its verification is simple and is relegated to the exercises. If $z(t)$ and $w(t)$ are two complex-valued functions on an interval $[a, b]$, both of which are differentiable, then so is their product and

$$(zw)'(t) = z'(t)w(t) + z(t)w'(t). \tag{4}$$

Repeated applications of (4), or mathematical induction, yield

$$(z^m)'(t) = m(z(t))^{m-1}z'(t), \quad m = 1, 2, \ldots \tag{5}$$

Example 10   Let γ be a piecewise smooth closed curve. Show that

$$\int_\gamma z^m dz = 0, \quad m = 0, 1, 2, \ldots \tag{6}$$

Solution Let $[t_j, t_{j+1}]$ be a segment in $[a, b]$ on which $\gamma'(t)$ is continuous. From (5) we have

$$\frac{d}{dt} (\gamma(t))^{m+1} = (m + 1)\gamma^m(t)\gamma'(t).$$

This gives us

$$\int_{t_j}^{t_{j+1}} \gamma^m(t)\gamma'(t)dt = \frac{1}{m + 1} \{\gamma^{m+1}(t_{j+1}) - \gamma^{m+1}(t_j)\}$$

Hence,

$$\int_\gamma z^m dz = \sum_{j=0}^{n-1} \int_{t_j}^{t_{j+1}} \gamma^m(t)\gamma'(t)dt$$

$$= \sum_{j=0}^{n-1} \frac{1}{m + 1} \{\gamma^{m+1}(t_{j+1}) - \gamma^{m+1}(t_j)\}$$

$$= \frac{1}{m + 1} \{\gamma^{m+1}(b) - \gamma^{m+1}(a)\} = 0$$

because $\gamma(b) = \gamma(a)$ since γ is a closed curve. The result in (6) will be of critical importance in Chapter 2 when we prove Cauchy's Theorem. □

Let p and q be distinct points in the plane and suppose that γ_1 and γ_2 are two piecewise smooth curves from p to q. Then $\gamma = \gamma_1 - \gamma_2$ is a closed piecewise smooth curve so that for $m = 0, 1, 2, \ldots$, we have

$$0 = \int_\gamma z^m \, dz = \int_{\gamma_1} z^m \, dz - \int_{\gamma_2} z^m \, dz.$$

That is, the value of the integral of z^m along a curve joining p to q does not depend on the curve but only on p, q, and m. Indeed, a glance at the computations in Example 10 shows that

$$\int_\gamma z^m \, dz = \int_a^b \gamma^m(t)\gamma'(t)dt = \frac{1}{m + 1} \gamma^{m+1}(t) \Big|_{t=a}^{t=b}$$

$$= \frac{1}{m + 1} [\gamma^{m+1}(b) - \gamma^{m+1}(a)]$$

$$= \frac{1}{m + 1} [q^{m+1} - p^{m+1}]$$

for any curve γ joining p to q.

Green's* Theorem

The most important result on line integrals is Green's Theorem, whose statement will come after several more definitions. Green's Theorem is formulated

---

* George Green, 1793–1841.

for a domain Ω whose boundary Γ consists of a finite number of disjoint, piecewise smooth simple closed curves $\gamma_1, \ldots, \gamma_n$ (see Fig. 1.30). We orient the boundary Γ of Ω **positively** by requiring that Ω remain on the left as we walk along Γ. Thus, the "outer" piece of the boundary of Ω is oriented counterclockwise and *each* "inner" piece of Γ, if there are any, is oriented clockwise (see Fig. 1.31). For example, if Ω is $\{z : 1 < |z| < 2\}$, then the boundary of Ω is positively oriented if the circle $\{z : |z| = 1\}$ is traversed clockwise and the circle $\{z : |z| = 2\}$ counterclockwise.

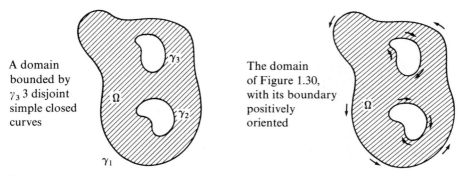

A domain bounded by γ_3 3 disjoint simple closed curves

The domain of Figure 1.30, with its boundary positively oriented

Figure 1.30   **Figure 1.31**

Henceforth *in this section* Ω is a domain whose boundary Γ consists of a finite number of disjoint, piecewise smooth simple closed curves $\gamma_1, \ldots, \gamma_n$ and Γ is positively oriented. If f is a continuous complex-valued function on Γ, then we define the line integral of f over Γ by

$$\int_\Gamma f(z)dz = \sum_{j=1}^n \int_{\gamma_j} f(z)dz,$$

where, of course, we already know how to compute each $\int_{\gamma_j} f(z)dz$ since each γ_j has a specific orientation.

Green's Theorem relates the line integral of a function f over Γ to the integral of a certain related function over Ω and, in order to properly formulate it, we assume that there is some open set D that contains both Ω and Γ and, on D, f has continuous partial derivatives with respect to both x and y. That is, if $f = p + iq$, then

$$\frac{\partial f}{\partial x} = \frac{\partial p}{\partial x} + i\frac{\partial q}{\partial x}, \quad \frac{\partial f}{\partial y} = \frac{\partial p}{\partial y} + i\frac{\partial q}{\partial y},$$

where all of $\partial p/\partial x$, $\partial q/\partial x$, $\partial p/\partial y$ and $\partial q/\partial y$ are continuous on D.

With all this background, then, we can state the theorem.

Theorem 1: Green's Theorem:

$$\int_\Gamma f(z)dz = i\iint_\Omega \left\{\frac{\partial f}{\partial x} + i\frac{\partial f}{\partial y}\right\}dx\,dy \qquad (7)$$

The reader who has studied Green's Theorem in calculus will probably *not* recognize this formulation, although it is one of the easiest for us to work with. Let us recast Green's Theorem in what may be more familiar terms. If u and v are two real-valued functions on an open set D containing Ω and Γ and if u and v both have continuous partial derivatives with respect to x and y on D and if on Γ we set $dx = (\operatorname{Re} \gamma'(t))dt$ and $dy = (\operatorname{Im} \gamma'(t))dt$, then Green's Theorem can also be stated in this way:

$$\int_\Gamma \{u\,dx + v\,dy\} = \iint_\Omega \left(\frac{\partial v}{\partial x} - \frac{\partial u}{\partial y}\right)dx\,dy. \tag{8}$$

The proof of equivalence of these two versions of Green's Theorem is left to the exercises.

The first example below is a verification of Green's Theorem for a triangle and some related figures. The following three examples derive some consequences of Green's Theorem.

Example 11   Verify Green's Theorem for a triangle.

Solution   We shall suppose first that the triangle Γ has one horizontal side (see Fig. 1.32).

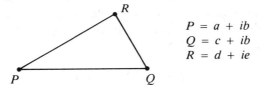

$$P = a + ib$$
$$Q = c + ib$$
$$R = d + ie$$

Figure 1.32

Let v be a function that has continuous partial derivatives on some open set containing the triangle Γ and its inside Ω. To compute $\int_\Gamma v\,dy$ we parametrize the edges of Γ in the following way.

On PQ: $x = t,\, y = b;\quad a \leqslant t \leqslant c$

On QR: $y = t,\, x = c + (d - c)\dfrac{t - b}{e - b};\quad b \leqslant t \leqslant e$

On RP: $y = t,\, x = d + (a - d)\dfrac{t - e}{b - e};\quad e \geqslant t \geqslant b$

(We have assumed that $b < e$; if $e < b$, then just reverse the range of t and the order of the sides.) Hence,

$$dy = \begin{cases} 0 & \text{on}\quad PQ \\ dt & \text{on}\quad QR \\ dt & \text{on}\quad RP \end{cases}$$

This gives

$$\int_\Gamma v \, dy = \int_{PQ} v \, dy + \int_{QR} v \, dy + \int_{RP} v \, dy$$

$$= 0 + \int_b^e v(At + B, t)dt + \int_e^b v(Ct + D, t)dt,$$

where for simplicity we have written

$$A = \frac{d - c}{e - b}, \quad B = c - b\frac{d - c}{e - b}, \quad C = \frac{a - d}{b - e}, \quad D = d - e\frac{a - d}{b - e}.$$

On the other hand the area integral of $\partial v/\partial x$ over the inside Ω of the triangle is computed in this way. Ω is described by

$$\Omega = \{(x, y) : b \leqslant y \leqslant e, Cy + D \leqslant x \leqslant Ay + B\}.$$

Consequently,

$$\iint_\Omega \frac{\partial v}{\partial x} \, dx \, dy = \int_b^e \left\{ \int_{Ct + D}^{At + B} \frac{\partial v}{\partial x} \, dx \right\} dt$$

$$= \int_b^e \{v(At + B, t) - v(Ct + D, t)\} dt$$

$$= \int_\Gamma v \, dy.$$

Next, any triangle Γ can be divided in two by a single horizontal line segment from one of its vertices to the opposite side so that each of the resulting triangles has one horizontal edge (Figs. 1.33a and 1.33b). Let Γ_1 and Γ_2 be the two resulting triangles, each positively oriented, and let Ω_1 and Ω_1 be their respective insides. Then

$$\int_{\Gamma_j} v \, dy = \iint_{\Omega_j} \frac{\partial v}{\partial x} \, dx \, dy, \quad j = 1, 2$$

from the computations above. The integral over the segment PS appears in

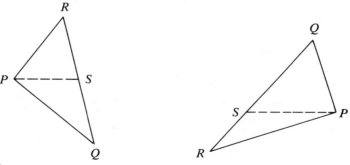

Figure 1.33

both $\int_{\Gamma_1} v \, dy$ and $\int_{\Gamma_2} v \, dy$ but in both cases it is zero since PS is horizontal. Hence,

$$\int_{\Gamma} v \, dy = \int_{\Gamma_1} v \, dy + \int_{\Gamma_2} v \, dy$$

$$= \iint_{\Omega_1} \frac{\partial v}{\partial x} \, dx \, dy + \iint_{\Omega_2} \frac{\partial v}{\partial x} \, dx \, dy$$

$$= \iint_{\Omega} \frac{\partial v}{\partial x} \, dx \, dy.$$

This establishes the formula

$$\iint_{\Omega} \frac{\partial v}{\partial x} \, dx \, dy = \int_{\Gamma} v \, dy$$

for any triangle Γ and its inside Ω. A similar argument, using triangles with a vertical edge, establishes the formula

$$\iint_{\Omega} \frac{\partial u}{\partial y} \, dx \, dy = -\int_{\Gamma} u \, dx.$$

Thus, Green's Theorem is verified for a triangle.   □

Some further comments: Once Green's Theorem is proved for all triangles, it also follows for many other regions. For instance, let Ω be a bounded convex domain whose boundary consists of straight line segments (Fig. 1.34a). Select a point P in Ω and draw the line segments from P to each vertex of Ω (Fig. 1.34b). This breaks Ω up into non-overlapping solid triangles $\Omega_1, \ldots, \Omega_r$ with boundaries $\Gamma_1, \ldots, \Gamma_r$, respectively. We then apply Green's Theorem to each of these triangles, obtaining

$$\iint_{\Omega_j} \left(\frac{\partial v}{\partial x} - \frac{\partial u}{\partial y} \right) dx \, dy = \int_{\Gamma_j} (v \, dy + u \, dx), \quad j = 1, \ldots, r.$$

Each line segment joining P to a vertex of Ω is integrated over twice, once from P to the vertex and once from the vertex to P. These line integrals then cancel each other and we find that

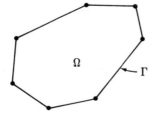

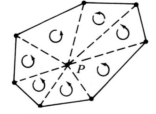

Figure 1.34

$$\int_{\Gamma} (v\ dy + u\ dx) = \sum_{j=1}^{r} \int_{\Gamma_j} (v\ dy + u\ dx)$$

$$= \sum_{j=1}^{r} \iint_{\Omega_j} \left(\frac{\partial v}{\partial x} - \frac{\partial u}{\partial y} \right) dx\ dy$$

$$= \iint_{\Omega} \left(\frac{\partial v}{\partial x} - \frac{\partial u}{\partial y} \right) dx\ dy.$$

Thus, Green's Theorem holds for Ω. A similar sort of "triangulation" argument can be applied to show that Green's Theorem is valid for a bounded domain Ω whose boundary Γ is composed of straight line segments (Figs. 1.35a and 1.35b show representative types).

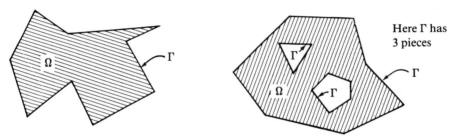

Here Γ has
3 pieces

Figure 1.35

Example 12   Suppose that γ is a piecewise smooth positively oriented simple closed curve. The values of the integrals

$$\frac{1}{2\pi i} \int_{\gamma} \frac{dz}{z - p}, \quad p \text{ not in } \gamma,$$

are crucial factors in many of the theorems of complex variables. We shall now use Green's Theorem to show that

$$\frac{1}{2\pi i} \int_{\gamma} \frac{dz}{z - p} = \begin{cases} 0 \text{ if } p \text{ is outside } \gamma, \\ 1 \text{ if } p \text{ is inside } \gamma. \end{cases}$$

Let Ω be the domain inside γ. Suppose first that p is outside γ. Let $f(z) = (z - p)^{-1}$; then f has continuous partial derivatives everywhere except at p and, indeed, it is elementary to find that

$$\frac{\partial f}{\partial x} = \frac{-1}{(z - p)^2}; \quad \frac{\partial f}{\partial y} = \frac{-i}{(z - p)^2}$$

Hence, $(\partial f/\partial x) + i\ (\partial f/\partial y) = 0$ in Ω and so

$$\int_{\gamma} f(z)dz = i \iint_{\Omega} \left\{ \frac{\partial f}{\partial x} + i\ \frac{\partial f}{\partial y} \right\} dx\ dy = 0.$$

Suppose now that p is inside γ; then, of course, the function $f(z) = (z - p)^{-1}$ does *not* have continuous partials on all of Ω. However, let γ_1 be a circle centered at p of radius ε, where ε is so small that the disc of radius 2ε centered at p lies within Ω. Take Ω_1 to be the region inside γ but outside γ_1 and orient the boundary of Ω_1 positively, so that γ is already correctly oriented and γ_1 is oriented clockwise. Then, just as above, we have

$$\frac{\partial f}{\partial x} + i\frac{\partial f}{\partial y} = 0 \text{ on } \Omega_1$$

and so

$$\int_\gamma \frac{dz}{z - p} + \int_{\gamma_1} \frac{dz}{z - p} = 0.$$

Reversing the orientation of γ_1 so that it is now traversed counterclockwise, we find that

$$\int_\gamma \frac{dz}{z - p} = \int_{-\gamma_1} \frac{dz}{z - p}.$$

To evaluate the integral over $-\gamma_1$, set $z = p + \varepsilon e^{it}$, $0 \leqslant t \leqslant 2\pi$; then $dz = i\varepsilon e^{it} dt$ so that

$$\int_{-\gamma_1} \frac{dz}{z - p} = \int_0^{2\pi} \frac{i\varepsilon e^{it}\, dt}{\varepsilon e^{it}} = 2\pi i. \qquad \square$$

Example 13   In this example we will derive from Green's Theorem an important formula with numerous applications; the formula is called, not surprisingly, **Green's formula**.

We begin by defining the normal derivative of a function h on the boundary of Ω. Recall from calculus that if $\lambda = \cos\theta + i\sin\theta$ is a complex number of modulus one then the **directional derivative** of a function h in the direction λ at a point z_0 is

$$\lim_{t\to 0} \frac{h(z_0 + t\lambda) - h(z_0)}{t}.$$

This is, in fact, equal to the dot product of the gradient of h with the vector $(\cos\theta, \sin\theta)$ and so has the value

$$\frac{\partial h}{\partial x}\cos\theta + \frac{\partial h}{\partial y}\sin\theta.$$

Let $z_0 = z(t_0)$ be a point of Γ, where $\gamma(t) = x(t) + iy(t)$ parameterizes a portion of Γ containing z_0. A unit vector pointing *out* of Ω and normal to Γ is

$$\lambda = (y'(t) - ix'(t))([x'(t)]^2 + [y'(t)]^2)^{-1/2}.$$

(We shall always assume that $[x'(t)]^2 + [y'(t)]^2$ is strictly positive.) Thus, the directional derivative of the function h in the direction normal to the boundary of Γ, with the normal vector pointing out of Γ, is

$$\frac{\partial h}{\partial n} = \left(\frac{\partial h}{\partial x} y'(t) - \frac{\partial h}{\partial y} x'(t)\right)([x'(t)]^2 + [y'(t)]^2)^{-1/2}.$$

Let f and g be two real-valued functions that have continuous partial derivatives on some open set containing both Ω and Γ. In the second formulation
of Green's Theorem, formula (8), take

$$v = \frac{\partial f}{\partial x} g - \frac{\partial g}{\partial x} f \quad \text{and} \quad u = -\frac{\partial f}{\partial y} g + \frac{\partial g}{\partial y} f.$$

We obtain

$$\int_\Gamma \left[g\left(\frac{\partial f}{\partial x} dy - \frac{\partial f}{\partial y} dx\right) - f\left(\frac{\partial g}{\partial x} dy - \frac{\partial g}{\partial y} dx\right)\right] = \iint_\Omega [(\Delta f)g - (\Delta g)f]dx\, dy,$$

where

$$\Delta f = \frac{\partial^2 f}{\partial x^2} + \frac{\partial^2 f}{\partial y^2}$$

is the **Laplacian*** of f and likewise Δg is the Laplacian of g. If, in the line
integral, we multiply and divide by the expression $[(x'(t))^2 + (y'(t))^2]^{-1/2}$ and
recall from calculus that the element of **arc-length** ds along Γ is given by

$$ds = [(x'(t))^2 + (y'(t))^2]^{1/2}\, dt,$$

we then obtain the formula we've been aiming for:

$$\int_\Gamma \left(g\frac{\partial f}{\partial n} - f\frac{\partial g}{\partial n}\right)ds = \iint_\Omega (g\Delta f - f\Delta g)\, dx\, dy. \qquad \square \quad (9)$$

Example 14   In a similar fashion, if we take

$$u = f\frac{\partial f}{\partial x} \quad \text{and} \quad v = -f\frac{\partial f}{\partial y}$$

in (8) we obtain the formula

$$\int_\Gamma f\frac{\partial f}{\partial n} ds = \iint_\Omega \left\{f\Delta f + \left(\frac{\partial f}{\partial x}\right)^2 + \left(\frac{\partial f}{\partial y}\right)^2\right\}dx\, dy. \qquad (10)$$

The formulas (9) and (10) will be of critical importance when we discuss
harmonic functions in Chapter 4. One application will suffice for now.

> **Definition**   A function $u(z) = u(x, y)$ with continuous first and second
> partial derivatives with respect to both x and y is **harmonic** on an
> open set D if

---

* Named after Pierre-Simon Laplace, 1749–1827.

$$\Delta u = \frac{\partial^2 u}{\partial x^2} + \frac{\partial^2 u}{\partial y^2} = 0 \text{ on } D.$$

Theorem 2: Suppose that u is harmonic on an open set D and D contains a domain Ω and the boundary Γ of Ω. Assume that Γ consists of a finite number of disjoint, piecewise smooth simple closed curves. If $u = 0$ on Γ, then $u = 0$ in Ω as well.

Proof. The proof comes directly from (10) with, of course, u in place of f. The integral $\int_\Gamma u \, \partial u/\partial n \, ds$ is zero since $u = 0$ on Γ. On Ω we have $\Delta u = 0$; hence we find that

$$0 = \iint_\Omega \left\{ \left(\frac{\partial u}{\partial x} \right)^2 + \left(\frac{\partial u}{\partial y} \right)^2 \right\} dx \, dy.$$

But then $\partial u/\partial x = \partial u/\partial y = 0$ on Ω and so $u(x, y)$ is constant on Ω. Since $u = 0$ on Γ it follows that $u = 0$ on Ω. □

Exercises for Section 1.6

Compute the following line integrals

1. $\int_\gamma z \, dz$, where γ is the semicircle from i to $-i$, which passes through -1.

2. $\int_\gamma e^z \, dz$, where γ is the line segment from 0 to z_0.

3. $\int_\gamma |z|^2 \, dz$, where γ is the line segment from 2 to $3 + i$.

4. $\int_\gamma 1/(z + 4)dz$, where γ is the circle of radius 1 centered at -4, oriented counterclockwise.

5. $\int_\gamma (\text{Re } z)dz$, where γ is the line segment from 1 to i.

6. $\int_\gamma (z^2 + 3z + 4)dz$, where γ is the circle $|z| = 2$ oriented counterclockwise.

7. Let γ_1 be the semicircle from 1 to -1 through i and γ_2 the semicircle from 1 to -1 through $-i$ (Fig. 1.36).

    Compute $\int_{\gamma_1} z \, dz$ and $\int_{\gamma_2} z \, dz$. Can you account for the fact that they are equal?
    Now compute $\int_{\gamma_1} \bar{z} \, dz$ and $\int_{\gamma_2} \bar{z} \, dz$. Can you account for the fact that they are *not* equal?

8. Explicitly verify the conclusion of Example 10 by computing $\int_\gamma z \, dz$ and $\int_\gamma z^2 \, dz$ when γ is the square with vertices at $\pm 1 \pm i$.

9. (a) Show that $\dfrac{1}{2\pi} \displaystyle\int_0^{2\pi} e^{ik\theta} \, d\theta = \begin{cases} 1, & k = 0 \\ 0, & k \neq 0 \end{cases}$.

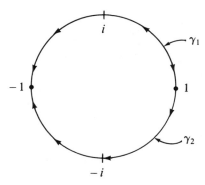

Figure 1.36

(b) Explicitly verify the conclusion of Example 11 for γ, the circle of radius 1, centered at 0. (**Hint**:

$$\frac{1}{z-p} = \sum_{k=0}^{\infty} \frac{p^k}{z^{k+1}} \text{ if } 0 \leqslant |p| < 1 \text{ and } |z| = 1$$

$$\frac{1}{z-p} = -\sum_{k=0}^{\infty} \frac{z^k}{p^{k+1}} \text{ if } |z| = 1 \text{ and } 1 < |p| < \infty.$$

Interchange summation and integration; then use (a).)

10.   Let $f = u + iv$ be a continuous function and $\gamma(t) = x(t) + iy(t)$ be a piecewise smooth curve. Show that

$$\text{Re}\left\{\int_{\gamma} f(z)dz\right\} = \int_{\gamma} (u\ dx - v\ dy)$$

and

$$\text{Im}\left\{\int_{\gamma} f(z)dz\right\} = \int_{\gamma} (v\ dx + u\ dy).$$

Here $dx = x'(t)dt$, $dy = y'(t)dt$.

11.   Use Exercise 10 to show the equivalence of the two formulations of Green's Theorem: (7) and (8).

12.   Use the result of (the extension of) Example 10 to compute the following integrals: (a) $\int_{\gamma}(z^3 - 6z^2 + 4)dz$, where γ is any curve joining $-1 + i$ to 1.
(b) $\int_{\gamma}(z^4 + z^2)dz$, where γ is any curve joining $-i$ to $2 + i$.

13.   Use Greens Theorem to derive **Green's Identity**:

$$\iint_{\Omega}\left\{\frac{\partial u}{\partial x}\frac{\partial v}{\partial x} + \frac{\partial u}{\partial y}\frac{\partial v}{\partial y}\right\}dx\ dy = \int_{\Gamma} v\frac{\partial u}{\partial n}\ ds - \iint_{\Omega} v\ \Delta u\ dx\ dy.$$

14.   Use Exercise 13 to derive the identity

$$\int_\Gamma \left(v \frac{\partial u}{\partial n} - u \frac{\partial v}{\partial n} \right) ds = \iint_\Omega (v \, \Delta u - u \, \Delta v) dx \, dy.$$

15.   Let u be a continuous function on the complex plane, which is bounded: $|u(z)| \leqslant C$ for all z. Let γ_R be the circle $|z| = R$. Show that

$$\lim_{R \to \infty} \int_{\gamma_R} \frac{u(z)}{(z - z_0)^2} \, dz = 0$$

for each z_0. (**Hint:** Use (3).)

16.   Let z_0 be outside a piecewise smooth simple closed curve γ. Continue Example 12 by showing that

$$\int_\gamma \frac{dz}{(z - z_0)^m} = 0, \quad m = 2, 3, 4, \dots.$$

Exact Differentials*

Definition   An expression $P(x, y)dx + Q(x, y)dy$ is an **exact differential** if there is a function $g(x, y)$ with

$$\frac{\partial g}{\partial x} = P \quad \text{and} \quad \frac{\partial g}{\partial y} = Q.$$

In Exercises 17 to 20, D is a disc and P and Q are functions on D with continuous partial derivatives with respect to x and y.

17.   If $P \, dx + Q \, dy$ is an exact differential, show that

$$\int_\gamma \{P \, dx + Q \, dy\} = 0 \qquad\qquad \textbf{(11)}$$

for each closed curve γ in D.

18.   Use Green's Theorem to show that

$$\frac{\partial Q}{\partial x} = \frac{\partial P}{\partial y} \qquad\qquad \textbf{(12)}$$

throughout D if and only if (11) holds.

19.   Suppose that (11) holds for each closed curve γ in D. Show that $P \, dx + Q \, dy$ is an exact differential. (**Hint:** Let $x_0 + iy_0$ be any point in D and set

$$g(x, y) = \int_{x_0}^x P(t, y_0)dt + \int_{y_0}^y Q(x, s)ds.$$

Use (12) to show that g is the desired function)

20.   Let f be a continuous complex-valued function on D. Show there is a complex-valued function F on D with

$$\frac{\partial F}{\partial x} = f, \qquad \frac{\partial F}{\partial y} = if \qquad\qquad (13)$$

if and only if

$$\int_\gamma f(z)dz = 0 \qquad\qquad (14)$$

for each closed curve γ in D. The functions f satisfying (14) are exactly the functions we will study in most of the remainder of the book.

21.   We found in Example 10 that $\int_\gamma z^m \, dz = 0$ for each closed curve γ, $m = 0, 1, 2, \ldots$. What is the function F in Exercise 20 that corresponds to $f(z) = z^m$?

Further reading

The paperback book by Konrad Knopp, *Elements of the Theory of Functions* (New York: Dover, 1952) is an excellent source for virtually all the material in this chapter. The sections there on linear fractional transformations can be skipped or read, depending on the reader's taste; these functions are covered in Chapter 3 of this book.

2

Basic properties of analytic functions

2.1 Analytic and harmonic functions; the Cauchy–Riemann equations

Analytic functions and their close relatives, harmonic functions, are the stuff of which the study of complex variables is built. This section introduces both of these types of functions and in this section and subsequent sections many of their significant properties are developed.

A function f defined for z in a domain D is **differentiable** at a point z_0 in D if

$$\lim_{z \to z_0} \frac{f(z) - f(z_0)}{z - z_0} = \lim_{h \to 0} \frac{f(z_0 + h) - f(z_0)}{h} \tag{1}$$

exists; the limit, if it exists, is denoted by $f'(z_0)$. If f is differentiable at each point of the domain D, then f is called **analytic** in D. A function analytic on the whole complex plane is called **entire**.

It is worth stressing here that the limit in (1) is required to exist *no matter how z approaches z_0*; equivalently, no matter how h approaches 0. That is, although $|h|$ must approach zero, the argument of h can change arbitrarily. This latitude is at the root of the difference between differentiable functions of a real variable and differentiable functions of a complex variable.

Example 1 $f(z) = z^n$, $n = 1, 2, \ldots$, is entire and $f'(z) = nz^{n-1}$. For

$$(z + h)^n - z^n = nz^{n-1}h + \frac{n(n-1)}{2} z^{n-2}h^2 + \cdots + h^n$$

$$= h\left(nz^{n-1} + \frac{n(n-1)}{2} z^{n-2}h + \cdots + h^{n-1} \right)$$

and thus

$$\frac{(z + h)^n - z^n}{h} \to nz^{n-1} \quad \text{as} \quad h \to 0.$$

Example 2 The sum and product of functions analytic on a common domain are again analytic and their derivatives follow the rules already familiar from calculus:

$$(f + g)' = f' + g' \tag{2}$$

and

$$(fg)' = f'g + fg'. \tag{3}$$

Example 3 The quotient of two analytic functions is differentiable at all points z_0 at which the denominator does not vanish and

$$\left(\frac{f}{g}\right)'(z_0) = \frac{g(z_0)f'(z_0) - f(z_0)g'(z_0)}{(g(z_0))^2}, \quad g(z_0) \neq 0. \tag{4}$$

Example 4 If f and g are differentiable and if the range of f lies within the domain of g, then $g(f(z))$ is differentiable with

$$[g(f(z))]' = g'(f(z))f'(z). \tag{5}$$

(The proofs of formulas 2 through 5 are straightforward and are left to the exercises at the end of this section.)

From Examples (1) to (4) we can immediately give two general types of analytic functions.

Example 5 Any polynomial $p(z) = a_0 + a_1 z + \cdots + a_n z^n$ is an entire function.

Example 6 A rational function $r = p/q$, where p and q are polynomials, is analytic on any domain containing no zero of q.

Example 7 The exponential function $f(z) = e^z$ is an entire function; this is not difficult to show directly. We know that

$$e^{z+h} - e^z = e^z(e^h - 1)$$

by the properties of the exponential function derived in Section 5 of Chapter 1. Furthermore, with $h = \sigma + i\tau$, we have

$$e^h - 1 - h = \{e^\sigma \cos \tau - 1 - \sigma\} + i\{e^\sigma \sin \tau - \tau\}$$
$$= \{e^\sigma(\cos \tau - 1) + e^\sigma - 1 - \sigma\} + i\{e^\sigma(\sin \tau - \tau) + \tau(e^\sigma - 1)\}.$$

Hence,

$$\left| \frac{e^h - 1}{h} - 1 \right| = \left| \frac{e^h - 1 - h}{h} \right|$$

$$\leqslant e^\sigma \left| \frac{1 - \cos \tau}{\tau} \right| + \left| \frac{e^\sigma - 1 - \sigma}{\sigma} \right| + e^\sigma \left| \frac{\sin \tau - \tau}{\tau} \right| + |e^\sigma - 1|.$$

To obtain this inequality we used the triangle inequality and the simple facts that

$$\frac{1}{|h|} \leqslant \frac{1}{|\tau|} \qquad \text{and} \qquad \frac{1}{|h|} \leqslant \frac{1}{|\sigma|}.$$

However, each of the four quantities within absolute value signs approaches zero as σ and τ independently approach zero (use l'Hôpital's Rule on each, if you like) so we find that

$$\lim_{h \to 0} \frac{e^h - 1}{h} = 1.$$

This finally gives

$$\lim_{h \to 0} \frac{e^{z+h} - e^z}{h} = e^z \lim_{h \to 0} \frac{e^h - 1}{h} = e^z.$$

Consequently, e^z is differentiable at all points z and

$$(e^z)' = e^z. \tag{6}$$

Example 8   It follows from (5) and (6) that if $f(z)$ is analytic on a domain D then

$$F(z) = \exp(f(z))$$

is also analytic on D and

$$F'(z) = f'(z) \exp(f(z)), \quad z \in D$$

Example 9   Let $f(z)$ be differentiable at a point z_0 in a domain D. Show that f is continuous at z_0.

Solution   For z near z_0, we have

$$|f(z) - f(z_0)| = \left| \frac{f(z) - f(z_0)}{z - z_0} \right| |z - z_0| \to |f'(z_0)| \cdot 0 = 0$$

as $z \to z_0$. Hence, $\lim_{z \to z_0} f(z) = f(z_0)$. In particular, both the real and the imaginary parts of f are continuous on any domain on which f is analytic.   □

The Cauchy*–Riemann† equations

The essential feature of an analytic function is the fact that the limit in (1) must exist *no matter how z approaches* z_0. This leads to the pair of famous partial differential equations, which connect the real and imaginary parts of an analytic function.

> **Theorem 1: Cauchy–Riemann equations.** Suppose $f = u + iv$ is analytic on a domain D. Then throughout D we have
>
> $$\frac{\partial u}{\partial x} = \frac{\partial v}{\partial y} \quad \text{and} \quad \frac{\partial u}{\partial y} = -\frac{\partial v}{\partial x}. \tag{7}$$

Proof. Let us employ the definition of the derivative of f at z_0 in two different ways. First take h to be real. Then

$$f'(z_0) = \lim_{h \to 0} \left[\frac{u(x_0 + h, y_0) - u(x_0, y_0)}{h} + i \frac{v(x_0 + h, y_0) - v(x_0, y_0)}{h} \right]$$

$$= \frac{\partial u}{\partial x}(x_0, y_0) + i \frac{\partial v}{\partial x}(x_0, y_0).$$

Next, take $h = ik$, where k is real. Then

$$f'(z_0) = \lim_{k \to 0} \left[\frac{u(x_0, y_0 + k) - u(x_0, y_0)}{ik} + i \frac{v(x_0, y_0 + k) - v(x_0, y_0)}{ik} \right]$$

$$= \frac{1}{i} \frac{\partial u}{\partial y}(x_0, y_0) + \frac{\partial v}{\partial y}(x_0, y_0).$$

All that remains is to equate real and imaginary parts of these two expressions for $f'(z_0)$. □

The Cauchy–Riemann equations imply that an analytic function $f = u + iv$ is actually determined by its real part u (or, equivalently, by its imaginary part v) apart from an additive constant. For if we know u we can (presumably) find v with

$$\frac{\partial v}{\partial x} = -\frac{\partial u}{\partial y} \quad \text{and} \quad \frac{\partial v}{\partial y} = \frac{\partial u}{\partial x}.$$

For example, if $u(x, y) = x^3 - 3xy^2$, then v is to be found from

---

* Augustine Louis Cauchy, 1789–1857.
† Bernhard Riemann, 1826–1866.

$$\frac{\partial v}{\partial x} = 6xy \qquad \text{and} \qquad \frac{\partial v}{\partial y} = 3x^2 - 3y^2.$$

We integrate $6xy$ with respect to x and find that $v(x, y) = 3x^2y + p(y)$, where p is a function of y alone. We then differentiate this expression for v with respect to y and compare the result with the known equality $\partial v/\partial y = 3x^2 - 3y^2$. This yields $3x^2 - 3y^2 = 3x^2 + p'(y)$. Hence, $p(y) = -y^3 + c$, where c is a constant. Consequently, $v(x, y) = 3x^2y - y^3 + c$. A word of warning is merited here: not every function $u(x, y)$ is the real part of an analytic function. Indeed, if $f = u + iv$ is analytic on a domain D and if we assume temporarily (and unnecessarily as it turns out) that both u and v have continuous partial derivatives of first and second order, then we find that

$$\frac{\partial^2 u}{\partial x^2} + \frac{\partial^2 u}{\partial y^2} = \frac{\partial}{\partial x}\left(\frac{\partial u}{\partial x}\right) + \frac{\partial}{\partial y}\left(\frac{\partial u}{\partial y}\right)$$

$$= \frac{\partial}{\partial x}\left(\frac{\partial v}{\partial y}\right) + \frac{\partial}{\partial y}\left(-\frac{\partial v}{\partial x}\right)$$

$$= \frac{\partial^2 v}{\partial x\,\partial y} - \frac{\partial^2 v}{\partial y\,\partial x} = 0.$$

Hence, if $u = \operatorname{Re} f$, where f is analytic on D then u must necessarily satisfy **Laplace's equation**

$$\Delta u = \frac{\partial^2 u}{\partial x^2} + \frac{\partial^2 u}{\partial y^2} = 0.$$

Recall from Section 6 of Chapter 1 that a continuous function u with continuous first and second partial derivatives on D is **harmonic** on D if it satisfies Laplace's equation on D. Thus, the real part u of an analytic function f is harmonic. Quite similarly, the imaginary part v of f is also harmonic. Harmonic functions arise with uncanny frequency in physical problems and provide one of the most important applications of complex variables. Many such applications are discussed in Chapter 4.

We note further that a real-valued harmonic function u determines many other harmonic functions v by the Cauchy–Riemann equations:

$$\frac{\partial v}{\partial y} = \frac{\partial u}{\partial x} \qquad \text{and} \qquad \frac{\partial v}{\partial x} = -\frac{\partial u}{\partial y}.$$

Such a v is called a **harmonic conjugate** of u; any two harmonic conjugates v and v_1 of u differ by a constant since

$$\frac{\partial}{\partial y}(v - v_1) = \frac{\partial u}{\partial x} - \frac{\partial u}{\partial x} = 0$$

and

$$\frac{\partial}{\partial x}(v - v_1) = -\frac{\partial u}{\partial y} + \frac{\partial u}{\partial y} = 0.$$

Moreover, if v is a harmonic conjugate of u in a disc Ω then $f = u + iv$ is analytic on Ω. (The verification that f is indeed analytic depends on Theorem 3 of this section.) We will study harmonic functions further in Chapter 4.

There are other consequences of the Cauchy–Riemann equations. In each case of the following theorem we shall draw the conclusion that the function in question is a constant.

Theorem 2: Suppose that $f = u + iv$ is analytic on a domain D. If either u is constant on D or $u^2 + v^2$ is constant on D, then f is constant on D.

Proof. Suppose that u is constant on D. Then $\partial u/\partial x = \partial u/\partial y = 0$ and so by the Cauchy–Riemann equations we also have $\partial v/\partial x = \partial v/\partial y = 0$ on D. Since $\partial v/\partial x$ is the rate of change of v on a horizontal line it follows that v is constant on each horizontal line segment in D. Likewise, v is constant on each vertical line segment in D since $\partial v/\partial y = 0$ in D. However, each pair of points in D can be joined by a polygonal curve consisting entirely of such segments (see Exercise 16, Section 3, of Chapter 1). Hence, v is constant in D.

If $u^2 + v^2$ is constant in D, then

$$u\,\frac{\partial u}{\partial x} + v\,\frac{\partial v}{\partial x} = 0$$

and

$$u\,\frac{\partial u}{\partial y} + v\,\frac{\partial v}{\partial y} = 0.$$

The Cauchy–Riemann equations then yield the matrix equation

$$\begin{pmatrix} \dfrac{\partial u}{\partial x} & -\dfrac{\partial u}{\partial y} \\[2mm] \dfrac{\partial u}{\partial y} & \dfrac{\partial u}{\partial x} \end{pmatrix}\begin{pmatrix} u \\ v \end{pmatrix} = \begin{pmatrix} 0 \\ 0 \end{pmatrix}.$$

Thus, either $u(z) = v(z) = 0$ or the determinant of the 2×2 matrix vanishes, giving

$$\left(\frac{\partial u}{\partial x}\right)^2 + \left(\frac{\partial u}{\partial y}\right)^2 = 0 \text{ at } z. \tag{8}$$

If u is not identically zero, then $u(z_0) \neq 0$ for some z_0 and so $u(z) \neq 0$ for all z with $|z - z_0| < \delta$ for some $\delta > 0$ since u is continuous. Hence, (8) implies that $u(z) = u(z_0)$ for all z with $|z - z_0| < \delta$. A repetition of the foregoing argument shows that u is constant on any horizontal or vertical line segment in D and thus u is constant throughout D. Consequently, f is also constant in D □

One interpretation of this theorem is that an analytic function f is constant if its range lies in either a vertical line or a circle centered at the origin. We shall see later that there are, in fact, even more severe restrictions than this on the range of a nonconstant analytic function.

The Cauchy–Riemann equations, which are implied by analyticity, themselves imply analyticity with an extra hypothesis of continuity on the partial derivatives.

Theorem 3: Suppose that $f = u + iv$ and that all of u, v, $\partial u/\partial x'$, $\partial v/\partial x'$, $\partial u/\partial y'$, and $\partial v/\partial y'$, are continuous in a disc centered at the point z_0. If u and v satisfy the Cauchy–Riemann equations at z_0, then f is differentiable at z_0.

Proof. The proof is basically a computation. We write $h = \delta + iv$. Then

$$f(z_0 + h) - f(z_0) = u(x_0 + \delta, y_0 + v) - u(x_0, y_0)$$

$$+ i[v(x_0 + \delta, y_0 + v) - v(x_0, y_0)].$$

We examine the terms with u first. We have

$$u(x_0 + \delta, y_0 + v) - u(x_0, y_0)$$

$$= u(x_0 + \delta, y_0 + v) - u(x_0, y_0 + v) + u(x_0, y_0 + v) - u(x_0, y_0)$$

$$= \delta \frac{\partial u}{\partial x}(x_0 + \delta', y_0 + v) + v \frac{\partial u}{\partial y}(x_0, y_0 + v')$$

$$= \delta \frac{\partial u}{\partial x}(x_0, y_0) + v \frac{\partial u}{\partial y}(x_0, y_0) + \delta \varepsilon_1 + v\varepsilon_2$$

by two applications of the mean-value theorem for real-valued functions of a real variable and the continuity of $\partial u/\partial x$ and $\partial u/\partial y$; ε_1 and ε_2 depend on δ and v and approach zero as δ and v approach zero.

Likewise,

$$v(x_0 + \delta, y_0 + v) - v(x_0, y_0) = \delta \frac{\partial v}{\partial x}(x_0, y_0) + v \frac{\partial v}{\partial y}(x_0, y_0) + \delta\varepsilon_3 + v\varepsilon_4 .$$

The Cauchy–Riemann equations are $\partial u/\partial x = \partial v/\partial y$ and $\partial u/\partial y = -\partial v/\partial x$ and so, all in all, we have

$$f(z_0 + h) - f(z_0) = \frac{\partial u}{\partial x}(x_0, y_0)\{\delta + iv\} + i\frac{\partial v}{\partial x}(x_0, y_0)\{\delta + iv\}$$

$$+ \delta(\varepsilon_1 + i\varepsilon_3) + v(\varepsilon_2 + i\varepsilon_4).$$

Hence,

$$\frac{f(z_0 + h) - f(z_0)}{h} = \frac{\partial u}{\partial x}(x_0, y_0) + i\frac{\partial v}{\partial x}(x_0, y_0) + \frac{\delta}{h}(\varepsilon_1 + i\varepsilon_3) + \frac{v}{h}(\varepsilon_2 + i\varepsilon_4).$$

However, $|\delta/h| \leqslant 1$ and $|v/h| \leqslant 1$ so that

$$\frac{f(z_0 + h) - f(z_0)}{h} \to \frac{\partial u}{\partial x}(x_0, y_0) + i\frac{\partial v}{\partial x}(x_0, y_0), \quad \text{as} \quad h \to 0.$$

Thus, f is differentiable at z_0 and

$$f'(z_0) = \frac{\partial u}{\partial x}(x_0, y_0) + i\frac{\partial v}{\partial x}(x_0, y_0). \qquad \square$$

Example 10 Let us employ Theorem 3 to show that $f(z) = \log z$ is differentiable and to find the derivative of $\log z$. Let D be a domain on which there is a single-valued branch of $\log z$; for example, we can take D to be the complex plane with any ray from the origin to ∞ deleted. We have

$$\begin{aligned} \log z &= \ln|z| + i \arg z \\ &= \tfrac{1}{2}\ln(x^2 + y^2) + i\arctan(y/x), \end{aligned}$$

where we have specified that the values of $\arctan(y/x)$ lie in an interval of the form $(\theta_0, \theta_0 + 2\pi)$. Thus,

$$u(x, y) = \tfrac{1}{2}\ln(x^2 + y^2), \qquad v(x, y) = \arctan(y/x)$$

and so

$$\frac{\partial u}{\partial x} = \frac{x}{x^2 + y^2}, \qquad \frac{\partial u}{\partial y} = \frac{y}{x^2 + y^2}, \qquad \frac{\partial v}{\partial x} = \frac{-y}{x^2 + y^2}, \qquad \frac{\partial v}{\partial y} = \frac{x}{x^2 + y^2}.$$

All the partial derivatives are continuous on D and they clearly satisfy the Cauchy–Riemann equations at all points of D:

$$\frac{\partial u}{\partial x} = \frac{\partial v}{\partial y}, \qquad \frac{\partial u}{\partial y} = -\frac{\partial v}{\partial x}.$$

Consequently, Theorem 3 implies that $\log z$ is differentiable with

$$(\log z)' = \frac{\partial u}{\partial x} + i\frac{\partial v}{\partial x} = \frac{x}{x^2 + y^2} + i\frac{-y}{x^2 + y^2}$$

$$= \frac{\bar{z}}{|z|^2} = \frac{1}{z}.$$

Thus,

$$(\log z)' = 1/z. \tag{9}$$

The reader should note that $1/z$ is analytic on the whole complex plane except at $z = 0$ but the function $\log z$, whose derivative is $1/z$, is *not* analytic on the punctured plane; indeed, it is not analytic on any domain D that contains a simple closed curve that surrounds the origin. (The reader is asked to prove this in a problem later in this chapter.)

Example 11   Let α be a real number. Show that the function

$$f(z) = z^\alpha$$

can be defined to be analytic on any domain D in the plane that omits a ray from the origin to infinity.

Solution   The domain D omits a ray from 0 to infinity so that we may choose a branch of $\log z$ that is well-defined and analytic in D. For instance, if D omits the negative real axis and 0, then $\text{Log } z$ is analytic on D. With this branch of $\log z$ fixed, we define

$$z^\alpha = e^{\alpha \log z}$$

The results earlier in this section, particularly Example 8, show that z^α is analytic on D.   □

Exercises for Section 2.1

1.   Establish the following differentiation formulas
       (a) $(\sin z)' = \cos z$ (b) $(\cos z)' = -\sin z$
       (c) $(\sinh z)' = \cosh z$ (d) $(\cosh z)' = \sinh z$
       (e) $(\tan z)' = \sec^2 z$ (f) $(\arcsin z)' = (1 - z^2)^{-1/2}$
       (g) $(\arctan z)' = (1 + z^2)^{-1}$
       Refer to Section 5 of Chapter 1 for the definitions of these trigonometric functions.

Use the rules for differentiation, formulas (2) to (6), to find the derivative of each of the functions in Exercises 2 to 7.

2. $z^2 + 10z$ 3. $\exp(z^3 - z)$ 4. $[\cos(z^2)]^3$ 5. $(z^3 + 100)^{-4}$

6. $(\text{Log } z)^3$ on the plane minus the negative reals   7. $\sinh(e^z)$

For each function f listed in Exercises 8 to 11 find an analytic function F with $F' = f$.

8. $f(z) = z - 2$ 9. $f(z) = \dfrac{z^4 + 1}{z^2}$ 10. $f(z) = \sin z \cos z$

11. $f(z) = \cosh(2z)$.

12.   Let f and g be analytic on a domain. Show (a) $f + g$ is analytic on D and $(f + g)' = f' + g'$; (b) fg is analytic on D and $(fg)' = f'g + fg'$; (c) f/g is analytic at all points z_0, where $g(z_0) \neq 0$ and $(f/g)' = (f'g - fg')/g^2$.

13.   Show that if f is analytic on a domain D and g is analytic on a domain Ω containing the range of f, then $g(f(z))$ is analytic on D and the chain rule holds: $(g(f(z)))' = g'(f(z))f'(z)$.

14. Let $P(z) = A(z - z_1) \cdots (z - z_n)$, where A and $z_1, \ldots, z_n$ are complex numbers and $A \neq 0$. Show that

$$\frac{P'(z)}{P(z)} = \sum_{j=1}^{n} \frac{1}{z - z_j}, \qquad z \neq z_1, \ldots, z_n.$$

15. Let f be analytic on a domain D and suppose that $f'(z) = 0$ for all $z \in D$. Show that f is constant on D.

16. Find the derivative of the **linear fractional transformation** $T(z) = (az + b)/(cz + d)$, $ad \neq bc$. In what way does the condition $ad - bc \neq 0$ enter? Conclude that $T'(z)$ is never zero, $z \neq -d/c$.

17. Suppose that f is analytic on a domain D and $f'(z) = \alpha f(z)$, $z \in D$, where α is a constant. Show that $f(z) = C \exp(\alpha z)$, C a constant. (**Hint:** consider $g(z) = e^{-\alpha z} f(z)$ and use Exercise 15 for g.)

18. Show $h(z) = \bar{z}$ is not analytic on any domain. (**Hint:** check the Cauchy–Riemann equations.)

19. Fill in the details to make the following argument correct: Let $w = g(z) = \text{Log } z$ and $w_0 = \text{Log } z_0$. Then

$$g'(z_0) = \lim_{z \to z_0} \frac{\text{Log } z - \text{Log } z_0}{z - z_0} = \lim_{z \to z_0} \frac{w - w_0}{z - z_0}$$

$$= \lim_{z \to z_0} \frac{w - w_0}{e^w - e^{w_0}} = \frac{1}{(e^w)'} \text{ at } w = w_0$$

$$= \frac{1}{e^{w_0}} = \frac{1}{z_0}.$$

20. Let $f = u + iv$ be analytic. In each of the following, find v given u.
    (a) $u = x^2 - y^2$ (b) $u = x/(x^2 + y^2)$
    (c) $u = 2x^2 + 2x + 1 - 2y^2$ (d) $u = \cosh y \sin x$
    (e) $u = \cosh x \cos y$.

21. Let γ be a piecewise smooth simple closed curve and suppose that F is analytic on some domain containing γ. Show that

$$\int_\gamma F'(z)dz = 0.$$

22. Show that if f is analytic on a domain D and if the range of f lies in either a straight line or a circle, then f is a constant.

23. If $z(t)$ is a differentiable function of the real variable t, $a \leqslant t \leqslant b$, and if g is an analytic function on a domain D that contains the range of $z(t)$, $a \leqslant t \leqslant b$, show that $g(z(t)) = w(t)$ is a differentiable function of t and

$$w'(t) = g'(z(t))z'(t)$$

Cauchy–Riemann equations in polar coordinates★

24.   Suppose $f = u + iv$ is analytic in a domain D. Show that the Cauchy–Riemann equations in polar coordinates are

$$r\frac{\partial u}{\partial r} = \frac{\partial v}{\partial \theta} \quad \text{and} \quad r\frac{\partial v}{\partial r} = -\frac{\partial u}{\partial \theta}.$$

2.1.1   Flows, fields, and analytic functions★

There are two physical phenomena each of which provides a physical setting that helps us to understand and motivate analytic functions. An analysis shows that these phenomena are really the same and that they differ only in terminology. However, the terminology is now fixed by tradition. Thus, it is worthwhile to begin by describing each of these situations separately; later we shall concentrate on just one of them.

Flows

Imagine a thin layer of incompressible liquid flowing smoothly across a domain D in the complex plane (Fig. 2.1). At each point $z \in D$, we can find the direction and speed of the liquid, which we assume does *not* vary with time; equivalently, we watch the path taken by a particle in the liquid and we find the tangent vector to this path. This correspondence between points of D and the value of the velocity (speed + direction) of the liquid is nothing more than a complex-valued function $f(z) = u(z) + iv(z)$, $z \in D$. Conversely, each complex-valued function f gives a flow on D by assigning to the point $z \in D$ the value $f(z)$ as the velocity of the flow at that point z. Hence, each complex-valued function on the domain D corresponds to a (time independent) flow on D and vice versa.

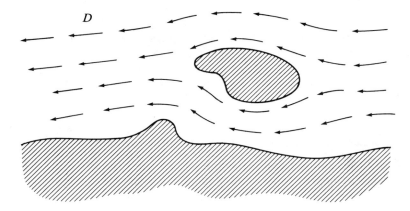

Figure 2.1

We assume now that the function f is continuous on D. Let γ be a smooth curve in D; we examine a small segment of γ. If the segment is sufficiently small it can be thought of as a straight line segment and the flow is virtually constant near the segment. The flow f has a component of magnitude $f \cdot n$ perpendicular (or normal) to the segment and so the amount of the flow that crosses this small segment in a unit time is just $f \cdot n \, ds$, where ds is the length of the small segment. Hence, $\int_\gamma f \cdot n \, ds$ is the total amount of the flow that crosses γ within a unit time. Likewise, the magnitude of the component of the flow that is parallel to the small segment is $f \cdot \tau$, where τ is a unit vector tangent to (or parallel with) the small segment and so the term $f \cdot \tau \, ds$ represents the amount of the flow that is tangential to the small segment. Thus, $\int_\gamma f \cdot \tau \, ds$ is a total amount of the flow that is tangential to γ.

A flow $f = u + iv$ is termed **sourceless** in D if

$$\int_\gamma (f \cdot n) ds = 0, \qquad ds = \text{arc length}$$

for all smooth simple closed curves γ in D, where $f \cdot n$ is the component of f normal to γ. Thus, for example, there is no *net* flow of the fluid in or out of any disc in D.

A flow $f = u + iv$ is termed **irrotational** in D if

$$\int_\gamma (f \cdot \tau) ds = 0, \qquad ds = \text{arc length}$$

for all smooth simple closed curves γ in D, where $f \cdot \tau$ is the component of f tangential to γ. Thus, if the flow is irrotational there is no net rotation of the fluid about any point of D.

Fields

In physics a complex-valued function f on a domain D is thought of as representing a **field**. The term **vector-field** is also frequently used because the value $f(z)$ can be viewed as a vector whose length $|f(z)|$ gives the strength of the field at z and whose direction $\arg f(z)$ gives the orientation of the field at z. For instance, f could be the force field induced by the gravitational attraction of a collection of masses or f could be the electrostatic field induced by a collection of charges. We assume, as with flows, that the field is in **steady state**.

Let γ be a smooth curve in D. The **work** done by the field in moving a particle along the curve γ is

$$\int_\gamma f \cdot \tau \, ds,$$

where $f \cdot \tau$ is the component of f tangent to the curve and ds is arc length along the curve. The field is termed **conservative** if the net work done when moving a particle around any closed curve is zero; that is, when

$$0 = \int_\gamma f \cdot \tau \, ds$$

for each smooth closed curve γ in D.

Again let γ be a smooth curve in D; the **flux** of f across γ is defined to be

$$\int_\gamma f \cdot n \, ds,$$

where $f \cdot n$ is the component of f normal to γ and again ds is arc length. The field f is termed **fluxless** if the net flux of f across any smooth closed curve γ in D is zero; that is, if $\int_\gamma f \cdot n \, ds = 0$ for any smooth closed curve γ in D.

> *The reader will have noticed by now that there is a difference only in terminology between fields and flows: what is termed "irrotational" for flows is called "conservative" for fields and what is termed "sourceless" for flows is called "fluxless" for fields. For this reason, we will discuss only flows in the remainder of the book.*

We now explore the relation between a sourceless/irrotational flow f and the analyticity of f. We shall assume that $f = u + iv$, where u and v have continuous partial derivatives throughout D with respect to both x and y.

Let $\gamma(t) = x(t) + iy(t)$, $a \leqslant t \leqslant b$, be a smooth curve in D; then

$$ds = \sqrt{(x'(t))^2 + (y'(t))^2} \, dt$$

and

$$f \cdot \tau = \frac{ux' + vy'}{\sqrt{(x')^2 + (y')^2}},$$

since a unit tangent vector is $\dfrac{x' + iy'}{\sqrt{(x')^2 + (y')^2}}$.

Thus,

$$\int_\gamma f \cdot \tau \, ds = \int_\gamma u \, dx + v \, dy = \mathrm{Re}\left\{ \int_\gamma \overline{f(z)} \, dz \right\}.$$

Consequently, the flow f is irrotational if and only if

$$0 = \mathrm{Re}\left\{ \int_\gamma \overline{f(z)} \, dz \right\}$$

for all smooth closed curves γ in D.

Just as above we find that

$$\int_\gamma f \cdot n \, ds = \int_\gamma u \, dy - v \, dx = \mathrm{Im}\left\{ \int_\gamma \overline{f(z)} \, dz \right\}.$$

Hence, the flow f is sourceless exactly when

$$0 = \text{Im}\left\{\int_\gamma \overline{f(z)}\, dz\right\}$$

for all smooth closed curves γ in D.

Let γ be a small circle in D whose inside Ω also lies in D. Green's Theorem then gives

$$\iint_\Omega \left[\frac{\partial v}{\partial x} - \frac{\partial u}{\partial y}\right] dx\, dy = \int_\gamma u\, dx + v\, dy$$

and

$$\iint_\Omega \left[\frac{\partial u}{\partial x} + \frac{\partial v}{\partial y}\right] dx\, dy = \int_\gamma u\, dy - v\, dx.$$

Thus, for a flow f, which is either irrotational or sourceless, either the first or the second of these area integrals must be zero. By letting γ shrink down to a point, we conclude that

if a flow f $=$ u $+$ iv *is irrotational on* D, *then* $(\partial v/\partial x) - (\partial u/\partial y) = 0$
on D

and

if a flow f $=$ u $+$ iv *is sourceless on* D, *then* $(\partial u/\partial x) + (\partial v/\partial y) = 0$ *on* D.

Consequently, if a flow $f = u + iv$ is *both* sourceless and irrotational then it must satisfy

$$\frac{\partial u}{\partial x} = -\frac{\partial v}{\partial y} \qquad \text{and} \qquad \frac{\partial u}{\partial y} = \frac{\partial v}{\partial x}.$$

But these are precisely the Cauchy–Riemann equations for the function $\bar{f} = u - iv$, the complex conjugate of $f = u + iv$. In other words when the flow is both sourceless and irrotational on D, $\bar{f}$ must be analytic. In fact, we have actually shown somewhat more.

A flow f *on a domain* D *is both sourceless and irrotational on* D *if and only if*

$$\int_\gamma \overline{f(z)}\, dz = 0$$

for all smooth closed curves γ *in* D.

On the other hand, the use of Green's Theorem as above shows us that if $\bar{f}$ is analytic on D, then the flow f is **locally sourceless** and **locally irrotational**. That is, each point of D is the center of some disc Ω in D and in Ω the flow is sourceless and irrotational. The reader is cautioned, however, that a flow can be locally sourceless/irrotational in D but not globally sourceless/irrotational in D; see Example 2 below.

Example 1   The function $f(z) = \bar{z} = x - iy$ gives a sourceless and irrotational flow on any domain. Let us find the path followed by a particle in this flow. If $x(t) + iy(t)$ is a parametrization of the path, then the tangent to the path is just $(dx/dt) + i(dy/dt)$ and so

$$\frac{dx}{dt} + i\frac{dy}{dt} = \bar{z} = x(t) - iy(t).$$

Hence, $(dx/dt) = x$, $(dy/dt) = -y$ and we have $x(t) = c_1 e^t$, $y(t) = c_2 e^{-t}$ for constants c_1 and c_2. In particular, $xy = c_1 c_2 = $ constant. The curves $xy = $ constant are thus the paths traced out by the flow. In general a **streamline** of the flow is the path followed by a particle in the flow. Here, then, the streamlines are the curves $xy = $ constant. The direction of the flow is determined by the signs of c_1 and c_2 (Fig. 2.2). The origin is a **stagnation point** of the flow since once there it is impossible to leave (of course, nothing actually gets there).

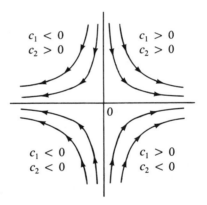

Figure 2.2

Example 2   The function $f(z) = 1/\bar{z}$ gives a *locally* sourceless and irrotational flow on any domain D that does not include the origin (where f is not defined). Note, however, that on the circle $|z| = 1$, we have

$$f \cdot n = \cos\theta\cos\theta + \sin\theta\sin\theta = 1,$$

and so

$$\int_{|z|=1} f \cdot n \, ds = 1 \int_{|z|=1} ds = 2\pi.$$

Thus, the flow $f(z) = 1/\bar{z}$ although locally sourceless, is not globally sourceless (Fig. 2.3).

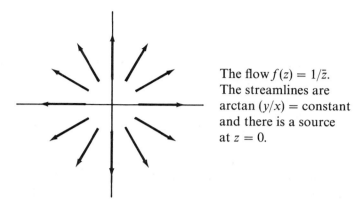

The flow $f(z) = 1/\bar{z}$.
The streamlines are
arctan $(y/x) =$ constant
and there is a source
at $z = 0$.

Figure 2.3

Example 3 The function $f(z) = z$ is not both sourceless and irrotational on *any* domain D, since $\bar{f}(z) = \bar{z}$ is not analytic on any domain D. More specifically, if γ is the circle of radius r_0 centered at $z_0 = x_0 + iy_0$, then

$$f \cdot n = (x_0 + r_0 \cos \theta)\cos \theta + (y_0 + r_0 \sin \theta)\sin \theta$$

$$= r_0 + x_0 \cos \theta + y_0 \sin \theta,$$

so that

$$\int_\gamma f \cdot n \, ds = \int_0^{2\pi} (r_0 + x_0 \cos \theta + y_0 \sin \theta)r_0 \, d\theta$$

$$= 2\pi r_0^2.$$

Hence, there is a source of strength $2(= \lim_{r_0 \to 0} 2\pi r_0^2 / \pi r_0^2)$ at the point z_0 (more flows out of the disc of radius r_0 centered at z_0 than flows into it).

We shall return to the ideas of this section briefly later in this chapter after we have developed some more techniques and again in Chapter 3, in the context of conformal mapping.

Exercises for Section 2.1.1

In Exercises 1 to 8 decide whether or not the given function represents a locally sourceless and/or irrotational flow. For those that do, decide if the flow is globally sourceless and/or irrotational. Sketch some of the streamlines.

1. $x^3 - 3xy^2 + i(y^3 - 3yx^2)$

2. $\dfrac{x}{x^2 + y^2} + i\dfrac{y}{x^2 + y^2}$, $x^2 + y^2 > 0$

3. $\cos x \cosh y + i \sin x \sinh y$

4. $x^2 - y^2 + 2ixy$

5. $x^2 - y^2 - 2ixy$ 6. $e^y \cos x + ie^y \sin x$

7. $e^x \cos y + ie^x \sin y$ 8. $\cosh x \cos y + i \sinh x \sin y$

9.   Suppose that G is analytic on a domain D and $f(z) = \overline{G'(z)}$ Show that f
       represents a sourceless/irrotational flow on D.

2.2   Power series

This section is devoted to examining functions of a very special nature: power
series. These functions represent a fusion of the ideas on infinite series introduced in Section 4 of Chapter 1 with the concept of an analytic function
introduced and studied in Section 1 of this chapter. We shall see in Section 4
of this chapter that there is a full and complete relationship between analytic
functions and power series.

A **power series** in z is an infinite series of the special form

$$\sum_{n=0}^{\infty} a_n(z - z_0)^n, \tag{1}$$

where $a_0, a_1, \ldots$ are complex numbers, called the **coefficients** of the series; z_0 is
fixed and is called the **center** of the series. We have already seen one example
of such a series,

$$f(z) = \sum_{n=0}^{\infty} z^n = \frac{1}{1 - z}, \quad |z| < 1,$$

in Section 4 of Chapter 1. Of course, any polynomial,

$$p(z) = \sum_{n=0}^{N} b_n z^n,$$

is also an example of such a series. Other examples are the power series

$$\sum_{n=0}^{\infty} \frac{z^n}{n!}, \quad \sum_{n=0}^{\infty} (-1)^n \frac{z^{2n}}{(2n)!}, \quad \sum_{n=0}^{\infty} (-1)^n \frac{z^{2n+1}}{(2n+1)!},$$

all three of which converge for all z by use of the ratio test (see Section 4 of
Chapter 1).

The first major result on power series is shown below.

Theorem 1:   Suppose there is some $z_1 \neq z_0$ such that $\sum a_n(z_1 - z_0)^n$
converges. Then for each z with $|z - z_0| < |z_1 - z_0|$, the series
$\sum a_n(z - z_0)^n$ is absolutely convergent.

Proof. Suppose that $|z - z_0| \leqslant r < |z_1 - z_0|$. Since $\sum_0^{\infty} a_n(z_1 - z_0)^n$ is con-
vergent we know that

$$\lim_{n \to \infty} a_n(z_1 - z_0)^n = 0.$$

In particular, there is a constant M with

$$|a_n| |z_1 - z_0|^n \leqslant M$$

for all n. Hence,

$$|a_n| |z - z_0|^n = |a_n| |z_1 - z_0|^n \left(\frac{|z - z_0|}{|z_1 - z_0|} \right)^n$$

$$\leqslant M\rho^n$$

where $\rho = r/|z_1 - z_0|$ is less than 1. Hence, the series $\sum a_n(z - z_0)^n$ is absolutely convergent. $\qquad\square$

The radius of convergence

For any power series $\sum_0^\infty a_n(z - z_0)^n$ there are always only three mutually exclusive possibilities:

The series $\sum a_n(z - z_0)^n$ converges only for $z = z_0$. (2)

The series $\sum a_n(z - z_0)^n$ converges for all z. (3)

The series $\sum a_n(z - z_0)^n$ converges for some $z \neq z_0$, but not for all z. (4)

Let us examine the third possibility (4) more thoroughly. Suppose z' and z'' are two points with

$$\sum_{n=0}^{\infty} a_n(z' - z_0)^n \text{ convergent}, \qquad \sum_{n=0}^{\infty} a_n(z'' - z_0)^n \text{ divergent}.$$

By Theorem 1 we then know that

$$|z - z_0| < |z' - z_0| \text{ implies that } \sum_0^{\infty} a_n(z - z_0)^n \text{ converges}$$

and

$$|z - z_0| > |z'' - z_0| \text{ implies that } \sum_0^{\infty} a_n(z - z_0)^n \text{ diverges}.$$

In particular, $|z' - z_0| < |z'' - z_0|$. Continuing in this fashion to examine the convergence of $\sum_0^\infty a_n(z - z_0)^n$ we see that we can define a number R by this rule: R is the unique number such that

$$|z - z_0| < R \text{ implies that } \sum_0^{\infty} a_n(z - z_0)^n \text{ converges}$$

and (5)

$$|z - z_0| > R \text{ implies that } \sum_0^{\infty} a_n(z - z_0)^n \text{ diverges}.$$

When $|z - z_0| = R$, the series $\sum_0^\infty a_n(z - z_0)^n$ may converge or it may diverge. The number R is the **radius of convergence** of the power series $\sum_0^\infty a_n(z - z_0)^n$. To make the picture complete, if the series satisfies (2) we define $R = 0$ and if (3) holds we define $R = \infty$. To reiterate then, a power series $\sum_0^\infty a_n(z - z_0)^n$, which converges for some $z \neq z_0$, always converges within a disc $\{z : |z - z_0| < R\}$, where R is positive or R is infinity.

The radius of convergence R of the power series $\sum_0^\infty a_n(z - z_0)^n$ depends in a very explicit way on the coefficients $\{a_n\}$. The general formula is a bit complicated but two simple and useful cases are given below.

Theorem 2: Suppose that $\sum_0^\infty a_n(z - z_0)^n$ is a power series with a positive or infinite radius of convergence R.

(a) If $\lim\limits_{n \to \infty} |a_{n+1}/a_n|$ exists, then

$$1/R = \lim_{n \to \infty} |a_{n+1}/a_n|. \tag{6}$$

(b) If $\lim\limits_{n \to \infty} \sqrt[n]{|a_n|}$ exists, then

$$1/R = \lim_{n \to \infty} \sqrt[n]{|a_n|}. \tag{7}$$

Proof. The proofs of (a) and (b) are nothing but the ratio and root tests, respectively. To see why (a) is true, let L be the limit of $|a_{n+1}/a_n|$. Then

$$\lim_{n \to \infty} \left| \frac{a_{n+1}(z - z_0)^{n+1}}{a_n(z - z_0)^n} \right| = |z - z_0| \lim_{n \to \infty} \left| \frac{a_{n+1}}{a_n} \right| = |z - z_0|L.$$

When $|z - z_0|L$ is less than one the series $\sum_0^\infty a_n(z - z_0)^n$ is absolutely convergent; when $|z - z_0|L$ is greater than one the series $\sum a_n(z - z_0)^n$ diverges; see Exercise 40, Section 4 of Chapter 1. By the definition of R we must have $R = 1/L$ or $1/R = L$.

The proof of (b) is virtually the same with the root test replacing the ratio test; see Exercise 39 of Section 4 of Chapter 1.   ☐

Example 1   The power series $\sum_0^\infty 5^n(z - 1)^n$ has radius of convergence $R = 1/5$, since

$$1/R = \lim_{n \to \infty} \sqrt[n]{5^n} = 5.$$

Hence, the series converges within the disc $|z - 1| < 1/5$ and diverges for all z with $|z - 1| > 1/5$.

Example 2   The power series $\sum_0^\infty (-1)^n z^n/n!$ converges for all z, since

$$\lim_{n \to \infty} \frac{|a_{n+1}|}{|a_n|} = \lim_{n \to \infty} \frac{|(-1)^{n+1}/(n+1)!|}{|(-1)^n/n!|} = \lim_{n \to \infty} \frac{1}{n+1} = 0,$$

and so $1/R = 0$, which means $R = \infty$.

Example 3   Find the radius of convergence of the power series

$$\sum_{n=0}^{\infty} n^2 \frac{1}{2^{2n}} z^n.$$

Solution   We use (7):

$$1/R = \lim_{n \to \infty} \sqrt[n]{n^2/2^{2n}} = \lim_{n \to \infty} (\sqrt[n]{n})^2 (\sqrt[n]{4^{-n}})$$

$$= (1^2)(1/4) = 1/4$$

because $\lim_{n \to \infty} \sqrt[n]{n} = 1$ (see Exercise 21(a) of this section). Hence, $R = 4$.   □

Example 4   Find the radius of convergence of the power series $\sum_{n=0}^{\infty} 4^n z^{3n}$

Solution   Here we have

$$a_k = \begin{cases} 4^{k/3} & \text{if } k = 0, 3, 6, \ldots \\ 0 & \text{otherwise} \end{cases}$$

so technically speaking we cannot apply either (6) or (7). However, when we write $w = z^3$, the power series becomes $\sum_0^{\infty} 4^n w^n$. This power series has radius of convergence $1/4$; that is, the given series

$$\text{converges if } 1/4 > |w| = |z|^3$$
$$\text{diverges if } 1/4 < |w| = |z|^3$$

Hence, the first series $\sum_{n=0}^{\infty} 4^n z^{3n}$ has radius of convergence $R = 4^{-1/3}$.   □

Example 5   Here is another example that shows the limitations of formulas (6) and (7). Let f be the power series

$$f(z) = \sum_{n=0}^{\infty} 4^{n(-1)^n} z^n.$$

That is,

$$a_n = \begin{cases} 4^n & \text{if } n = 0, 2, 4, \ldots \\ \dfrac{1}{4^n} & \text{if } n = 1, 3, 5, \ldots. \end{cases}$$

The two power series

$$f_1(z) = \sum_{n=0}^{\infty} 4^{2n} z^{2n}$$

and

$$f_2(z) = \sum_{n=0}^{\infty} \frac{1}{4^{2n+1}} z^{2n+1}$$

have radii of convergence $R_1 = 1/4$ and $R_2 = 4$, respectively, and

$$f(z) = f_1(z) + f_2(z) \quad \text{if } |z| < 1/4.$$

Hence, the radius of convergence of f is at least $1/4$. But it cannot be bigger than $1/4$ since the series for f_2 diverges whenever $|z| > 1/4$. Hence, $R = 1/4$. But note that neither (6) nor (7) was available to us to reach this conclusion. The general formula for R, which *does* cover this case, depends on some special knowledge from real analysis and is presented in the exercises.

The derivative of a power series

We now set out on the chief task of this section. We shall show that a function f given by a power series

$$f(z) = \sum_{0}^{\infty} a_n(z - z_0)^n \tag{8}$$

with a positive or infinite radius of convergence R is actually *analytic* within the disc $|z - z_0| < R$ and its derivative is given by

$$f'(z) = \sum_{1}^{\infty} na_n(z - z_0)^{n-1}. \tag{9}$$

We begin by showing that the power series on the right side of (9) has radius of convergence at least R. Let $|z - z_0| = r < R$ and let s be between r and R, $r < s < R$. Then for all integers n greater than some N, we have

$$nr^{n-1} \leqslant s^n, \quad n \geqslant N \tag{10}$$

since $\lim_{n \to \infty} n(r/s)^n = 0$; see Exercise 21(b) at the end of this section. Thus,

$$n|a_n||z - z_0|^{n-1} \leqslant |a_n|s^n, \quad n \geqslant N$$

and since $s < R$, the series $\sum_{0}^{\infty} |a_n|s^n$ converges. Hence, the series $\sum na_n(z - z_0)^{n-1}$ also converges whenever $|z - z_0| = r < R$.

Let g be the series on the right side of (9):

$$g(z) = \sum_{n=1}^{\infty} na_n(z - z_0)^{n-1}.$$

It simplifies the notation to assume $z_0 = 0$; this can be done with no loss of generality. Suppose z is in the disc of radius R centered at 0; let $\delta = \frac{1}{2}(R - |z|)$ and suppose that $|h| < \delta$. We now compute

$$f(z + h) - f(z) = \sum_{1}^{\infty} a_n[(z + h)^n - z^n]$$

$$= a_1 h + \sum_{2}^{\infty} a_n[(z + h)^n - z^n].$$

This gives

$$\frac{f(z + h) - f(z)}{h} - g(z) = \sum_{2}^{\infty} a_n \left\{ \frac{(z + h)^n - z^n}{h} - nz^{n-1} \right\}.$$

Now the binomial theorem (Exercise 20, this section) gives us

$$\frac{(z + h)^n - z^n}{h} - nz^{n-1} = \sum_{j=2}^{n} \binom{n}{j} z^{n-j} h^{j-1},$$

where $\binom{n}{j} = \dfrac{n!}{j!(n-j)!}$. Hence,

$$\left| \frac{(z + h)^n - z^n}{h} - nz^{n-1} \right| \leqslant \sum_{j=2}^{n} \binom{n}{j} |z|^{n-j} |h|^{j-1}$$

$$\leqslant |h| \sum_{j=2}^{n} \binom{n}{j} (R - 2\delta)^{n-j} \delta^{j-2}$$

$$= |h| \delta^{-2} \sum_{j=2}^{n} \binom{n}{j} (R - 2\delta)^{n-j} \delta^{j}$$

$$< |h| \delta^{-2} \sum_{j=0}^{n} \binom{n}{j} (R - 2\delta)^{n-j} \delta^{j}$$

$$= |h| \delta^{-2} [(R - 2\delta) + \delta]^n = |h| \delta^{-2} (R - \delta)^n.$$

This leads us to the estimate

$$\left| \frac{f(z + h) - f(z)}{h} - g(z) \right| \leqslant |h| \delta^{-2} \sum_{n=2}^{\infty} |a_n| (R - \delta)^n.$$

This last quantity goes to zero as $h \to 0$ (recall that δ is fixed) and so we have established that f is differentiable with derivative equal to g:

$$\left(\sum_{0}^{\infty} a_n z^n \right)' = \sum_{1}^{\infty} n a_n z^{n-1}.$$

This is what we set out to accomplish. However, even more can now be said. The series $\sum_{1}^{\infty} n a_n z^{n-1}$ is also a power series with radius of convergence R so that it, too, is differentiable with

$$\left(\sum_{1}^{\infty} n a_n z^{n-1} \right)' = \sum_{2}^{\infty} n(n - 1) a_n z^{n-2}.$$

We can continue this process indefinitely, of course, and we thereby obtain the following theorem.

Theorem 3: If $f(z) = \sum_0^\infty a_n(z - z_0)^n$ has a positive or infinite radius of convergence R, then within the disc $|z - z_0| < R$, f is infinitely differentiable, each derivative being again given by a power series

$$f^{(k)}(z) = \sum_{n=k}^\infty n(n - 1) \cdots (n - k + 1)a_n(z - z_0)^{n-k}, \; k = 1, 2, \ldots. \tag{11}$$

In particular, by setting $z = z_0$, we have

$$\frac{f^{(n)}(z_0)}{n!} = a_n, \; n = 0, 1, 2, \ldots. \tag{12}$$

Example 6 The power series $\sum_{n=0}^\infty z^n/n!$ has radius of convergence infinity by the ratio test and hence is an entire function. According to Theorem 3 its derivative is

$$\left(\sum_{n=0}^\infty \frac{z^n}{n!} \right)' = \sum_{n=1}^\infty \frac{n}{n!} z^{n-1} = \sum_{n=1}^\infty \frac{z^{n-1}}{(n-1)!}$$

$$= \sum_{j=0}^\infty \frac{z^j}{j!}.$$

That is, the function is its own derivative. Let us denote the power series by F:

$$F(z) = \sum_{n=0}^\infty \frac{z^n}{n!}.$$

Then

$$(e^{-z}F(z))' = -e^{-z}F(z) + e^{-z}F'(z)$$
$$= 0,$$

since $F' = F$. Thus, $F(z)e^{-z} = \lambda$ for some constant λ by Exercise 15 of Section 1. But $1 = F(0) = \lambda$, so we have found that $F(z) = e^z$ and thus,

$$e^z = \sum_{n=0}^\infty \frac{z^n}{n!}. \tag{13}$$

Example 7 Show that $\sin z$ and $\cos z$ have power series expansions valid for all z; find the series explicitly.

Solution From Example 6 we know that $e^z = \sum_{n=0}^\infty (z^n/n!)$. Hence,

$$\cos z = \frac{1}{2} (e^{iz} + e^{-iz}) = \frac{1}{2} \left(\sum_{n=0}^\infty \frac{(iz)^n}{n!} + \sum_{n=0}^\infty \frac{(-iz)^n}{n!} \right)$$

$$= \sum_{n=0}^\infty (-1)^n \frac{z^{2n}}{(2n)!}.$$

Likewise,

$$\sin z = \sum_{n=0}^{\infty} (-1)^n \frac{z^{2n+1}}{(2n+1)!}.$$ □

Multiplication and division of power series★

Suppose that the two power series $f(z) = \sum_0^{\infty} a_n z^n$ and $g(z) = \sum_0^{\infty} b_n z^n$ both have radius of convergence at least R for some positive number R; the product $f(z)g(z)$ is then defined and analytic in the disc $\{z : |z| < R\}$. We can formally multiply out the series for $f(z)$ and that for $g(z)$ and collect equal powers of z. The result is this.

$$f(z)g(z) = (a_0 + a_1 z + a_2 z^2 + \cdots)(b_0 + b_1 z + b_2 z^2 + \cdots)$$
$$= a_0 b_0 + (a_0 b_1 + a_1 b_0)z + (a_0 b_2 + a_1 b_1 + a_2 b_0)z^2 + \cdots$$
$$= \sum_{n=0}^{\infty} c_n z^n,$$

where

$$c_n = \sum_{k=0}^{n} a_k b_{n-k}.$$

These manipulations are, in fact, correct and the power series $\sum c_n z^n$ has radius of convergence at least R; the formal statement of this result follows and the proof is presented in the exercises at the end of this section.

Theorem 4: Suppose that $f(z) = \sum_{n=0}^{\infty} a_n z^n$ and $g(z) = \sum_{n=0}^{\infty} b_n z^n$ are two power series each with radius of convergence R or more, $R > 0$. Then their product fg can also be expressed as a power series in the disc $|z| < R$:

$$(fg)(z) = \sum_{n=0}^{\infty} c_n z^n, \text{ where } c_n = \sum_{k=0}^{n} a_k b_{n-k}.$$

Example 8   Find the power series expansion of $e^z/(1-z)$ about $z_0 = 0$ in $|z| < 1$.

Solution   We know that

$$e^z = 1 + z + z^2/2! + z^3/3! + \cdots, \text{ all } z$$

and that

$$1/(1-z) = 1 + z + z^2 + z^3 + \cdots, \quad |z| < 1.$$

Hence,

$$e^z/(1 - z) = 1 + 2z + \frac{5}{2} z^2 + \frac{8}{3} z^3 + \frac{65}{24} z^4 + \cdots, \quad |z| < 1. \qquad \square$$

Example 9   Find the power series for $(1 - z^2)\sin z$.

Solution   Again we just formally multiply out the two series

$$(1 - z^2)(\sin z) = (1 - z^2)(z - z^3/6 + z^5/120 - \cdots)$$

$$= z - \frac{7}{6} z^3 + \frac{21}{120} z^5 - \cdots. \qquad \square$$

Example 10   Later in this chapter we will show that $z \csc z$ has a power series expansion near $z = 0$. Assuming this for the moment, find the first few terms of this series.

Solution   We have $z \csc z = z/\sin z$, so here we are faced with a problem of *division* of series. However, every division problem can be replaced by an equivalent multiplication problem and that is what we will do here. Write

$$z \csc z = b_0 + b_1 z + b_2 z^2 + \cdots.$$

Then

$$z = (z \csc z)\sin z$$

$$= (b_0 + b_1 z + b_2 z^2 + \cdots)(z - z^3/6 + z^5/120 - \cdots)$$

$$= b_0 z + b_1 z^2 + \left(b_2 - \frac{b_0}{6} \right)z^3 + \left(b_3 - \frac{b_1}{6} \right)z^4 + \left(b_4 - \frac{b_2}{6} + \frac{b_0}{120} \right)z^5 + \cdots.$$

We equate coefficients of equal powers of z and obtain a series of equations for the b_js:

$$1 = b_0, \quad 0 = b_1, \quad 0 = b_2 - b_0/6, \quad 0 = -b_1/6 + b_3,$$

$$0 = b_0/120 - b_2/6 + b_4$$

Hence, $b_0 = 1$, $b_1 = 0$, $b_2 = 1/6$, $b_3 = 0$, $b_4 = 7/360$. Consequently,

$$z \csc z = 1 + z^2/6 + (7/360) z^4 + \cdots. \qquad \square$$

You should note that this technique of dividing series works all the time provided that it is known that the quotient does actually have a series representation. We shall look further into this problem in Section 5 of this chapter.

Exercises for Section 2.2

In Exercises 1 to 6 find the radius of convergence of the given power series

1. $\displaystyle\sum_{k=1}^{\infty} k(z-1)^k$

2. $\displaystyle\sum_{k=0}^{\infty} \frac{(k!)^2}{(2k)!} (z-2)^k$

3. $\displaystyle\sum_{j=0}^{\infty} \frac{z^{3j}}{2^j}$

4. $\displaystyle\sum_{k=0}^{\infty} (-1)^k z^{2k}$

5. $\displaystyle\sum_{n=0}^{\infty} 5^{(-1)^n} z^n$

6. $\displaystyle\sum_{k=1}^{\infty} \frac{(2k)(2k-2)\cdots 4\cdot 2}{(2k-1)(2k-3)\cdots 3\cdot 1} z^k$

In exercises 7 to 13 find the power series about the origin for the given function

7. e^{-z}

8. $z^2 \cos z$

9. $\displaystyle\frac{z^3}{1-z^3}, |z| < 1$

10. $\displaystyle\frac{1+z}{1-z}, |z| < 1$

11. $\displaystyle\frac{z^2}{(4-z)^2}, |z| < 4 \left(\textbf{Hint:} \frac{1}{(a-z)^2} = \frac{d}{dz}[(a-z)^{-1}].\right)$

12. $\sinh(z^2)$

13. $(\sin z)^2$

In Exercises 14 to 18 find a "closed form" (that is, a simple expression) for each of the given power series

14. $\displaystyle\sum_{n=0}^{\infty} \frac{z^{2n}}{n!}$

15. $\displaystyle\sum_{n=1}^{\infty} z^{3n}$

16. $\displaystyle\sum_{n=1}^{\infty} n(z-1)^{n-1}$

17. $\displaystyle\sum_{n=0}^{\infty} (-1)^n \frac{(z-2\pi i)^n}{n!}$

18. $\displaystyle\sum_{n=2}^{\infty} n(n-1)z^n$ **(Hint: divide by z^2.)**

19. (a) Suppose that the radius of convergence of the power series $f(z) = \sum_{n=0}^{\infty} a_n(z-z_0)^n$ is R, $R > 0$. Show that the radius of convergence of

$$F(z) = \sum_{n=0}^{\infty} \frac{a_n}{n+1} (z-z_0)^{n+1}$$

is at least R

(b) Show that $F'(z) = f(z)$ for all z with $|z - z_0| < R$.

(c) Use parts (a) and (b) to show that

$$\text{Log}(1-z) = -z - \frac{z^2}{2} - \frac{z^3}{3} - \frac{z^4}{4} - \cdots, \text{ if } |z| < 1.$$

20. Establish the **binomial formula** for any pair (a, b) of complex numbers

$$(a + b)^n = \sum_{j=0}^{n} \binom{n}{j} a^j b^{n-j}, \qquad \binom{n}{j} = \frac{n!}{j!(n-j)!}$$

(Hint: One technique is to prove the formula by induction on n.)

21.   Show that (a) $\lim_{n \to \infty} \sqrt[n]{n} = 1$ (b) $\lim_{n \to \infty} n\rho^n = 0$ if $|\rho| < 1$

        (c) $\lim_{n \to \infty} \sqrt[n]{M} = 1$ if M is positive.

(Hint: For (a) write

$$\sqrt[n]{n} = 1 + \varepsilon_n, \ \varepsilon_n > 0; \text{ then } n = (1 + \varepsilon_n)^n > 1 + \frac{n(n-1)}{2} \varepsilon_n^2$$

[by Exercise 20].)

22.   (a) If $f(z) = \sum_{n=0}^{\infty} a_n(z - z_0)^n$ has radius of convergence $R > 0$ and if
        $f(z) = 0$ for all z, $|z - z_0| < r \leqslant R$, show that $a_0 = a_1 = \cdots = 0$.
        (b) If $F(z) = \sum_{0}^{\infty} a_n(z - z_0)^n$ and $G(z) = \sum_{0}^{\infty} b_n(z - z_0)^n$ are equal on
        some disc $|z - z_0| < r$, show that $a_n = b_n$ for all n.

23.   Let $a_0, a_1, a_2, \ldots$ be complex numbers with either

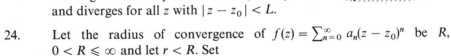

$$\lim_{n \to \infty} \sqrt[n]{|a_n|} = L \text{ or } \lim_{n \to \infty} \frac{|a_{n+1}|}{|a_n|} = L.$$

        Show that the series

$$\sum_{n=0}^{\infty} \frac{a_n}{(z - z_0)^n}$$

        converges absolutely for all z with $|z - z_0| > L$
        and diverges for all z with $|z - z_0| < L$.

24.   Let the radius of convergence of $f(z) = \sum_{n=0}^{\infty} a_n(z - z_0)^n$ be R,
        $0 < R \leqslant \infty$ and let $r < R$. Set

$$f_n(z) = \sum_{j=0}^{n} a_j(z - z_0)^j, \ n = 1, 2, \ldots.$$

        Given $\varepsilon > 0$, show that there is an N such that $|f_n(z) - f(z)| < \varepsilon$ for all
        z with $|z - z_0| \leqslant r$ and all $n \geqslant N$.

(Hint: The series $\sum_{j=0}^{\infty} |a_j| r^j$ converges so there is an N such that

$$\sum_{N}^{\infty} |a_j| r^j < \varepsilon.$$

        Hence, for $n \geqslant N$ and $|z - z_0| \leqslant r$, we have

$$|f_n(z) - f(z)| = \left| \sum_{n+1}^{\infty} a_j(z - z_0)^j \right| \leqslant \sum_{n+1}^{\infty} |a_j| r^j$$

$$\leqslant \sum_{N}^{\infty} |a_j| r^j < \varepsilon.)$$

Multiplication of power series⋆

Let $f(z) = \sum_{n=0}^{\infty} a_n z^n$ and $g(z) = \sum_{n=0}^{\infty} b_n z^n$ both have radius of convergence at least R, $R > 0$. Set $c_n = \sum_{k=0}^{n} a_k b_{n-k}$ for $n = 0, 1, 2, \ldots$ and fix numbers r and s with $0 < r < s < R$.

25. Show there are constants M and M' with $|a_k| s^k \leqslant M$ and $|b_k| s^k \leqslant M'$ for all $k = 0, 1, 2, \ldots$.

26. Show that $|c_n| \leqslant (n + 1) M M' s^{-n}$. (**Hint:** $|c_n| \leqslant \sum_{k=0}^{n} |a_k| |b_{n-k}|$; now apply Exercise 25 to $|a_k|$ and $|b_{n-k}|$.)

27. Use the ratio test to show that the series $\sum_{0}^{\infty} (n + 1)(r/s)^n$ is convergent.

28. Apply the conclusion from Exercise 26 to show that $\sum_{0}^{\infty} |c_n| r^n \leqslant M M' \sum_{0}^{\infty} (n + 1)(r/s)^n$.

29. Show that the radius of convergence of $\sum_{0}^{\infty} c_n z^n$ is at least R. (**Hint:** Apply the conclusion of Exercise 28 with $|z| = r$.)

30. Let $|z| = r < R$ and let $r < s < R$. Supply justification for each step in the following argument

$$|(a_N b_1 + \cdots + a_0 b_{N+1}) z^{N+1} + (a_N b_2 + \cdots + a_0 b_{N+2}) z^{N+2}$$

$$+ \cdots + (a_N b_N + \cdots + a_0 b_{2N}) z^{2N} + \cdots|$$

$$\leqslant \sum_{n=N+1}^{\infty} r^n \sum_{k=0}^{N} |a_k| |b_{n-k}| \leqslant M M' \sum_{n=N+1}^{\infty} (n + 1)(r/s)^n < \varepsilon$$

if N is sufficiently large.

31. Show that $f(z)g(z) = \sum_{0}^{\infty} c_n z^n$ by supplying the justification for each of the following steps

$$f(z)g(z) = \lim_{N \to \infty} \left(\sum_{0}^{N} a_k z^k \right) \left(\sum_{0}^{\infty} b_k z^k \right)$$

$$= \lim_{N \to \infty} \left\{ \sum_{0}^{N} c_n z^n + \sum_{n=N+1}^{\infty} z^n \left(\sum_{k=0}^{N} a_k b_{n-k} \right) \right\}$$

$$= \sum_{0}^{\infty} c_n z^n.$$

(**Hint:** The term $\sum_{n=N+1}^{\infty} z^n (\sum_{k=0}^{N} a_k b_{n-k})$ is very small when N is large by Exercise 30.)

General formula for the radius of convergence of a power series⋆

Let $\{x_n\}$ be a bounded sequence of nonnegative real numbers. Define

$$\limsup_{n \to \infty} x_n = \rho$$

to be the *largest* number ρ such that *each* interval $(\rho - \varepsilon, \rho + \varepsilon)$ contains x_n for infinitely many indices n. (The fact that such a ρ exists is a fundamental property of the real numbers.)

If $\{x_n\}$ is unbounded, define $\lim\sup\limits_{n\to\infty} x_n$ to be infinity.

32.   (a) Show that there are integers $\{n_j\}_{j=1}^{\infty}$ increasing to ∞ such that

$$\lim_{j\to\infty} x_{n_j} = \rho.$$

(b) If $\{n_k\}_{k=1}^{\infty}$ is a sequence of integers increasing to ∞ and if

$$\lim_{k\to\infty} x_{n_k} = \sigma, \text{ show that } \sigma \leqslant \rho.$$

33.   Let $\sum_0^{\infty} a_n(z - z_0)^n$ be a power series with radius of convergence R. Show that

$$1/R = \lim_{n\to\infty}\sup \sqrt[n]{|a_n|}.$$

Outline: Let $L = \lim\sup \sqrt[n]{|a_n|}$. Given $\varepsilon > 0$ there is an N such that $\sqrt[n]{|a_n|} < L + \varepsilon$ for all $n \geqslant N$. (Why?) Suppose that $|z - z_0| < (1 - \varepsilon)/(L + \varepsilon)$; then $|a_n||z - z_0|^n < (1 - \varepsilon)^n$ for all $n \geqslant N$ and so $\sum_0^{\infty} a_n(z - z_0)^n$ is absolutely convergent. Hence, $R \geqslant (1 - \varepsilon)/(L + \varepsilon)$ for each $\varepsilon > 0$. (Why?) Therefore, $R \geqslant 1/L$. Conversely, suppose that $|z - z_0| > 1/(L - \varepsilon)$ for some $\varepsilon > 0$. There are infinitely many indices n with $\sqrt[n]{|a_n|} > L - \varepsilon$. (Why?) Hence, $|a_n||z - z_0|^n > 1$ for infinitely many indices n and so $\sum_0^{\infty} a_n(z - z_0)^n$ diverges. (Why?) Hence, $R \leqslant 1/(L - \varepsilon)$ for each $\varepsilon > 0$ and so $R \leqslant 1/L$. Hence, $R = 1/L$.

The binomial theorem for fractional exponents★

34.   Suppose that α is a real number but not a nonnegative integer. Define $\binom{\alpha}{j}$ for each nonnegative integer j by

$$\binom{\alpha}{0} = 1, \qquad \binom{\alpha}{1} = \alpha,$$

and in general

$$\binom{\alpha}{j} = \frac{\alpha(\alpha - 1) \cdots (\alpha - j + 1)}{j!}, \quad j = 1, 2, \ldots$$

Show that $\lim\limits_{j\to\infty} \dfrac{\binom{\alpha}{j+1}}{\binom{\alpha}{j}} = 1$.

$$\left(\text{Hint: } \binom{\alpha}{j} = \frac{(j - 1 - \alpha)}{j} \frac{(j - 2 - \alpha)}{j - 1} \cdots \frac{(1 - \alpha)}{2} \frac{(\alpha)}{1} . \right)$$

35.   Show that the radius of convergence of the series

$$F(z) = \sum_{k=0}^{\infty} \binom{\alpha}{k} z^k$$

is 1.

36. (a) Show that the function F defined in Exercise 35 satisfies the differential equation

$$(1 + z)F'(z) = \alpha F(z).$$

(**Hint:** Compute F'; then multiply it by $(1 + z)$ and add terms with equal powers of z.)

(b) Conclude from (a) that $F(z) = (1 + z)^{\alpha}$; that is

$$(1 + z)^{\alpha} = \sum_{k=0}^{\infty} \binom{\alpha}{k} z^k, \quad |z| < 1.$$

2.3 Cauchy's Theorem and Cauchy's Formula

Cauchy's Theorem is the lynchpin of complex analysis. We shall give two proofs of the result, the first uses Green's Theorem and therefore requires that f' be continuous. This restriction of f' is actually unnecessary as the second proof shows.

Theorem 1: Cauchy's Theorem (*version 1*). Suppose f is analytic in a domain D and that f' is continuous on D. If Γ is a triangle in D whose inside Ω also lies in D, then

$$\int_{\Gamma} f(z)\, dz = 0. \tag{1}$$

Proof. We know from Green's Theorem that

$$\int_{\Gamma} f(z)\, dz = i \iint_{\Omega} \left\{ \frac{\partial f}{\partial x} + i \frac{\partial f}{\partial y} \right\} dx\, dy.$$

However, because $f = u + iv$ is analytic, u and v satisfy the Cauchy–Riemann equations, and so

$$\frac{\partial f}{\partial x} + i \frac{\partial f}{\partial y} = \frac{\partial u}{\partial x} + i \frac{\partial v}{\partial x} + i\left(\frac{\partial u}{\partial y} + i \frac{\partial v}{\partial y}\right)$$

$$= \frac{\partial u}{\partial x} - \frac{\partial v}{\partial y} + i\left(\frac{\partial v}{\partial x} + \frac{\partial u}{\partial y}\right)$$

$$= 0 + i(0) = 0. \qquad \square$$

Theorem 2: Cauchy's Theorem (*version 2*)⋆ Suppose f is analytic in a domain D. If Γ is a triangle in D whose inside Ω is also in D, then

$$\int_{\Gamma} f(z)\, dz = 0. \tag{2}$$

Proof. Without the hypothesis of continuity on f' we cannot apply Green's Theorem so we must use a more sophisticated technique. Let

$$I = \left| \int_{\Gamma} f(z)\, dz \right|$$

so that $I \geqslant 0$. We wish, of course, to show that $I = 0$. Divide the (solid) triangle $\Gamma \cup \Omega$ into four triangles by joining the midpoints of the three sides of Γ (Fig. 2.4). Orient the boundaries of all the triangles positively, call the four smaller solid triangles $\Delta_1, \ldots, \Delta_4$, and let $\Gamma_1, \ldots, \Gamma_4$ be the boundaries of $\Delta_1, \ldots, \Delta_4$, respectively. Then

$$\int_{\Gamma} f(z)\, dz = \sum_{j=1}^{4} \int_{\Gamma_j} f(z)\, dz,$$

so that at least one of the small triangles Δ_m satisfies

$$I = \left| \int_{\Gamma} f(z)\, dz \right| \leqslant 4 \left| \int_{\Gamma_m} f(z)\, dz \right| = 4 I_m.$$

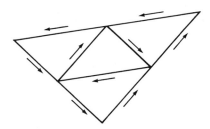

Figure 2.4

Rename this triangle Δ_1. Note that the diameter of Δ_1, the length of its longest side, is one half the diameter of Δ. Proceed as above to divide Δ_1 into four smaller triangles by joining the midpoints of the three sides of Δ_1. As above we will obtain a second triangle, named Δ_2, with boundary Γ_2, such that

$$I_1 = \left| \int_{\Gamma_1} f(z)\, dz \right| \leqslant 4 \left| \int_{\Gamma_2} f(z)\, dz \right| = 4 I_2$$

and diameter $(\Delta_2) = \frac{1}{2}$ diameter $(\Delta_1) = \frac{1}{4}$ diameter (Δ). Continuing in this manner we obtain a sequence of triangles, $\Delta_1, \Delta_2, \Delta_3, \ldots$ with respective boundaries $\Gamma_1, \Gamma_2, \Gamma_3, \ldots$ such that

  (i) Δ_{j+1} is a subset of Δ_j
  (ii)   length $(\Gamma_{j+1}) = \frac{1}{2}$ length (Γ_j)
  (iii)   the diameter of $\Delta_{j+1} = \frac{1}{2}$ {diameter of Δ_j}

  (iv)   if $I_j = \left| \int_{\Gamma_j} f(z)\, dz \right|$, then $|I_j| \leqslant 4\,|I_{j+1}|$.

In particular, we have for $j = 1, 2, \ldots$

  (ii)′   length $(\Gamma_j) = \dfrac{1}{2^j}$ length (Γ)

  (iii)′   diameter $(\Gamma_j) = \dfrac{1}{2^j}$ diameter (Γ)

  (iv)′ $I \leqslant 4^j I_j$.

   Since the diameters of the triangles $\{\Delta_j\}$ decrease to zero and since (i) holds, there is a unique point z_0 within all the Δ_j and z_0 lies in D. Thus, f is differentiable at z_0 so, given $\varepsilon > 0$, there is a (small) $\delta > 0$ such that

$$\left| \frac{f(z) - f(z_0)}{z - z_0} - f'(z_0) \right| < \varepsilon \text{ if } |z - z_0| < \delta.$$

Equivalently,

$$|f(z) - [f(z_0) + f'(z_0)(z - z_0)]| < \varepsilon |z - z_0| \text{ if } |z - z_0| < \delta. \tag{3}$$

We know that for $j \geqslant j_0$, the triangle Δ_j lies within the disc $\{z : |z - z_0| < \delta\}$ because diameter $(\Delta_j) \to 0$ as $j \to \infty$. Further, we know that

$$\int_{\Gamma_j} dz = \int_{\Gamma_j} z\, dz = 0 \tag{4}$$

by version 1 of Cauchy's Theorem. The inequality given in (3) together with (4) and inequality (3) in Section 6 of Chapter 1 give

$$I_j = \left| \int_{\Gamma_j} f(z)\, dz \right| = \left| \int_{\Gamma_j} \{f(z) - [f(z_0) + f'(z_0)(z - z_0)]\}\, dz \right|$$

$$< \varepsilon \left(\max_{z \in \Gamma_j} |z - z_0| \right) (\text{length } (\Gamma_j))$$

$$\leqslant \varepsilon \text{ diameter } (\Delta_j) \text{ length } (\Gamma_j)$$

$$\leqslant \varepsilon \frac{1}{4^j} \text{ diameter } (\Delta) \text{ length } (\Gamma).$$

Finally, employing (iv)′ above, we find that

$$0 \leqslant I \leqslant 4^j I_j \leqslant 4^j \left\{ \varepsilon \frac{1}{4^j} \text{ diameter } (\Delta) \text{ length } (\Gamma) \right\}$$

and so

$$0 \leqslant I \leqslant \varepsilon \text{ diameter } (\Delta) \text{ length } (\Gamma).$$

Since ε is an arbitrary positive number, we see that the number I must be zero.

□

Although the result looks very specialized at this point, we can build on it to obtain much more general results. We begin by considering a convex open set D and an analytic function f on D. Fix a point z_0 in D and define

$$F(z) = \int_{z_0}^{z} f(w) \, dw, \tag{5}$$

where the path of integration is the straight line segment joining z_0 to z. (This segment lies in D because D is convex.) We will now show that F is analytic on D and, indeed, $F' = f$ throughout D.

If $z_1 \in D$, then

$$F(z_1 + h) - F(z_1) = \int_{z_1}^{z_1 + h} f(w) \, dw$$

by Cauchy's Theorem applied to the triangle with vertices at z_0, z_1, and $z_1 + h$ (Fig. 2.5). Hence,

$$\frac{F(z_1 + h) - F(z_1)}{h} - f(z_1) = \int_{z_1}^{z_1 + h} \frac{1}{h} \{f(w) - f(z_1)\} \, dw. \tag{6}$$

Given $\varepsilon > 0$, choose δ so small that

$$|f(w) - f(z_1)| < \varepsilon \text{ whenever } |w - z_1| < \delta.$$

This is possible because f is continuous at z_1. Then for $|h| < \delta$, we have from

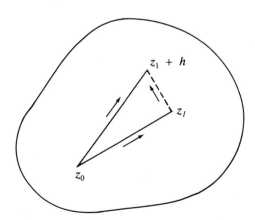

Figure 2.5

(6) and inequality (3) in Section 6 of Chapter 1

$$\left| \frac{F(z_1 + h) - F(z_1)}{h} - f(z_1) \right| \leqslant \frac{1}{|h|} \left\{ \max_{|w - z_1| \leqslant |h|} |f(w) - f(z_1)| \right\} |h| < \varepsilon.$$

Consequently, the derivative, $F'(z_1)$ exists and equals $f(z_1)$.

Actually, a review of the foregoing calculation shows that it is not necessary that D be convex; all that is needed is that D be open and that there be *some* point z_0 in D with the property that for each $z \in D$, the line segment from z_0 to z lies in D. Such domains are called **star-shaped** (Fig. 2.6). To summarize, then,

> *if f is analytic in a star-shaped domain* D, *then there is an*
> *analytic function* F *on* D *with* $F' = f$ *throughout* D   (7)

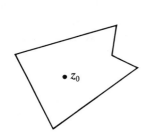

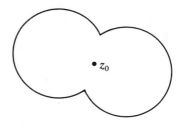

Two star-shaped
domains that
are not convex

Figure 2.6

With (7) in hand we can obtain further results. Let γ be a piecewise smooth curve in a star-shaped domain D. Then

$$\frac{d}{dt} \{F(\gamma(t))\} = F'(\gamma(t))\gamma'(t) = f(\gamma(t))\gamma'(t). \tag{8}$$

Hence,

$$\int_\gamma f(z) \, dz = \int_a^b f(\gamma(t))\gamma'(t) \, dt = \int_a^b \frac{d}{dt} F(\gamma(t)) \, dt$$

$$= F(\gamma(b)) - F(\gamma(a)).$$

In particular, if the curve γ is closed, then $\gamma(b) = \gamma(a)$, and so the line integral of f over γ is zero. Thus, we obtain the result:

> *If f is analytic in a star-shaped domain* D *and if* γ *is a*
> *piecewise smooth closed curve in* D, *then* $\int_\gamma f(z) \, dz = 0$.   (9)

One goal of Section 4 of this chapter is to prove (7) and (9) with the star-shaped hypothesis on D weakened.

Cauchy's formula (first version)

Theorem 3: Suppose f is analytic in a domain D and γ is a positively oriented circle in D whose inside Ω is also in D. Then,

$$f(z) = \frac{1}{2\pi i} \int_\gamma \frac{f(\zeta)}{\zeta - z} \, d\zeta \qquad \text{for all } z \in \Omega.$$

Proof. Let z_1 be a point inside γ (that is, in Ω) and let r be a positive number so small that the closed disc $\{z : |z - z_1| \leqslant r\}$ lies within Ω. Let γ_r be the circle of radius r centered at z_1; we orient γ counterclockwise and γ_r clockwise. The horizontal line through z_1 meets γ at two points and γ_r at two points and has two segments L_1 and L_2 within γ but exterior to γ_r (Fig. 2.7). Let Γ_1 be the curve consisting of the bottom part of γ, then L_2 oriented right to left, then the bottom half of γ_r, and finally L_1 oriented right to left. Let Γ_2 be the top part of γ, then L_1 oriented left to right, then the top half of γ_r, and then finally, L_2 oriented left to right. Suppose now that g is analytic on D except possibly at z_1; by considering an appropriate star-shaped region we can employ (9) and find that

$$\int_{\Gamma_1} g(z) \, dz = \int_{\Gamma_2} g(z) \, dz = 0.$$

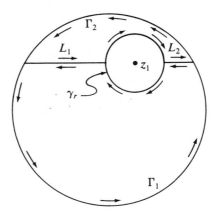

Figure 2.7

Thus, because the line integrals on L_1 and L_2 cancel when we consider $\Gamma_1 + \Gamma_2$, we have

$$\int_\gamma g(z) \, dz + \int_{\gamma_r} g(z) \, dz = \int_{\Gamma_1 + \Gamma_2} g(z) \, dz = 0. \tag{10}$$

Now choose $g(z) = f(z)/(z - z_1)$, $z \neq z_1$. From (10) we find that

$$\int_\gamma \frac{f(z)}{z - z_1}\, dz = -\int_{\gamma_r} \frac{f(z)}{z - z_1}\, dz$$

$$= \int_0^{2\pi} f(z_1 + re^{it}) \frac{1}{re^{it}} ire^{it}\, dt$$

$$= i\int_0^{2\pi} f(z_1 + re^{it})\, dt.$$

Let r decrease to 0 and make use of the continuity of f at z_1 (see Example 9 in Section 6 of Chapter 1). We find that

$$\int_\gamma \frac{f(z)}{z - z_1}\, dz = i2\pi f(z_1),$$

which is just what we wished to show.   □

We first make use of Theorem 3 to evaluate some definite integrals over the circle $|z| = 1$.

Example 1   Find the value of $\displaystyle\int_0^{2\pi} \frac{d\theta}{2 + \sin\theta}$.

Solution   The idea in this example, and others like it, is to use the substitution $z = e^{i\theta}$ effectively. We have

$$\cos\theta = \frac{1}{2}(z + 1/z), \quad z = e^{i\theta}$$

$$\sin\theta = \frac{1}{2i}(z - 1/z), \quad z = e^{i\theta}$$

$$d\theta = \frac{1}{i}\, dz/z.$$

Thus, $2 + \sin\theta = 2 + (1/2i)(z - 1/z) = (1/2)(4 - iz + i/z)$ and

$$\frac{d\theta}{2 + \sin\theta} = \frac{2\, dz}{iz(4 - iz + i/z)} = \frac{2\, dz}{z^2 + 4iz - 1}.$$

Now,

$$z^2 + 4iz - 1 = [z - i(\sqrt{3} - 2)][z + i(\sqrt{3} + 2)].$$

Set $p = i(\sqrt{3} - 2)$ and $q = -i(\sqrt{3} + 2)$; p lies within the circle $|z| = 1$ while q lies outside the circle $|z| = 1$, since $|q| = \sqrt{3} + 2$. The function $(z - q)^{-1}$ is analytic in the disc $|z| < \sqrt{3} + 2$ and so Cauchy's formula (Theorem 3) gives us

$$\frac{1}{2\pi i} \int_{|z|=1} \frac{dz}{(z-q)(z-p)} = \frac{1}{p-q} = \frac{1}{2\sqrt{3}i}.$$

But the integral is just

$$\frac{1}{2\pi i} \int_{|z|=1} \frac{dz}{(z-q)(z-p)} = \frac{1}{2\pi i} \int_0^{2\pi} \frac{ie^{i\theta}\, d\theta}{2ie^{i\theta}(2+\sin\theta)} = \frac{1}{4\pi i} \int_0^{2\pi} \frac{d\theta}{2+\sin\theta}.$$

This yields

$$\int_0^{2\pi} \frac{d\theta}{2+\sin\theta} = \frac{2\pi}{\sqrt{3}}. \qquad \qquad \square$$

Example 2   Evaluate the integral

$$\frac{1}{2\pi} \int_0^{2\pi} \frac{d\theta}{1 - 2a\cos\theta + a^2}, \quad 0 < a < 1.$$

Solution   Again we use the substitution $z = e^{i\theta}$. This yields

$$1 - 2a\cos\theta + a^2 = 1 + a^2 - a(z + 1/z),$$

so

$$\frac{d\theta}{1 - 2a\cos\theta + a^2} = \frac{dz}{iz(1 + a^2 - a(z + 1/z))} = \frac{dz}{i(-az^2 + (1 + a^2)z - a)}.$$

Now

$$-az^2 + (1 + a^2)z - a = -a(z - 1/a)(z - a).$$

The point $1/a$ is outside the circle $|z| = 1$ and the point a is inside the circle $|z| = 1$. Hence,

$$\frac{1}{2\pi i} \int_{|z|=1} \frac{dz}{-az^2 + (1 + a^2)z - a} = \frac{1}{2\pi i} \int_{|z|=1} \frac{1}{-a(z - 1/a)} \frac{1}{z - a}\, dz$$

$$= \frac{1}{-a(a - 1/a)} = \frac{1}{1 - a^2}.$$

But

$$\frac{1}{2\pi i} \int_{|z|=1} \frac{dz}{-az^2 + (1 + a^2)z - a} = \frac{1}{2\pi i} \int_0^{2\pi} \frac{ie^{i\theta}\, d\theta}{e^{i\theta}(1 - 2a\cos\theta + a^2)}$$

$$= \frac{1}{2\pi} \int_0^{2\pi} \frac{d\theta}{1 - 2a\cos\theta + a^2}.$$

Thus, the integral has the value

$$\frac{1}{2\pi} \int_0^{2\pi} \frac{d\theta}{1 - 2a\cos\theta + a^2} = \frac{1}{1 - a^2}, \quad 0 < a < 1. \qquad (11)$$

The function

$$P_a(\theta) = \frac{1 - a^2}{1 - 2a \cos \theta + a^2}$$

is called the **Poisson kernel** and plays an important part in the subject of harmonic functions; we will return to $P_a(\theta)$ in Chapter 4. ☐

Example 3 The following is a more difficult application, this time of (9). The function $f(z) = \exp(-z^2)$ is entire. Let β be fixed, $\beta > 0$, and let R be a large positive number; let γ be the contour in Figure 2.8. Hence, $\int_\gamma f(z) \, dz = 0$ by (9). Parametrize the contour γ as indicated in Figure 2.8. Thus,

$$0 = \int_0^R e^{-x^2} \, dx + \int_0^\beta e^{-(R+it)^2} i \, dt + \int_R^0 e^{-(i\beta+x)^2} \, dx + \int_\beta^0 e^{-(it)^2} i \, dt$$

$$= \int_0^R e^{-x^2} \, dx + ie^{-R^2} \int_0^\beta e^{-2iRt} e^{t^2} \, dt - e^{\beta^2} \int_0^R e^{-x^2} e^{-2i\beta x} \, dx - i \int_0^\beta e^{t^2} \, dt.$$

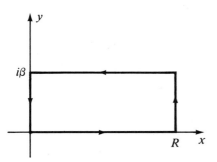

$$\gamma : \begin{array}{ll} z = x, & 0 \leqslant x \leqslant R \\ z = R + it, & 0 \leqslant t \leqslant \beta \\ z = x + i\beta, & R \geqslant x \geqslant 0 \\ z = it, & \beta \geqslant t \geqslant 0 \end{array}$$

Figure 2.8

As $R \to \infty$, the first integral approaches $\frac{1}{2}\sqrt{\pi}$, since

$$\int_{-\infty}^{\infty} e^{-x^2} \, dx = \sqrt{\pi}.$$

(This should be familiar from calculus; also see Exercise 20 in this section.)
The second integral can be estimated by

$$\left| \int_0^\beta e^{-2iRt} e^{t^2} \, dt \right| \leqslant \int_0^\beta e^{t^2} \, dt \leqslant \beta e^{\beta^2},$$

so that

$$e^{-R^2} \int_0^\beta e^{-2iRt} e^{t^2} \, dt \to 0$$

as $R \to \infty$. (Recall that β is fixed.) The third integral converges to

$$\int_0^\infty e^{-x^2}[\cos(2\beta x) - i \sin(2\beta x)] \, dx$$

as $R \to \infty$. Setting the real and imaginary parts of the resulting expression separately equal to zero we find that

$$e^{\beta^2} \int_0^\infty e^{-x^2} \cos(2\beta x) \, dx = \frac{\sqrt{\pi}}{2} \tag{12}$$

and

$$e^{\beta^2} \int_0^\infty e^{-x^2} \sin(2\beta x) \, dx = \int_0^\beta e^{t^2} \, dt. \tag{13}$$

The term on the right in (13) is called the "error function" at β, abbreviated erf(β), and its values for various β can be found in standard tables of mathematical functions.

Exercises for Section 2.3

In Exercises 1 to 4 evaluate the given integral using Cauchy's formula.

1. $\displaystyle\int_{|z|=1} \frac{z}{(z-2)^2} \, dz$

2. $\displaystyle\int_{|z|=2} \frac{e^z}{z(z-3)} \, dz$

3. $\displaystyle\int_{|z+1|=2} \frac{z^2}{4-z^2} \, dz$

4. $\displaystyle\int_{|z|=1} \cot z \, dz$

In Exercises 5 to 8 evaluate the definite trigonometric integral making use of the technique of Examples 1 and 2 in this section.

5. $\displaystyle\int_0^{2\pi} \frac{d\theta}{2 + \cos \theta}$

6. $\displaystyle\int_0^{2\pi} \frac{d\theta}{a + b \sin \theta}, \quad a > b > 0$

7. $\displaystyle\int_0^{2\pi} \frac{d\theta}{3 + \sin \theta + \cos \theta}$

8. $\displaystyle\int_0^{\pi} \frac{d\theta}{1 + \sin^2 \theta}$

In Exercises 9 to 12 evaluate the given integral; indicate which theorem or device you used to obtain your answer.

9. $\displaystyle\int_\gamma \frac{dz}{z^2}$, where γ is any curve in Re $z > 0$ joining $1 - i$ to $1 + i$.

10. $\displaystyle\int_\gamma \left(z + \frac{1}{z}\right) dz$, where γ is any curve in Im $z > 0$ joining $-4 + i$ to $6 + 2i$.

11. $\displaystyle\int_\gamma e^z \, dz$, where γ is the semicircle from -1 to 1 passing through i.

12. $\int_\gamma \sin z \; dz$, where γ is any curve joining i to π.

13.   Integrate e^{iz^2} around the contour γ shown in Figure 2.9 to obtain the **Fresnel integrals**:

$$\int_0^\infty \cos(x^2) \; dx = \int_0^\infty \sin(x^2) \; dx = \frac{\sqrt{2\pi}}{4}. \tag{14}$$

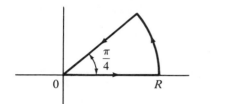

$$
\begin{aligned}
z &= x, & 0 \leqslant x \leqslant R \\
z &= Re^{i\theta}, & 0 \leqslant \theta \leqslant \pi/4 \\
z &= te^{i\pi/4}, & R \geqslant t \geqslant 0
\end{aligned}
$$

Figure 2.9

(**Hint:** On the circular arc $z = Re^{i\theta}$, $0 \leqslant \theta \leqslant \pi/4$, you will need to know that $\sin \psi \geqslant (2/\pi)\psi$, $0 \leqslant \psi \leqslant \pi/2$, and hence, $e^{-R^2 \sin \psi} \leqslant e^{-2R^2\psi/\pi}$. The inequality $\sin \psi \geqslant (2/\pi)\psi$ is proved by calculus.)

14.   Specialize Theorem 3 to the case when ζ is the center of the circle and show that

$$f(z_0) = \frac{1}{2\pi} \int_0^{2\pi} f(z_0 + re^{it}) \; dt. \tag{15}$$

15.   (a) Use the estimate (3), Section 6, Chapter 1, and (15) above to conclude that

$$|f(z_0)| \leqslant \max_{0 \leqslant t \leqslant 2\pi} |f(z_0 + re^{it})| \tag{16}$$

for all sufficiently small r.
       (b) Conclude from (a) that $|f|$ cannot have a strict local maximum within its domain of analyticity. That is, the graph in three-dimensional space of $(x, y, |f(x + iy)|)$ has no "mountain peaks."

16.   Let $f(z) = \sum_{j=0}^\infty a_j(z - z_0)^j$ have a radius of convergence of $R > 0$. Show that for each $r \in (0, R)$ we have

$$\frac{f^{(j)}(z_0)}{j!} = a_j = \frac{1}{2\pi i} \int_{|z - z_0| = r} f(z)(z - z_0)^{-j-1} \; dz, \quad j = 0, 1, 2, \ldots$$

(**Hint:** Substitute the series into the integral and interchange summation and integration.)

17.   Use the result of Exercise 16 to establish these formulas:

(a) $\dfrac{1}{2\pi}\displaystyle\int_0^{2\pi} e^{-in\theta}\sin(e^{i\theta})\,d\theta = \begin{cases} 0, & n = 0, 2, 4, \ldots \\ \dfrac{(-1)^{(n-1)/2}}{n!}, & n = 1, 3, 5, \ldots \end{cases}$

(b) $\dfrac{1}{2\pi}\displaystyle\int_0^{2\pi} e^{-in\theta}e^{e^{i\theta}}\,d\theta = \dfrac{1}{n!}, \quad n = 0, 1, 2, \ldots$

(c) $\dfrac{1}{2\pi}\displaystyle\int_0^{2\pi} e^{im\theta}(2e^{in\theta} - e^{i(n+1)\theta})^{-1}\,d\theta = \begin{cases} 0, & n < m \\ 2^{m-n-1}, & n \geqslant m \end{cases}$

18.   Use the conclusion of Exercise 16 to establish the **Cauchy estimates**:

$$|f^{(n)}(z_0)| \leqslant \frac{n!}{r^n}\ \max_{|z-z_0|=r}\ |f(z)|, \ n = 0, 1, 2, \ldots,$$

whenever f is analytic on a domain containing the set $\{z : |z - z_0| \leqslant r\}$.

19.   The function $g(z) = 1/z$ is analytic on the annulus $D = \{z : r < |z| < R\}$ for each $r, R, 0 \leqslant r < R$. Show, however, that there is no analytic function G on D with $G' = g$. This example shows that some hypothesis on a domain D is needed if each analytic function on D is to be the derivative of some other analytic function on D.

The integral $\int_{-\infty}^{\infty} e^{-x^2}\,dx\star$

20.   Let $I = \int_{-\infty}^{\infty} e^{-x^2}\,dx$; then

$$I^2 = \int_{-\infty}^{\infty} e^{-x^2}\,dx \int_{-\infty}^{\infty} e^{-y^2}\,dy = \int_{-\infty}^{\infty}\int_{-\infty}^{\infty} e^{-(x^2+y^2)}\,dx\,dy$$

$$= \int_0^{2\pi}\int_0^{\infty} e^{-r^2}r\,dr\,d\theta = 2\pi\int_0^{\infty} re^{-r^2}\,dr = \pi.$$

Hence, $I = \sqrt{\pi}$.

The Poisson Kernel$\star$

21.   Verify each of the statements (a) through (e) below for $z = re^{i\theta}$ and $r < 1$.

(a) $\mathrm{Re}\left(\dfrac{e^{it} + z}{e^{it} - z}\right) = \dfrac{1 - r^2}{1 - 2r\cos(\theta - t) + r^2}.$

(b) $\dfrac{e^{it}}{e^{it} - z} + \dfrac{\bar{z}e^{it}}{1 - e^{it}\bar{z}} = \mathrm{Re}\left(\dfrac{e^{it} + z}{e^{it} - z}\right).$

(c) $\dfrac{1}{2\pi}\displaystyle\int_0^{2\pi} f(e^{it})\dfrac{e^{it}\bar{z}}{1 - \bar{z}e^{it}}\,dt = 0,$ if $f(w)$ is analytic for $|w| < 1 + \varepsilon.$

(d) $\dfrac{1}{2\pi} \displaystyle\int_0^{2\pi} f(e^{it}) \dfrac{e^{it}}{e^{it} - z}\, dt = f(z)$.

(e) Add (c) and (d); then use (a) and (b) to conclude that

$$f(z) = \frac{1}{2\pi} \int_0^{2\pi} f(e^{it}) \frac{1 - r^2}{1 - 2r \cos(\theta - t) + r^2}\, dt, \; z = re^{i\theta}.$$

2.4   Consequences of Cauchy's Theorem

Cauchy's formula, Theorem 3, Section 3 of this chapter, has far-reaching impli-
cations, allowing us to draw numerous non-obvious conclusions about the
properties of analytic functions. We begin with one of the most important.

Theorem 1: Suppose f is analytic in a domain D and the disc $\Delta = \{z : |z - z_0| \leqslant r\}$ lies in D. Then within the disc Δ, f is given by a
convergent power series

$$f(z) = \sum_{n=0}^{\infty} a_n(z - z_0)^n, \; |z - z_0| < r. \tag{1}$$

Furthermore,

$$a_n = \frac{1}{2\pi i} \int_\gamma \frac{f(\zeta)}{(\zeta - z_0)^{n+1}}\, d\zeta, \; n = 0, 1, 2, \ldots, \tag{2}$$

where γ is the circle $\{\zeta : |\zeta - z_0| = r\}$ oriented positively.

Proof. Fix z with $|z - z_0| = s < r$; then for any ζ with $|\zeta - z_0| = r$, we have

$$\frac{1}{\zeta - z} = \frac{1}{(\zeta - z_0) - (z - z_0)} = \frac{1}{\zeta - z_0} \frac{1}{1 - \left(\dfrac{z - z_0}{\zeta - z_0}\right)}.$$

Now $\left| \dfrac{z - z_0}{\zeta - z_0} \right| = \dfrac{s}{r} < 1$ and so we can represent $\left(1 - \left(\dfrac{z - z_0}{\zeta - z_0}\right)\right)^{-1}$ by the
geometric series

$$\left(1 - \left(\frac{z - z_0}{\zeta - z_0}\right)\right)^{-1} = 1 + \left(\frac{z - z_0}{\zeta - z_0}\right) + \left(\frac{z - z_0}{\zeta - z_0}\right)^2 + \cdots$$

$$= \sum_{k=0}^{\infty} \left(\frac{z - z_0}{\zeta - z_0}\right)^k.$$

The series $\sum_{k=0}^{\infty} (s/r)^k$ converges, since $0 < s < r$. Consequently, given $\varepsilon > 0$, we can choose N so big that $\sum_{k=N}^{\infty} (s/r)^k < \varepsilon$. Hence, for all ζ with $|\zeta - z_0| = r$ and all $n \geqslant N$, we have

$$\left| \sum_{k=n}^{\infty} \left(\frac{z - z_0}{\zeta - z_0} \right)^k \right| \leqslant \sum_{k=n}^{\infty} (s/r)^k \leqslant \sum_{k=N}^{\infty} (s/r)^k < \varepsilon. \tag{3}$$

Now for any n we can write

$$\int_{\gamma} \frac{f(\zeta)}{\zeta - z} \, d\zeta = \int_{\gamma} \frac{f(\zeta)}{\zeta - z_0} \left\{ \sum_{k=0}^{\infty} \left(\frac{z - z_0}{\zeta - z_0} \right)^k \right\} \, d\zeta$$

$$= \sum_{k=0}^{n} (z - z_0)^k \left\{ \int_{\gamma} \frac{f(\zeta)}{(\zeta - z_0)^{k+1}} \, d\zeta \right\}$$

$$+ \int_{\gamma} \frac{f(\zeta)}{\zeta - z_0} \left\{ \sum_{k=n+1}^{\infty} \left(\frac{z - z_0}{\zeta - z_0} \right)^k \right\} \, d\zeta.$$

When we employ (3) and inequality (3) from Section 6 in Chapter 1 we see that

$$\left| \int_{\gamma} \left\{ \frac{f(\zeta)}{\zeta - z_0} \right\} \left\{ \sum_{k=n+1}^{\infty} \left(\frac{z - z_0}{\zeta - z_0} \right)^k \right\} \, d\zeta \right| \leqslant \frac{M}{r} \varepsilon 2\pi r = 2\pi \varepsilon M,$$

where $M = \max\{|f(\zeta)| : |\zeta - z_0| = r\}$. From Theorem 3 we see that for $n \geqslant N$

$$f(z) = \frac{1}{2\pi i} \int_{\gamma} \frac{f(\zeta)}{\zeta - z} \, d\zeta = \sum_{k=0}^{n} (z - z_0)^k \left\{ \frac{1}{2\pi i} \int_{\gamma} \frac{f(\zeta)}{(\zeta - z_0)^{k+1}} \, d\zeta \right\} + E(z; z_0),$$

where $|E(z; z_0)| < \varepsilon M$. This implies that the infinite series

$$\sum_{k=0}^{\infty} (z - z_0)^k \left\{ \frac{1}{2\pi i} \int_{\gamma} \frac{f(\zeta)}{(\zeta - z_0)^{k+1}} \, d\zeta \right\}$$

converges and its sum is $f(z)$. $\square$

Theorem 1 has an immediate, and important, corollary.

If f is analytic on a domain D, *then so is* f'. *Hence, f has derivatives of all orders and each derivative is analytic on* D.   (4)

This follows because in each disc within D, f is given by a power series and, by Theorem 3, Section 2 of this chapter, each power series is infinitely differentiable. It is critical to stress here that the result expressed in (4) requires only that f have a derivative at each point of D *and nothing more*. From the existence of *one* derivative we are able to infer the existence of *infinitely many* derivatives. The reader should contrast this with the case of real-valued functions of a real variable. For instance, the function

$$u(t) = \begin{cases} t^2, & t \geqslant 0 \\ -t^2, & t < 0 \end{cases}$$

is differentiable at all points t, $-\infty < t < \infty$, with derivative

$$u'(t) = \begin{cases} 2t, & t \geqslant 0 \\ -2t, & t < 0 \end{cases}$$

and the derivative is even continuous. However, u' is itself *not* differentiable at $t = 0$.

Let us find the power series for some of the elementary analytic functions discussed in Section 1. This is accomplished by use of Theorem 1 and formula (12), Section 2, of this chapter.

Example 1   The function e^z is entire and is its own derivative. Hence, $a_n = 1/n!$ for $n = 0, 1, 2, \ldots$ and so

$$e^z = \sum_{k=0}^{\infty} \frac{z^k}{k!}$$

as we saw already in formula (13), Section 2.

Example 2   The functions $\cos z$ and $\sin z$ are entire and satisfy $(\sin z)' = \cos z$ and $(\cos z)' = -\sin z$. Since $\sin(0) = 0$ and $\cos(0) = 1$, we obtain

$$\sin z = \sum_{j=0}^{\infty} (-1)^j \frac{z^{2j+1}}{(2j+1)!},$$

$$\cos z = \sum_{j=0}^{\infty} (-1)^j \frac{z^{2j}}{(2j)!}.$$

Example 3   The function $\text{Log}(1 - z)$ is analytic in the disc $|z| < 1$ (see Example 10, Section 1 of this chapter) and its derivatives are

$$\frac{d^n}{dz^n}(\text{Log}(1 - z)) = -\frac{(n-1)!}{(1-z)^n}, \quad n = 1, 2, \ldots$$

Hence, for $|z| < 1$

$$\text{Log}(1 - z) = -\sum_{n=1}^{\infty} \frac{z^n}{n}.$$

Example 4   The functions $\sinh(z)$ and $\cosh(z)$ are defined by

$$\sinh(z) = \tfrac{1}{2}(e^z - e^{-z}), \qquad \cosh(z) = \tfrac{1}{2}(e^z + e^{-z}).$$

From the expansion of e^z above, we find that

$$\cosh(z) = \sum_{k=0}^{\infty} \frac{z^{2k}}{(2k)!}$$

and

$$\sinh(z) = \sum_{k=0}^{\infty} \frac{z^{2k+1}}{(2k+1)!}.$$

Both series converge for all z by an application of the ratio test.

Example 5   The   function $f(z) = 1/z^3$ is   analytic   in   the   disc
$D = \{z : |z - 1| < 1\}$; at 1, the value of the nth derivative of f is $\frac{1}{2}(n + 2)!$
$(-1)^n$, $n = 1, 2, 3, \ldots$; thus, in D we have

$$\frac{1}{z^3} = 1 + \sum_{1}^{\infty} \frac{1}{2}(n + 2)!(-1)^n \frac{(z - 1)^n}{n!}$$

$$= 1 + \frac{1}{2} \sum_{n=1}^{\infty} (-1)^n (n + 2)(n + 1)(z - 1)^n.$$

The series is valid for $|z - 1| < 1$.

Theorem 1 has another significant implication

*Suppose that f is analytic in a domain D and further at
some point $z_0 \in D$, we have $f^{(k)}(z_0) = 0$, $k = 0, 1, 2, \ldots$.*   **(5)**
Then $f(z) = 0$ for all $z \in D$.

The proof of (5) lies a little deeper than that of (4). To begin, let Δ be
any disc in D with center at z_0; then by Theorem 1 and (12) of Section 2, $f(z)$ is
given by a power series within Δ:

$$f(z) = \sum_{k=0}^{\infty} a_k(z - z_0)^k; \qquad a_k = \frac{f^{(k)}(z_0)}{k!}, \quad k = 0, 1, \ldots.$$

Thus the hypothesis that $f^{(k)}(z_0) = 0$ for all k implies that $f(z) = 0$ for
all $z \in \Delta$. Next, let L be a line segment in D from z_0 to another point z_1 in D.
(It is important that all of L lies within D.) We shall show that f and, in fact, all
the derivatives of f vanish at all points of L. We take note of the fact that there
is some positive number δ such that each disc of radius δ centered at a point of
L lies entirely within D. Let $\Delta_0, \Delta_1, \Delta_2, \ldots, \Delta_N$ be a "chain" of discs of radius
$\delta/2$, centered at points $\zeta_0 = z_0, \zeta_1, \ldots, \zeta_N$ in L, and with the point ζ_j in the disc
Δ_{j-1} for $j = 1, 2, \ldots, N$, and $z_1 \in \Delta_N$ (Fig. 2.10).
We know already that $f = 0$ at all points of the disc Δ_0; hence,
$f^{(k)}(\zeta_1) = 0$ for $k = 0, 1, \ldots$, since $\zeta_1 \in \Delta_0$. Thus, the power-series representation for $f(z)$, which is valid in Δ_1 shows that $f = 0$ at all points of the disc Δ_1.
Consequently, $f^{(k)}(\zeta_2) = 0$ for $k = 0, 1, \ldots$ since $\zeta_2 \in \Delta_1$ and we learn $f \equiv 0$ in
the disc Δ_2. Continuing in this way, after N steps we learn that $f^{(k)}(z_1) = 0$ for
$k = 0, 1, \ldots$, which is what we wished to show. Hence, if $f^{(k)}(z_0) = 0$ for $k = 0$,
$1, 2, \ldots$, then $f^{(k)}(z_1) = 0$ for $k = 0, 1, \ldots$ at any point z_1 that can be reached
from z_0 by a straight line segment. Since *all* points of D can be reached from z_0

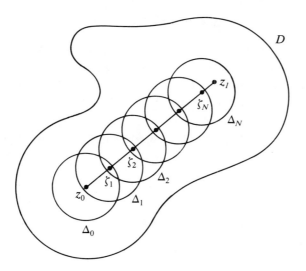

Figure 2.10

by a polygonal curve consisting of a finite number of straight line segments, we
see that $f(z) = 0$ for all $z \in D$.

The order of a zero

Suppose that f is analytic and not identically zero on a domain D and $f(z_0) = 0$ for some $z_0 \in D$. Thus, the power series for f centered at z_0 is

$$f(z) = a_1(z - z_0) + a_2(z - z_0)^2 + \cdots.$$

We know from (5) that not all of the coefficients a_k can vanish and hence there
is an integer $m \geqslant 1$ such that

$$a_1 = \cdots = a_{m-1} = 0 \ \text{ but } \ a_m \neq 0.$$

Thus,

$$f(z) = a_m(z - z_0)^m + a_{m+1}(z - z_0)^{m+1} + \cdots, \ a_m \neq 0.$$

This means that

$$f^{(k)}(z_0) = 0 \ \text{ for } \ k = 0, \ldots, m-1 \text{ but } f^{(m)}(z_0) \neq 0.$$

In this case we say that f **has a zero of order** m at z_0. Furthermore, if we set

$$g(z) = \frac{f(z)}{(z - z_0)^m},$$

then g is actually analytic in D and $g(z_0) = a_m \neq 0$. Consequently,

> *if f has a zero of order m at z_0, then f(z) = $(z - z_0)^m$g(z),
> where g is analytic in D and g(z_0) $\neq$ 0.*   (6)

Morera's* Theorem

Cauchy's Theorem, Theorem 1 and/or 2 of Section 3, on which so much of the development of complex variables depends, has a converse. This theorem, given precisely below, states that if the integral of a continuous function f over every triangle in some domain is zero, then f must be analytic in that domain.

> **Theorem 2: Morera's Theorem:** If f is a continuous function on a domain D and if
>
> $$\int_\gamma f(z)\, dz = 0$$
>
> for *every* triangle γ that lies, together with its interior, in D, then f is analytic on D.

Proof. The proof can be adapted almost word for word from parts of Section 3. Let z_0 be a point of D and Ω be the disc $\{z : |z - z_0| < r\}$, where $r > 0$ is so small that Ω is in D. We shall show that f is analytic on the disc Ω. Define

$$F(z) = \int_{z_0}^{z} f(\zeta)\, d\zeta,$$

where the integration is along the radius joining z_0 to z. We shall show that F is analytic in Ω and that $F' = f$. It will then follow from (4) of this section that f is analytic in Ω, as well.

Let h be a small complex number; then

$$F(z + h) - F(z) = \int_{z}^{z+h} f(\zeta)\, d\zeta,$$

because the integral of f around the triangle with vertices $\{z_0, z, z + h\}$ is zero (Fig. 2.11). Hence

$$\frac{F(z + h) - F(z)}{h} - f(z) = \int_{z}^{z+h} \left\{ \frac{f(\zeta) - f(z)}{h} \right\} d\zeta.$$

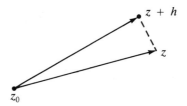

Figure 2.11

* Ciacinto Morera, 1856–1909.

Suppose now that $\varepsilon > 0$ is given; choose δ so small that

$$|f(\zeta) - f(z)| < \varepsilon$$

if $|\zeta - z| < \delta$; this is possible by the continuity of f. Then for $|h| < \delta$, we have

$$\left| \int_z^{z+h} \{f(\zeta) - f(z)\} \, d\zeta \right| < \varepsilon |h|$$

by the estimate (3) from Section 6 of Chapter 1. Putting all this together we find that

$$\left| \frac{F(z+h) - F(z)}{h} - f(z) \right| < \varepsilon \text{ if } |h| < \delta,$$

which is precisely the statement that $F'(z) = f(z)$.   □

Cauchy's Formula (again)

The final objective of this section is to extend formulas (7) and (9) of Section 3 to a more general context by weakening the restriction on the domain D. As a byproduct we will also obtain a more general form of Cauchy's formula than that given in Theorem 3 of Section 3.

> **Definition** A domain D is **simply-connected** if whenever γ is a simple closed curve in D, then the inside of γ is also a subset of D.

A few examples will help set the concept of simple connectivity.

Example 6 The disc $\{z: |z - z_0| < R\}$ is simply-connected, as is the horizontal strip $\{z: c < \text{Im } z < d\}$.

Example 7 The annulus $\{z: 0 < r_1 < |z - z_0| < r_2\}$ is not simply-connected, since for each r, $r_1 < r < r_2$, the circle $|z - z_0| = r$ lies in the annulus, but its inside, the disc $\{z: |z - z_0| < r\}$ is *not* a subset of the annulus.

Example 8 The punctured disc $\{z: 0 < |z - z_0| < r_2\}$ is not simply-connected for reasons identical to those in Example 7.

Example 9 The domain obtained by deleting the line segment $0 \leqslant x \leqslant 1$ from the open disc $\{z: |z| < 1\}$ is simply-connected while the domain obtained by deleting the segment $0 \leqslant x \leqslant 1/2$ is *not* simply-connected (Fig. 2.12(a) and 2.12(b)).

Example 10 The domain obtained from the annulus $\{z: 1 < |z| < 2\}$ by deleting those x with $1 < x < 2$ is simply-connected (Fig. 2.12(c)).

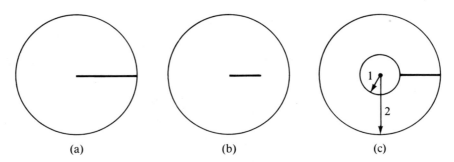

(a)   (b)   (c)

Figure 2.12

In a nutshell a domain is simply-connected if it has no "holes" in it; this is the intuitive idea behind the definition.

We need one fact relating analytic functions and a certain special type of closed curve in a simply-connected domain.

> Let D *be simply connected and let* Γ *be a closed curve*
> *in* D *consisting entirely of vertical and/or horizontal line*   (7)
> *segments. If* f *is analytic in* D*, then*

$$\int_\Gamma f(z)\ dz = 0.$$

This is actually rather difficult to prove completely, though it is reasonably clear from a picture (see, for instance, Figure 2.13). The reader should draw several more pictures of such closed curves to explore their other possible configurations. The critical issue is that any closed loops in Γ surround only points of D since D is simply-connected. With (7) in hand, we are ready to begin the extension of Cauchy's Theorem to simply-connected domains.

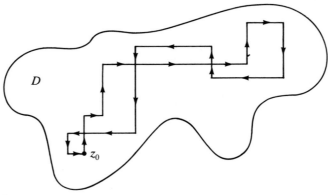

Figure 2.13

Theorem 3: Let f be analytic on a simply-connected domain D. Then there is an analytic function F on D with $F' = f$ on D.

Proof. Fix a point $z_0 \in D$ and define $F(z)$ by the rule

$$F(z) = \int_\gamma f(w) \, dw,$$

where γ is any polygonal curve from z_0 to z that is composed entirely of horizontal/vertical line segments. If γ_1 and γ_2 are two such paths, then $\Gamma = \gamma_1 - \gamma_2$ is a closed curve in D consisting entirely of horizontal/vertical line segments so that

$$0 = \int_\Gamma f(w) \, dw = \int_{\gamma_1} f(w) \, dw - \int_{\gamma_2} f(w) \, dw$$

by (7). Thus, the value of $F(z)$ does *not* depend on the choice of curve γ from z_0 to z; in mathematical language, $F(z)$ is "well defined." The proof that $F' = f$ is now identical to the proof of (7) of Section 3; the reader is urged to fill in the details.   □

Corollary: If f is analytic on a simply-connected domain D and if Γ is any piecewise smooth closed curve in D, then

$$\int_\Gamma f(z) \, dz = 0.$$

Proof. Just as in (8) of Section 3 note that

$$\frac{d}{dt}(F(\Gamma(t))) = F'(\Gamma(t))\Gamma'(t) = f(\Gamma(t))\Gamma'(t),$$

where F is the function from Theorem 3. Hence,

$$\int_\Gamma f(z) \, dz = \int_a^b f(\Gamma(t))\Gamma'(t) \, dt$$

$$= \int_a^b \frac{d}{dt}(F(\Gamma(t))) \, dt$$

$$= F(\Gamma(b)) - F(\Gamma(a)) = 0,$$

since $\Gamma(b) = \Gamma(a)$ because Γ is closed.   □

Finally, we are ready for a reasonably general version of Cauchy's Formula.

Theorem 4: Cauchy's Formula: If f is analytic on a simply-connected domain D and if γ is a piecewise smooth simple closed curve in D, then

$$f(\zeta) = \frac{1}{2\pi i} \int_\gamma \frac{f(w)}{w - \zeta}\, dw \tag{8}$$

for all ζ inside γ.

Proof. The function $g(z)$, defined by

$$g(z) = \begin{cases} \dfrac{f(z) - f(\zeta)}{z - \zeta}, & z \neq \zeta \\[2mm] f'(\zeta), & z = \zeta \end{cases}$$

is analytic in D. For in a small disc centered at ζ, we have

$$f(z) = \sum_0^\infty c_k(z - \zeta)^k, \qquad c_k = \frac{f^{(k)}(\zeta)}{k!}, \quad k = 0, 1, 2, \ldots,$$

so that

$$g(z) = \sum_1^\infty c_k(z - \zeta)^{k-1}$$

in this same disc. Hence, $\int_\gamma g(w)\, dw = 0$ by the corollary to Theorem 3. Therefore,

$$\frac{1}{2\pi i} \int_\gamma \frac{f(w)}{w - \zeta}\, dw = f(\zeta)\, \frac{1}{2\pi i} \int_\gamma \frac{dw}{w - \zeta}.$$

$$= f(\zeta). \qquad \square$$

We now give three applications of the theorems of this section.

Application 1   Liouville's Theorem

Here we shall use Cauchy's Formula to show that if an entire function is bounded, then it must be identically constant. This result is known as **Liouville's* Theorem.**

> **Theorem 5:** If F is entire and if there is a constant M such that $|F(z)| \leq M$ for all z, then F is identically constant.

Proof. Set $g(z) = (F(z) - F(0))/z$; g is then an entire function and for all z, $|z| = R > 0$, and we may make the simple estimate

---

* Joseph Liouville, 1809–1882.

$$|g(Re^{i\theta})| \leqslant \frac{|F(Re^{i\theta})| + |F(0)|}{R} \leqslant \frac{2M}{R}.$$

Let ζ be a point of the plane and assume R is so big that $R > |\zeta|$. By Cauchy's formula

$$g(\zeta) = \frac{1}{2\pi i} \int_{|z| = R} \frac{g(z)}{z - \zeta} \, dz,$$

so that

$$|g(\zeta)| \leqslant \frac{1}{2\pi} \frac{2M}{R} \frac{2\pi R}{R - |\zeta|} \to 0 \text{ as } R \to \infty.$$

Consequently $g = 0$. Thus, $F \equiv F(0)$. $\qquad\qquad\square$

Application 2   Analytic logarithms

Suppose f is analytic and zero-free in a simply-connected domain D. Fix $z_0 \in D$ and define

$$h(z) = \int_{z_0}^{z} \frac{f'(w)}{f(w)} \, dw,$$

where the integral is taken over any piecewise smooth curve in D from z_0 to z. Since the integrand f'/f is analytic in D, the function h is analytic in D. As we have seen several times already, the derivative of h is the integrand; in this case, $h' = f'/f$. Thus,

$$\begin{aligned}(e^{-h(z)}f(z))' &= -h'(z)e^{-h(z)}f(z) + e^{-h(z)}f'(z) \\ &= e^{-h(z)}[-f'(z) + f'(z)] \\ &= 0.\end{aligned}$$

Consequently, $e^{-h(z)}f(z) = c$ for all $z \in D$, where c is some nonzero constant. Setting $z = z_0$ yields $c = f(z_0)$, since $h(z_0) = 0$, and so

$$f(z) = f(z_0)e^{h(z)}.$$

That is to say, if f is zero-free and analytic on a simply connected domain D, then there is an analytic function g with

$$e^{g(z)} = f(z), \ z \in D.$$

[Just set $g = h - \text{Log}\, f(z_0)$.]

Application 3★   Flows continued

Suppose that f is a sourceless and irrotational flow in a domain D. We know from the discussion in Section 1.1 of this chapter that $\bar{f}$ is analytic on D. Even more, we learned there that

$$\int_\gamma \overline{f(z)} \, dz = 0 \tag{9}$$

for all smooth closed curves γ in D. It is simple to see from this that (9) also holds for all piecewise smooth closed curves γ, as well. Equation (9) actually implies that $\bar{f}$ is the derivative of some analytic function G on D. To see this, fix a point a in D and define

$$G(z) = \int_\Gamma \overline{f(w)} \, dw,$$

where Γ is any piecewise smooth curve from a to z. $G(z)$ is independent of the choice of Γ, since if Γ' is another curve from a to z, then $\gamma = \Gamma - \Gamma'$ is a piecewise smooth closed curve so that

$$0 = \int_\gamma \overline{f(w)} \, dw = \int_\Gamma \overline{f(w)} \, dw - \int_{\Gamma'} \overline{f(w)} \, dw.$$

Now an argument identical to the one employed in proving Theorem 2, Morera's Theorem, shows that $G' = \bar{f}$ on D. Any function G with $G' = \bar{f}$ is called a **complex potential** of the flow f. We shall see in Section 4.1 of Chapter 3 that the complex potential is intimately related to conformal mapping.

Exercises for Section 2.4

In Exercises 1 to 8, give the order of *each* of the zeros of the given function

1. $\dfrac{\sin z}{z}$

   2. $(e^z - 1)^2$

   3. $(z^2 + z - 2)^3$

4. $(z^2 - 4z + 4)^3$

   5. $z^2(1 - \cos z)$

   6. $\text{Log}(1 - z), \ |z| < 1$

7. $e^{2z} - 3e^z - 4$

   8. $\dfrac{z}{z^2 + 1}$

In Exercises 9 to 16, find the power-series expansion about the given point for each of the functions; find the largest disc in which the series is valid

9. $z(e^z - 1)$ about $z_0 = 0$ 10. e^z about $z_0 = \pi i$

11. $z^3 + 6z^2 - 4z - 3$ about $z_0 = 1$

12. $\dfrac{z^2}{1 - z}$ about $z_0 = 0$ 13. $\dfrac{z + 2}{z + 3}$ about $z_0 = -1$

14. $[\text{Log}(1 - z)]^2$ about $z_0 = 0$ (first four terms)

15. $\sin \pi z$ about $z_0 = 1/2$

16. $\tan z$ about $z_0 = 0$ (first four terms)

17.   Suppose that f is analytic on a domain D and has a zero of order m at z_0 in D. Show that (a) f' has a zero of order $m - 1$ at z_0; (b) f^2 has a zero of order $2m$ at z_0.

18.   Suppose that f has a zero of order m at z_0 and that g is analytic on a domain containing the range of f with a zero of order l at the origin. Show that $h(z) = h(f(z))$ has a zero of order ml at z_0.

19.   Use the Cauchy estimates, Exercise 18, Section 3 of this chapter, for $n = 1$ to give another proof of Liouville's Theorem by showing that the derivative of a bounded analytic function is identically zero.

20.   Suppose that f is an entire function and $\operatorname{Re} f(z) \leqslant c$ for all z. Show that f is constant. (**Hint:** Consider $\exp(f(z))$.)

21.   Suppose that f is an entire function and that there are positive constants A and m with $|f(z)| \leqslant A|z|^m$ if $|z| \geqslant R_0$. Show that f is a polynomial of degree m or less. (**Hint:** Use the Cauchy estimates (Exercise 18, Section 3) for $n > m$ and let $r \to \infty$.)

22.   Let D be a simply connected domain and f an analytic function on D that has no zeros in D. Let γ be a complex number, $\gamma \neq 0$. Show that there is an analytic function g on D with $f = g^\gamma$. (**Hint:** Use Application 2.)

23.   Suppose that F is analytic in the region $|z| > R$, including at ∞ (that is, $G(z) = F(1/z)$ is analytic for $|z| < 1/R$). Show that F can be expressed as a "power series" in $1/z$:

$$F(z) = \sum_{k=0}^{\infty} c_k \frac{1}{z^k}, \quad |z| > R$$

and derive a formula for c_k similar to that given in equation (2).

24.   Use Morera's Theorem and an interchange of the order of integration to show that each of the following functions is analytic on the indicated domain; find a power-series expansion for each function by using the known power series for the integrand and interchanging the summation and integration

(a) $\displaystyle\int_0^1 \frac{dt}{1 - tz}$ on $|z| < 1$ (b) $\displaystyle\int_0^{1/2} \operatorname{Log}(1 - tz)\, dt$ on $|z| < 2$

(c) $\displaystyle\int_{-\pi/2}^{\pi/2} \sin(z + t)\, dt,$ all z.

Differential equations in the complex plane⋆

25.   Find all solutions to the differential equation
$$f''(z) + \beta^2 f(z) = 0, \ f \text{ is an entire function.}$$

(Hint: Write $f(z) = \sum_0^\infty a_j z^j$ and solve for the coefficients a_2, a_3, ... in terms of a_0, a_1, and β.)

26.   Use the technique of Exercise 25 to give the solutions of these differential equations

(a) $f''(z) - 3f'(z) + 2f(z) = 0$; $a_0 = 1$, $a_1 = 2$

(b) $f''(z) - \dfrac{2}{z} f'(z) + f(z) = 0$; $z \neq 0$; $f(0) = 1$, $f'(0) = 0$

(c) $f''(z) + 2f'(z) + f(z) = 0$; $a_0 = 0$, $a_1 = 1$

27.   If f and A are analytic in a simply-connected domain D and

$$f'(z) = A(z)f(z), \quad z \in D,$$

show that $f(z) = C \exp[\int_{z_0}^z A(w)\, dw]$ for a constant C, where the integral is taken over any piecewise smooth curve joining (a fixed point) z_0 to z. **(Hint:** Let

$$g(z) = \exp\left[-\int_{z_0}^z A(w)\, dw \right].$$

Show that $(f(z)g(z))' = 0$ in D.)

28.   Use the result of Exercise 27 to give the solution to each of these differential equations:

(a) $f'(z) - 2zf(z) = 0$
(b) $f'(z) + e^z f(z) = 0$

(c) $f'(z) + \dfrac{1}{z} f(z) = 0$

Bessel functions★

29.   Let v be an integer, $v \geq 0$. Show that one solution of the differential equation

$$z^2 f''(z) + zf'(z) + (z^2 - v^2)f(z) = 0$$

is

$$J_v(z) = \sum_{n=0}^\infty (-1)^n \frac{z^{2n+v}}{2^{2n+v}(v+n)!\,n!}.$$

J_v is the **Bessel function of the first kind of order** v.

30.   Show that the Bessel functions satisfy the **recurrence relation**

$$J_{v-1} - J_{v+1} = 2J_v'.$$

2.5   Isolated singularities

An analytic function f has an **isolated singularity** at a point z_0 if f is analytic in the punctured disc $0 < |z - z_0| < r$, for some $r > 0$. The isolated singularities of an analytic function play a large role in the applications of complex variables, including analyzing flows and fields and computing definite integrals by means of the residue theorem. We shall analyze isolated singularities in this section; the applications will come in the next section with the residue theorem and in later chapters.

Three examples of isolated singularities are

(1) z_0 is an isolated singularity for $f(z) = (z^2 - z_0^2)/(z - z_0)$
(2) z_0 is an isolated singularity for $g(z) = 7(z - z_0)^{-4}$
(3) z_0 is an isolated singularity for $h(z) = \exp(1/(z - z_0))$.

As these examples show there are three possible modes of behavior for $|f|$ when $0 < |z - z_0| < r$:

$$|f(z)| \text{ remains bounded as } z \to z_0 \tag{1}$$

$$\lim_{z \to z_0} |f(z)| = \infty \tag{2}$$

$$\text{Neither (1) nor (2) holds.} \tag{3}$$

We will explore the first two of these possibilities in this section; the third we will not need and its development is in the exercises at the end of this section.

Removable singularities

Suppose first that (1) holds for f. Set

$$g(z) = \begin{cases} (z - z_0)^2 f(z), & 0 < |z - z_0| < r \\ 0, & z = z_0. \end{cases}$$

The function g is most certainly analytic for $0 < |z - z_0| < r$ but it is also differentiable at z_0, since

$$\lim_{z \to z_0} \frac{g(z) - g(z_0)}{z - z_0} = \lim_{z \to z_0} (z - z_0) f(z) = 0.$$

(The last equality is because $|f|$ remains bounded as $z \to z_0$.) Since g is analytic in $|z - z_0| < r$, we know that g has a power series valid for $|z - z_0| < r$:

$$g(z) = b_0 + b_1(z - z_0) + b_2(z - z_0)^2 + \cdots$$

However, $b_0 = g(z_0) = 0$ and $b_1 = g'(z_0) = 0$, so that

$$g(z) = b_2(z - z_0)^2 + b_3(z - z_0)^3 + \cdots,$$

and hence,

$$f(z) = b_2 + b_3(z - z_0) + \cdots$$

for $0 < |z - z_0| < r$. Set $f(z_0) = b_2$. Since the power series is valid for $|z - z_0| < r$ we see that the hypothesis (1) actually implies that f is analytic in $|z - z_0| < r$. When (1) holds, we then learn that f can be extended to be analytic in the disc $|z - z_0| < r$. In this case, z_0 is called a **removable singularity** for f.

Poles

Suppose next that (2) holds; there is no harm in assuming that r is so small that $|f(z)| > 1$ for $0 < |z - z_0| < r$. Consequently, $h(z) = 1/f(z)$ is analytic in $0 < |z - z_0| < r$ and is bounded there: $|h(z)| < 1$ if $0 < |z - z_0| < r$. But then the foregoing discussion shows that z_0 is a removable singularity for h and $h(z_0) = 0$ because (2) holds. Let $m \geqslant 1$ be the order of the zero of $h(z)$ at z_0 and write

$$h(z) = (z - z_0)^m H(z),$$

where $H(z)$ is analytic for $|z - z_0| < r$ and $H(z_0) \neq 0$. Because $h(z) \neq 0$ if $0 < |z - z_0| < r$ we also know that $H(z) \neq 0$ if $|z - z_0| < r$. Consequently, the function $F(z) = 1/H(z)$ is analytic for $|z - z_0| < r$. This leads to

$$f(z) = \frac{1}{h(z)} = \frac{1}{(z - z_0)^m} \frac{1}{H(z)} = \frac{F(z)}{(z - z_0)^m}, \qquad (4)$$

where $F(z)$ is analytic on $|z - z_0| < r$ and $F(z_0) \neq 0$. Thus, if (2) holds then f has predictable behavior near z_0 and, in particular, $|f(z)|$ grows to infinity just like some power of $1/|z - z_0|$. In this case, we say f has a **pole** at z_0; the **order of the pole** of f is just the order of the zero of $1/f$ at z_0.

Essential singularities

If (3) holds, then z_0 is termed an **essential singularity** of f. Information on the behavior of f near an essential singularity is contained in the exercises at the end of this section.

The residue at z_0

We note here the simple but significant fact that if f is a function analytic in the punctured disc $0 < |z - z_0| < r$, then the value of the integral

$$\frac{1}{2\pi i} \int_{|\zeta - z_0| = s} f(\zeta)\, d\zeta \qquad (5)$$

is the same for all s, $0 < s < r$. The proof is easy; if $0 < s_1 < s_2 < r$, then the region $0 < |z - z_0| < r$ is divided into four sectors by the horizontal and vertical lines through z_0 (Fig. 2.14).

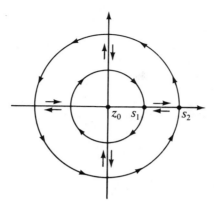

Figure 2.14

As indicated in Figure 2.14 the integral over $|\zeta - z_0| = s_1$ (oriented clockwise) plus the integral over $|\zeta - z_0| = s_2$ (oriented counterclockwise) can be broken up into the sum of four integrals (one in each sector); the line integrals over the interior line segments cancel. The integral in each sector is zero since f is analytic in a convex set containing this sector. Thus, the value of the integral in (5) does not depend on s. The value of the integral in (5) is called the **residue** of f at z_0, and we denote it by $\operatorname{Res}(f; z_0)$. Thus,

$$\operatorname{Res}(f; z_0) = \frac{1}{2\pi i} \int_{|\zeta - z_0| = s} f(\zeta)\, d\zeta. \tag{5}'$$

The residues of a function with only isolated singularities play an important role in complex variables.

Example 1   Find the residue of $f(z) = e^z/(z - 1)$ at $z_0 = 1$.

Solution   The residue is

$$\operatorname{Res}(f; 1) = \frac{1}{2\pi i} \int_{|\zeta - 1| = s} \frac{e^\zeta}{\zeta - 1}\, d\zeta = e^1 = e$$

by Cauchy's formula.   □

Example 2   Find the residue of $g(z) = e^z/(z - 1)^3$ at $z_0 = 1$.

Solution   We expand e^z in powers of $z - 1$:

$$e^z = e^1 e^{z-1} = e\left(1 + (z - 1) + \frac{(z - 1)^2}{2!} + \cdots\right)$$

$$= \sum_{j=0}^{\infty} \frac{e}{j!}(z - 1)^j.$$

Thus,

$$g(z) = \frac{e}{(z-1)^3} + \frac{e}{(z-1)^2} + \frac{e}{2(z-1)} + \frac{e}{6} + \frac{e}{24}(z-1) + \cdots,$$

so that

$$\text{Res}(g; 1) = \frac{1}{2\pi i} \int_{|\zeta-1|=s} g(\zeta) \, d\zeta = \frac{e}{2},$$

since the integrals of all the terms are zero except that of $e/[2(z-1)]$.   □

Example 3   Suppose f has a pole of order $m \geq 1$ at z_0 so that

$$f(z) = \frac{F(z)}{(z-z_0)^m}.$$

Then

$$\text{Res}(f; z_0) = \frac{1}{2\pi i} \int_{|\zeta-z_0|=s} f(\zeta) \, d\zeta = \frac{1}{2\pi i} \int_{|\zeta-z_0|=s} \frac{F(\zeta)}{(\zeta-z_0)^m} \, d\zeta.$$

Thus,

$$\text{Res}(f; z_0) = \frac{F^{(m-1)}(z_0)}{(m-1)!}, \tag{6}$$

according to Theorem 1, of Section 4 of this chapter, and, in particular, formulas (1) and (2) of Section 4. This is, in fact, what was used in Example 2.

Example 4   Find the residue of $\cot \alpha z$ at $z_0 = 0$.

Solution   We know that

$$\cot(\alpha z) = \frac{\cos(\alpha z)}{\sin(\alpha z)}$$

and

$$\sin(\alpha z) = \left(\frac{\sin(\alpha z)}{z}\right) z,$$

where $\lim_{z \to 0} (\sin \alpha z)/z = \alpha$. Hence, the residue of $\cot(\alpha z)$ at $z_0 = 0$ is

$$\text{Res}(\cot(\alpha z); 0) = \text{value of} \quad \frac{\cos(\alpha z)}{\frac{\sin(\alpha z)}{z}} \quad \text{at } z = 0$$

$$= \frac{1}{\alpha}.$$ □

Example 5 Let P and Q be polynomials with no common zero and suppose that Q has a zero of order 1 at z_0. Show that

$$\text{Res}(P/Q; z_0) = P(z_0)/Q'(z_0).$$

Solution We know that P/Q has a pole of order 1 at z_0 and hence,

$$\text{Res}(P/Q; z_0) = \lim_{z \to z_0} (z - z_0)P(z)/Q(z)$$

$$= \lim_{z \to z_0} \left[P(z) \middle/ \left(\frac{Q(z)}{z - z_0} \right) \right]$$

$$= P(z_0)/Q'(z_0),$$

as we wished to show. □

Laurent* series

We know that a function analytic in a disc $|z - z_0| < r$ can be expanded there in a power series and, conversely, if a function has a power series valid in $|z - z_0| < r$, then it is analytic on this disc. But what about a function that is only analytic in the punctured disc $0 < |z - z_0| < r$? We shall see that something almost as good as a power series can be given to represent the function in this punctured disc.

Suppose that f is analytic in the punctured disc $0 < |z - z_0| < r$ and has a pole of order m at z_0. We then know that

$$f(z) = \frac{F(z)}{(z - z_0)^m},$$

where F is analytic in the disc $|z - z_0| < r$ and $F(z_0) \neq 0$; see (4) in this section. Let

$$F(z) = b_0 + b_1(z - z_0) + b_2(z - z_0)^2 + \cdots, \quad b_0 \neq 0$$

be the power series for F valid in this disc. Simple division then yields

$$f(z) = \frac{b_0}{(z - z_0)^m} + \frac{b_1}{(z - z_0)^{m-1}} + \cdots + \frac{b_{m-1}}{z - z_0} + b_m + b_{m+1}(z - z_0) + \cdots.$$

The sum of the terms involving the negative powers of $z - z_0$ is the **principal part** of f at z_0; this is nothing but $P[1/(z - z_0)]$, where P is a polynomial of degree m. It is conventional to write the principal part using coefficients with negative subscripts and the remainder of the series with nonnegative subscripts in this way:

---

* Pierre Alphonse Laurent, 1813–1854.

$$f(z) = \frac{a_{-m}}{(z - z_0)^m} + \cdots + \frac{a_{-1}}{z - z_0} + a_0 + a_1(z - z_0) + \cdots.$$

This is the **Laurent series** for f at the point z_0. Note that the coefficient a_{-1} is exactly the residue of f at z_0, since for $0 < s < r$, we have

$$\frac{1}{2\pi i} \int_{|z - z_0| = s} f(z)\, dz = \sum_{j=1}^{m} a_{-j} \left(\frac{1}{2\pi i} \int_{|z - z_0| = s} (z - z_0)^{-j}\, dz \right)$$

$$+ \sum_{j=0}^{\infty} a_j \left(\frac{1}{2\pi i} \int_{|z - z_0| = s} (z - z_0)^j\, dz \right)$$

$$= a_{-1}.$$

Furthermore, the other coefficients in the Laurent series for f can also be expressed as integrals involving f. For $k = -m,\ -m + 1,\ \ldots,\ -2,\ -1,\ 0,\ 1,\ 2,\ \ldots$, we have

$$a_k = b_{m+k} = \frac{1}{2\pi i} \int_{|z - z_0| = s} \frac{F(z)}{(z - z_0)^{m + k + 1}}\, dz.$$

Therefore,

$$a_k = \frac{1}{2\pi i} \int_{|z - z_0| = s} \frac{f(z)}{(z - z_0)^{k+1}}\, dz, \quad k = -m,\ -m + 1,\ \ldots. \tag{7}$$

(The Laurent series developed above for functions with a pole at z_0 has several extensions. See the starred topic at the end of this section for the development of the Laurent series in an annulus.)

We note further that

$$f(z) = P\left(\frac{1}{z - z_0} \right) + g(z), \qquad |z - z_0| < r,$$

where g is analytic on the disc $|z - z_0| < r$. Indeed, suppose that f is analytic on some domain D except for a pole of order m at a point z_0 of D. We can then write f in the above form, where g is analytic on D because $P(1/(z - z_0))$ is analytic everywhere except at z_0. Moreover, by repeated applications of this procedure we can derive the following result.

If f is analytic on a domain D except for poles at points $z_1, \ldots, z_N$ of D of order $m_1, \ldots, m_N$, respectively, then we can write

$$f(z) = P_1\left(\frac{1}{z - z_1} \right) + \cdots + P_N\left(\frac{1}{z - z_N} \right) + g(z),\ z \in D \tag{8}$$

where $P_1, \ldots, P_N$ are polynomials of degrees $m_1, \ldots, m_N$, respectively, each with zero constant term, and g is analytic on D.

Example 6 Find the principal part and residue of $f(z) = (z^3 + z^2)/(z - 1)^2$ at $z_0 = 1$.

Solution   We   expand $z^3 + z^2$ in   powers   of $z - 1$: $z^3 + z^2 =$ $2 + 5(z - 1) + 4(z - 1)^2 + (z - 1)^3$. Hence,

$$f(z) = \frac{2}{(z - 1)^2} + \frac{5}{(z - 1)} + 4 + (z - 1).$$

The principal part is $2(z - 1)^{-2} + 5(z - 1)^{-1}$ and the residue is 5. $\square$

Example 7   Compute the Laurent series about $z_0 = 0$ for $f(z) = (\sin z)/z^2$.

Solution   We have

$$f(z) = \frac{\sin z}{z^2} = \frac{z - z^3/3! + z^5/5! - z^7/7! + \cdots}{z^2}$$

$$= \frac{1}{z} - \frac{z}{3!} + \frac{z^3}{5!} - \frac{z^5}{7!} + \cdots.$$

The series is valid for all $z \neq 0$. $\square$

Example 8   Find the Laurent series about $z_0 = 1$ and then about $z_0 = 0$ for $f(z) = 1/(z^2 - z^3)$.

Solution   We have

$$\frac{1}{z} = \frac{1}{1 - (1 - z)} = 1 + (1 - z) + (1 - z)^2 + (1 - z)^3 + \cdots$$

if $|1 - z| < 1$. Hence,

$$\frac{1}{z^2} = -\frac{d}{dz}\left(\frac{1}{z}\right) = -\frac{d}{dz}\{1 + (1 - z) + (1 - z)^2 + \cdots\}$$

$$= -\{-1 - 2(1 - z) - 3(1 - z)^2 - 4(1 - z)^3 - \cdots\}$$

$$= 1 + 2(1 - z) + 3(1 - z)^2 + 4(1 - z)^3 + \cdots.$$

Thus,

$$f(z) = \frac{1}{z^2(1 - z)} = \frac{1}{1 - z} + 2 + 3(1 - z) + 4(1 - z)^2 + \cdots.$$

The series is valid if $0 < |z - 1| < 1$.
    Around $z_0 = 0$, we use the expansion

$$f(z) = \frac{1}{z^2(1 - z)} = \frac{1}{z^2}\{1 + z + z^2 + \cdots\} \text{ if } |z| < 1$$

$$= \frac{1}{z^2} + \frac{1}{z} + 1 + z + z^2 + \cdots.$$

The series is valid for $|z| < 1$. The principal part of f is $(1/z^2) + (1/z)$ and the residue of f at z_0 is 1. $\qquad\qquad\Box$

Example 9 Find the first four terms of the Laurent series about $z_0 = 0$ of $g(z) = \cot(\alpha z)$, $\alpha \neq 0$.

Solution We have

$$g(z) = \frac{\cos(\alpha z)}{\sin(\alpha z)} = \frac{1}{z} \frac{z \cos(\alpha z)}{\sin(\alpha z)}.$$

Now $\cos(\alpha z)$ is an entire function and 0 is a removable singularity for $z/\sin(\alpha z)$, since

$$\lim_{z \to 0} \frac{z}{\sin(\alpha z)} = 1/\alpha.$$

Thus, $g(z)$ has a pole of order 1 at $z_0 = 0$ and consequently

$$g(z) = \frac{a_{-1}}{z} + a_0 + a_1 z + a_2 z^2 + \cdots.$$

Hence, $\cos(\alpha z) = g(z) \sin(\alpha z)$ and so

$$\cos(\alpha z) = 1 - \frac{\alpha^2 z^2}{2!} + \frac{\alpha^4 z^4}{4!} - \cdots = \left\{ \frac{a_{-1}}{z} + a_0 + a_1 z + a_2 z^2 + \cdots \right\}$$

$$\times \left\{ \alpha z - \frac{\alpha^3 z^3}{3!} + \cdots \right\}$$

$$= \alpha a_{-1} + \alpha a_0 z + \left(\alpha a_1 - \frac{\alpha^3}{6} a_{-1} \right) z^2 + \left(\alpha a_2 - \frac{\alpha^3}{6} a_0 \right) z^3 + \cdots.$$

Equating equal powers of z we obtain a collection of equations that can be solved for the unknown coefficients in the Laurent series of g. We only want a_{-1}, a_0, a_1, a_2 so we go out to the z^3 term and obtain the four equations

$$1 = \alpha a_{-1}, \qquad 0 = \alpha a_0, \qquad -\frac{\alpha^2}{2} = \alpha a_1 - \frac{\alpha^3}{6} a_{-1}, \qquad 0 = \alpha a_2 - \frac{\alpha^3}{6} a_0.$$

These yield $a_{-1} = 1/\alpha$, $a_0 = 0$, $a_1 = -\alpha/3$, and $a_2 = 0$, so

$$\cot(\alpha z) = \frac{1}{\alpha z} - \frac{\alpha}{3} z + \cdots. \qquad\qquad\Box$$

The method outlined in Example 9 is usually quite effective in determining the first few terms of a Laurent expansion; here is another illustration.

Example 10 Find the first five terms in the Laurent expansion of $h(z) = 4(e^z - 1)^{-2}$ about $z_0 = 0$.

Solution We have

$$4 = h(z)(e^z - 1)^2.$$

Now $(e^z - 1)^2$ has a zero at $z_0 = 0$ of order 2 and thus h has a pole at 0 of order 2:

$$h(z) = \frac{a_{-2}}{z^2} + \frac{a_{-1}}{z} + a_0 + a_1 z + a_2 z^2 + \cdots.$$

Furthermore,

$$(e^z - 1)^2 = (z + z^2/2 + z^3/6 + \cdots)^2$$

$$= z^2 + z^3 + \frac{7}{12} z^4 + \frac{1}{4} z^5 + \frac{31}{360} z^6 + \cdots.$$

Thus,

$$4 = \left(\frac{a_{-2}}{z^2} + \frac{a_{-1}}{z} + a_0 + \cdots \right)\left(z^2 + z^3 + \frac{7}{12} z^4 + \cdots \right)$$

$$= a_{-2} + (a_{-2} + a_{-1})z + \left(\frac{7}{12} a_{-2} + a_{-1} + a_0 \right) z^2$$

$$+ \left(\frac{1}{4} a_{-2} + \frac{7}{12} a_{-1} + a_0 + a_1 \right) z^3$$

$$+ \left(\frac{31}{360} a_{-2} + \frac{1}{4} a_{-1} + \frac{7}{12} a_0 + a_1 + a_2 \right) z^4 + \cdots.$$

By equating equal powers of z we obtain five equations for the coefficients of the first five terms in the Laurent series:

$$4 = a_{-2}, \qquad 0 = a_{-2} + a_{-1}, \qquad 0 = \frac{7}{12} a_{-2} + a_{-1} + a_0$$

$$0 = \frac{1}{4} a_{-2} + \frac{7}{12} a_{-1} + a_0 + a_1, \qquad 0 = \frac{31}{360} a_{-2} + \frac{1}{4} a_{-1} + \frac{7}{12} a_0 + a_1 + a_2.$$

These yield

$$a_{-2} = 4, \qquad a_{-1} = -4, \qquad a_0 = \frac{5}{3}, \qquad a_1 = -\frac{1}{3}, \qquad a_2 = \frac{1}{60},$$

and hence,

$$\frac{4}{(e^z - 1)^2} = \frac{4}{z^2} - \frac{4}{z} + \frac{5}{3} - \frac{z}{3} + \frac{z^2}{60} + \cdots. \qquad \square$$

The Laurent expansion of a rational function in powers of z and $1/z^\star$

The Laurent expansion of a rational function is important in applications, for instance, in connection with the Z-transform discussed in Section 5 of Chapter 5. Here we discuss one particularly simple but useful case. Let R be a rational function all of whose poles in the plane have order one and which has no pole at the origin. After an initial long division we can write

$$R(z) = S(z) + \frac{P(z)}{Q(z)},$$

where S, P, and Q are polynomials and the degree of P is strictly less than the degree of Q. Since S is already expressed as the sum of powers of z we concentrate on expanding $f = P/Q$. Here we shall show the following.

Let f be a rational function of all whose poles $z_1, \ldots, z_N$ in the plane have order one and which has no pole at the origin and which is zero at ∞. Suppose that no pole of f lies in the annular region $r < |z| < R$. Then

$$f(z) = \sum_{-\infty}^{\infty} a_k z^k \text{ for } r < |z| < R, \tag{9}$$

where

$$a_k = \begin{cases} \displaystyle\sum_{|z_j| < r} z_j^{-k-1} \, Res(f; z_j), & k \leqslant -1 \tag{10} \\[2ex] -\displaystyle\sum_{|z_j| > R} z_j^{-k-1} \, Res(f; z_j), & k \geqslant 0 \tag{11} \end{cases}$$

The chief tool to prove equations (9) through (11) is equation (8). Note first that the function g in (8) must be identically zero since it is a rational function with no poles and which is zero at ∞. Let $z_1, \ldots, z_m$ be the poles of f in the set $|z| < r$ and let $z_{m+1}, \ldots, z_N$ be the remaining poles of f, which must necessarily lie in the region $|z| > R$. Then from (8) we see that $f = F_1 + F_2$, where

$$F_1(z) = \sum_{j=1}^{m} P_j\left(\frac{1}{z - z_j}\right), \qquad F_2(z) = \sum_{j=m+1}^{N} P_j\left(\frac{1}{z - z_j}\right),$$

and consequently F_1 is analytic in the region $|z| > r$ including at ∞ and F_2 is analytic in the region $|z| < R$. This implies that F_2 can be expanded in a power series

$$F_2(z) = \sum_{k=0}^{\infty} a_k z^k,$$

where

$$a_k = \frac{1}{2\pi i} \int_{|z| = s} \frac{F_2(z)}{z^{k+1}} \, dz, \quad k = 0, 1, 2, \ldots$$

and F_1 can be expanded in a series in powers of $1/z$:

$$F_1(z) = \sum_{-\infty}^{-1} a_k z^k,$$

where

$$a_k = \frac{1}{2\pi i} \int_{|z|=s} \frac{F_1(z)}{z^{k+1}} \, dz, \quad k = -1, -2, \ldots;$$

(see Exercise 23, Section 4). Here s is any number between r and R. Furthermore, these series representations converge absolutely for each z with $r < |z| < R$. Consequently, we can write in a unified way

$$a_k = \frac{1}{2\pi i} \int_{|z|=s} \frac{f(z)}{z^{k+1}} \, dz, \quad k = 0, \pm 1, \pm 2, \ldots.$$

To obtain the formulas (10) and (11) we note that since z_j is a pole of order one we have

$$P_j\left(\frac{1}{z - z_j}\right) = \frac{b_j}{z - z_j}, \quad b_j = \mathrm{Res}(f; z_j)$$

and hence for $k \leqslant -1$, we have

$$z^{-k-1} P_j\left(\frac{1}{z - z_j}\right) = \frac{b_j z_j^{-k-1}}{z - z_j} + b_j \frac{z^{-k-1} - z_j^{-k-1}}{z - z_j}$$

$$= \frac{b_j z_j^{-k-1}}{z - z_j} + \text{polynomial}.$$

Furthermore, $F_2(z) z^{-k-1}$ is analytic for $|z| < R$ if $k \leqslant -1$ so that

$$a_k = \frac{1}{2\pi i} \int_{|z|=s} \frac{f(z)}{z^{k+1}} \, dz = \sum_{j=1}^{m} \frac{1}{2\pi i} \int_{|z|=s} z^{-k-1} P_j\left(\frac{1}{z - z_j}\right) dz$$

$$= \sum_{j=1}^{m} b_j z_j^{-k-1}, \quad k \leqslant -1$$

which is (10). Next, for $k \geqslant 0$, we make the change of variables $z = 1/w$. With this change we find that

$$\frac{1}{2\pi i} \int_{|z|=s} \frac{b_j z^{-k-1}}{z - z_j} \, dz = \frac{-1}{z_j} \frac{1}{2\pi i} \int_{|w|=1/s} \frac{w^k b_j}{w - (1/z_j)} \, dw$$

$$= \begin{cases} 0, & j = 1, \ldots, m \\ -\dfrac{b_j}{z_j^{k+1}}, & j = m+1, \ldots, N. \end{cases}$$

This gives (11) after summing on j.

Example 11 Find the expansion of $R(z) = (z^3 - 3z^2 + 3)/(z - 1)(z - 3)$ in powers of z and $1/z$ in the regions (a) $|z| < 1$; (b) $1 < |z| < 3$; (c) $|z| > 3$.

Solution   The initial long division gives $R(z) = z + 1 + z/(z - 1)(z - 3)$. We concentrate on expanding $f(z) = z/(z - 1)(z - 3)$.

(a) In this case, $r = 0$, $R = 1$, and there are no terms to use in (10) so that only (11) is used and we obtain

$$\frac{z}{(z - 1)(z - 3)} = \sum_{k=0}^{\infty} a_k z^k, \qquad a_k = -\left[-\frac{1}{2} + \frac{3}{2}\frac{1}{3^{k+1}}\right], \quad k \geqslant 0.$$

(b) Here, $r = 1$ and $R = 3$ and so directly from (10) and (11)

$$\frac{z}{(z - 1)(z - 3)} = \sum_{-\infty}^{\infty} a_k z^k, \qquad a_k = \begin{cases} -\dfrac{1}{2}, & k \leqslant -1 \\[2mm] -\dfrac{3}{2}\dfrac{1}{3^{k+1}}, & k \geqslant 0. \end{cases}$$

(c) Here, $r = 3$ and R does not enter the picture (or, if you prefer, is ∞) so that there are no terms to use in (11). Thus,

$$\frac{z}{(z - 1)(z - 3)} = \sum_{-\infty}^{-1} a_k z^k, \qquad a_k = -\frac{1}{2} + \frac{3}{2}\frac{1}{3^{k+1}}, \quad k \leqslant -1. \qquad \square$$

Laurent series in an annulus★

Here we shall show that a function g that is analytic on the annulus $\{z : m < |z - z_0| < M\}$ can be written as

$$g(z) = g_1(z) + g_2(z), \quad m < |z - z_0| < M, \tag{12}$$

where g_1 is analytic on the disc $|z - z_0| < M$ and g_2 is analytic on the region $|z - z_0| > m$. g_1 has a power series in $(z - z_0)$ valid for $|z - z_0| < M$, while g_2 has a power series in the variable $(z - z_0)^{-1}$ valid for $|z - z_0| > m$ (Exercise 23, Section 4). Consequently, it follows from (12) that

$$g(z) = \sum_{0}^{\infty} a_k(z - z_0)^k + \sum_{1}^{\infty} b_k(z - z_0)^{-k}$$

$$= \sum_{-\infty}^{\infty} a_k(z - z_0)^k, \qquad a_{-k} = b_k, \ k = 1, 2, \ldots .$$

This representation of g is called the **Laurent series** for g.
   Fix a z in the annulus and choose r, R with

$$m < r < |z - z_0| < R < M.$$

Let Γ be the circle $|w - z_0| = R$ oriented counterclockwise and let γ be the circle $|w - z_0| = r$ oriented clockwise. The closed curve beginning at **P** and comprised of Γ followed by the horizontal segment **PQ**, followed by γ, and ended by the segment **QP** surrounds z (Figure 2.15). We may apply Cauchy's formula, therefore, and write

$$g(z) = \frac{1}{2\pi i} \int_\Gamma \frac{g(w)}{w - z} \, dw + \frac{1}{2\pi i} \int_\gamma \frac{g(\zeta)}{\zeta - z} \, d\zeta.$$

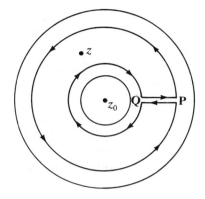

Figure 2.15

(Note that the line integral over **PQ** is cancelled by that over **QP**.) For $w \in \Gamma$, we may write

$$\frac{1}{w - z} = \frac{1}{(w - z_0) - (z - z_0)} = \frac{1}{w - z_0} \left(\frac{1}{1 - \dfrac{z - z_0}{w - z_0}} \right)$$

$$= \sum_{k=0}^{\infty} \frac{(z - z_0)^k}{(w - z_0)^{k+1}}.$$

The series is absolutely convergent since $|(z - z_0)/(w - z_0)| < 1$. In an entirely similar way we obtain

$$\frac{1}{\zeta - z} = -\sum_{j=0}^{\infty} \frac{(\zeta - z_0)^j}{(z - z_0)^{j+1}}$$

for each $\zeta \in \gamma$; here again, the series is absolutely convergent since $|(\zeta - z_0)/(z - z_0)| < 1$. In the integral representation of g given before, replace $1/(w - z)$ and $1/(\zeta - z)$ by the series obtained above. After interchanging the summation and integration we find that

$$g(z) = \sum_{k=0}^{\infty} (z - z_0)^k \left\{ \frac{1}{2\pi i} \int_\Gamma \frac{g(w)}{(w - z_0)^{k+1}} \, dw \right\}$$

$$+ \sum_{j=0}^{\infty} (z - z_0)^{-j-1} \left\{ \frac{-1}{2\pi i} \int_\gamma g(\zeta)(\zeta - z_0)^j \, d\zeta \right\}$$

$$= g_1(z) + g_2(z).$$

The integral $\int_{|w|=s} g(w) (w - z_0)^k \, dw$, $k = 0, \pm 1, \pm 2, \dots$ is independent of s for $m < s < M$, since the integrand is analytic on the annulus $m < |w - z_0| < M$.

Let

$$a_k = \frac{1}{2\pi i} \int_{|w|=s} \frac{g(w)}{(w - z_0)^{k+1}} \, dw, \quad k = 0, \pm 1, \ldots . \tag{13}$$

Then

$$g_1(z) = \sum_{k=0}^{\infty} a_k(z - z_0)^k, \qquad g_2(z) = \sum_{k=-\infty}^{-1} a_k(z - z_0)^k$$

are analytic on the sets $\{z : |z - z_0| < M\}$ and $\{z : |z - z_0| > m\}$, respectively, and we obtain the desired formula:

$$g(z) = g_1(z) + g_2(z) = \sum_{-\infty}^{\infty} a_k(z - z_0)^k, \quad m < |z - z_0| < M. \tag{14}$$

Exercises for Section 2.5

In Exercises 1 to 6 locate *each* of the isolated singularities of the given function and tell whether it is a removable singularity, a pole, or an essential singularity (case (3)). If the singularity is removable, give the value of the function at the point; if the singularity is a pole, give the order of the pole.

1. $\dfrac{e^z - 1}{z}$ 2. $\dfrac{z^2}{\sin z}$ 3. $\dfrac{z^4 - 2z^2 + 1}{(z - 1)^2}$

4. $\pi \cot \pi z$ 5. $\dfrac{2z + 1}{z + 2}$ 6. $\dfrac{e^z - 1}{e^{2z} - 1}$

In exercises 7 to 13, find the Laurent series for the given function about the indicated point. Also, give the residue of the function at the point.

7. $\dfrac{e^z - 1}{z^2}$; $z_0 = 0$ 8. $\dfrac{z^2}{z^2 - 1}$; $z_0 = 1$ 9. $\dfrac{\sin z}{(z - \pi)^2}$; $z_0 = \pi$

10. $\dfrac{z}{(\sin z)^2}$; $z_0 = 0$ (four terms of the Laurent series)

11. $\dfrac{az + b}{cz + d}$; $z_0 = -d/c$, $c \neq 0$

12. $\dfrac{1}{e^z - 1}$; $z_0 = 0$ (four terms of the Laurent series)

13. $\dfrac{1}{1 - \cos z}$; $z_0 = 0$ (four terms of the Laurent series)

14. If f is analytic in $|z - z_0| < R$ and has a zero of order m at z_0, show that

$$\text{Res}(f'/f; z_0) = m.$$

15.   If f is analytic in $0 < |z - z_0| < R$ and has a pole of order l at z_0, show that

$$\text{Res}(f'/f; z_0) = -l.$$

(**Hint:** Write $f(z) = g(z)/(z - z_0)^l$, where g is analytic in $|z - z_0| < R$ and $g(z_0) \neq 0$.)

16.   Suppose f is analytic in $|z - z_0| < R$ and has a zero of order $m \geq 1$ at z_0. If g is analytic in $0 < |z - z_0| < R$ and has a pole of order $l \leq m$ at z_0, show that fg has a removable singularity at z_0.

17.   Let f be analytic in $0 < |z - z_0| < r$ and suppose that f has an essential singularity at z_0. Let w be any complex number. Show that

$$g(z) = 1/(f(z) - w), \; z \in D$$

is *not* bounded in any punctured disc $0 < |z - z_0| < \varepsilon$ (**Hint:** if g is bounded, then show f has a pole or a removable singularity at z_0.) Conclude that the range of f gets arbitrarily close to all points in the complex plane. How does this differ from the situation in (1) and (2)?

18.   Here is an alternate proof that

$$\frac{1}{2\pi i} \int_{|z - z_0| = s} f(z) \, dz = \text{Res}(f; z_0)$$

is independent of s. Assume $z_0 = 0$ with no loss of generality.
(a) Show that

$$\frac{\partial}{\partial t} \{f(se^{it})e^{it}\} = ie^{it}f(se^{it}) + ise^{2it}f'(se^{it}).$$

(b) Show that

$$\frac{d}{ds}\left\{\frac{1}{2\pi} \int_0^{2\pi} sf(se^{it})e^{it} \, dt\right\} = \frac{1}{2\pi} \int_0^{2\pi} [f(se^{it}) + sf'(se^{it})e^{it}]e^{it} \, dt;$$

(c) Conclude from (a) and (b) that

$$\frac{d}{ds}\left\{\frac{1}{2\pi i} \int_{|z| = s} f(z) \, dz\right\} = \frac{1}{2\pi i} \int_0^{2\pi} \frac{\partial}{\partial t}\{e^{it}f(se^{it})\} \, dt.$$

Finally, calculate that the second integral is 0.

19.   Suppose that the Laurent series $\sum_{-\infty}^{\infty} a_n(z - z_0)^n$ converges for $0 < |z - z_0| < r$ and

$$\sum_{-\infty}^{\infty} a_n(z - z_0)^n = 0, \qquad 0 < |z - z_0| < r.$$

Show that $a_n = 0$, $n = 0, \pm 1, \pm 2, \ldots$. (**Hint:** Multiply the series by $(z - z_0)^{-m}$ and integrate around the circle $|z - z_0| = s$, $0 < s < r$ with respect to z. The result must be zero, but it is also a_{m-1}.

20.   If f is analytic in $0 < |z - z_0| < r$, show that its Laurent series is uniquely determined.

21.   If F is analytic on $0 < |z - z_0| < R$ and if there is an analytic function G on $0 < |z - z_0| < R$ with $G' = F$, then

$$\operatorname{Res}(F; z_0) = 0$$

(Hint: Use (5).)

22.   Find the Laurent series about $z_0 = 0$ for the following functions, valid in the indicated regions
(a) $e^{1/z}$ in $0 < |z| < \infty$
(b) $z^4 \sin(1/z)$ in $0 < |z| < \infty$

(c) $\dfrac{1}{z - 1} - \dfrac{1}{z + 1}$ in $2 < |z| < \infty$

(d) $\exp(z + 1/z)$ in $0 < |z| < \infty$
(e) $z \cos(1/z)$ in $0 < |z| < \infty$

23.   Use equations (9) to (11) to find the Laurent expansions of the following rational functions in powers of z and $1/z$ in the indicated region(s).

(a) $f(z) = \dfrac{z + 2}{z^2 - z - 2}$ in $1 < |z| < 2$ and then in $2 < |z| < \infty$

(b) $f(z) = \dfrac{z^2 + 2z - 4}{(z^2 - 9)(z + 1)}$ in $|z| < 1$ and then in $1 < |z| < 3$

(c) $f(z) = \dfrac{z + 1}{(z - 2)(z - 3)(z - 5)}$ in $2 < |z| < 3$ and then in $5 < |z| < \infty$

24.   Let f be analytic in $0 < |z - z_0| < r$ and let

$$f(z) = \sum_{n = -\infty}^{\infty} a_n(z - z_0)^n$$

be its Laurent series. Show that
(a) z_0 is a removable singularity for f if and only if $a_n = 0$ for $n = -1, -2, \ldots$.
(b) z_0 is a pole of order $m \geq 1$ for f if and only if $a_{-m} \neq 0$ but $a_{-n} = 0$ for all $n \geq m + 1$.
(c) z_0 is an essential singularity for f if and only if there are infinitely many a_{-n}, $n > 0$, that are *not* zero.

Singularities at ∞⋆

Suppose that f is analytic for $|z| > R$; set $g(z) = f(1/z)$ so that g is analytic for $0 < |z| < 1/R$. We classify the nature of the singularity of f at ∞ by that of g at $z = 0$; for instance, if g has a pole of order m at 0 then we say f has a pole of order m at ∞. The Laurent series for f about is just the Laurent series for

$g(z) = f(1/z)$ about 0 with z replaced by $1/z$.

25. Show that f has a removable singularity at ∞ if and only if $|f(z)|$ is bounded for $|z| > R_0$.

26. Show that f has a pole of order m at ∞ if and only if $h(z) = f(z)/z^m$ has a removable singularity at ∞ *and* $h(\infty) \neq 0$.

27. Classify the nature of the singularity at ∞ of each of the following functions. If the singularity is removable, give the value at ∞. If the singularity is a pole, give its order; in each case find the first few terms in the Laurent series about ∞.

    (a) $3z^2 + 4 - 1/z$ (b) $(1 - z)(z - 4)$ (c) $z^2/(z - 4)$

    (d) $(-z^2 - 2z + 3)^{-1}$ (e) $e^{1/z}$ (f) e^{-z^2}

    (g) $\left(\dfrac{1}{z} + z\right)^3$ (h) $\sin(1/z)$ (i) $\displaystyle\sum_{n=0}^{\infty} (-1)^n \frac{z^{-2n}}{(2n)!}$.

Bessel functions⋆

28. Let u be a complex number and let

$$G(z; u) = \exp[(u/2)(z - 1/z)], \quad z \neq 0.$$

Show that $G(z; u)$ is an analytic function of z for $z \neq 0$ and has an essential singularity at $z = 0$ (unless $u = 0$).

29. Let

$$G(z; u) = \sum_{-\infty}^{\infty} J_n(u) z^n, \tag{15}$$

be the Laurent series of $G(z; u)$ about the origin, where the coefficients $\{J_n(u)\}$ are given by (7). In (7) choose $s = 1$ and conclude that

$$J_n(u) = \frac{1}{2\pi} \int_0^{2\pi} \cos(u \sin \theta - n\theta)\, d\theta. \tag{16}$$

(**Hint:** Prove (16) for u real by showing that $J_n(u)$ is real if u is real.)

30. Multiply out the series for $\exp(uz/2)$ and the series for $\exp(-u/2z)$ and then collect equal powers of z. Conclude that

$$J_n(u) = \frac{u^n}{2^n} \sum_{k=0}^{\infty} \frac{(-1)^k u^{2k}}{2^k k!(n + k)!}$$

for $n \geq 0$. Compare this to the conclusion of Exercise 29, Section 4.

31. Show that $J_{-n}(u) = (-1)^n J_n(u)$. (**Hint:** Replace z by $-1/z$ in the Laurent series; then change the summation index n to $-n$ and compare series.)

2.6   The residue theorem and its application to the evaluation of definite integrals

The residue theorem is a result of both great theoretical and practical importance. It allows us to compute with ease what seem to be very difficult definite integrals and, at the same time, provides the key in unlocking the secret to counting the zeros of an analytic function. In this section we concentrate on the application to the evaluation of definite integrals. Later sections will contain other applications.

Suppose that f is analytic on a star-shaped or, more generally, a simply-connected domain D except for a finite number of poles, say at $z_1, \dots, z_N$ in D. Let γ be a piecewise smooth positively-oriented simple closed curve in D that does not pass through any of the points $z_1, \dots, z_N$. We then have this result:

Theorem 1: The Residue Theorem. Let f, D, and γ be as above. Then

$$\frac{1}{2\pi i} \int_\gamma f(z)\, dz = \sum_{z_j \text{ inside } \gamma} \text{Res}(f; z_j), \tag{1}$$

where the sum is over all those poles z_j of f that lie inside γ.

Proof   According to (8) of Section 5 we can write

$$f(z) = P_1\!\left(\frac{1}{z - z_1}\right) + \cdots + P_N\!\left(\frac{1}{z - z_N}\right) + g(z), \quad z \in D,$$

where $P_1, \dots, P_N$ are polynomials and g is analytic on D. Thus,

$$\frac{1}{2\pi i} \int_\gamma f(z)\, dz = \sum_{j=1}^{N} \frac{1}{2\pi i} \int_\gamma P_j\!\left(\frac{1}{z - z_j}\right) dz + \frac{1}{2\pi i} \int_\gamma g(z)\, dz.$$

The integral of g over γ is zero by Cauchy's Theorem (Section 4). Furthermore,

$$\frac{1}{2\pi i} \int_\gamma P_j\!\left(\frac{1}{z - z_j}\right) dz = \text{Res}(f; z_j),$$

since

$$\frac{1}{2\pi i} \int_\gamma \frac{1}{(z - z_j)^m}\, dz = \begin{cases} 1 & \text{if } m = 1 \\ 0 & \text{if } m = 2, 3, \dots \end{cases};$$

see Exercise 16, Section 6, Chapter 1. $\square$

Remark: The residue theorem is valid for functions with essential singularities; see Exercise 25 at the end of this section.

We now turn to the application of the residue theorem to the evaluation of definite integrals.

Integrals of rational functions

It is convenient to isolate here several simple facts about a polynomial

$$p(z) = a_n z^n + a_{n-1} z^{n-1} + \cdots + a_1 z + a_0, \ a_n \neq 0.$$

> If $p(\alpha) = 0$, then $p(z) = (z - \alpha)q(z)$, where q is
> a polynomial of degree $n - 1$ (2)
>
> There are at most n distinct points $\alpha_1, \ldots, \alpha_s$ in the
> plane at which $p(\alpha_j) = 0$ (3)
>
> For $|z| = R$, R large, we have the estimate

$$\tfrac{1}{2}|a_n| R^n \leqslant |p(z)| \leqslant 2|a_n| R^n. \tag{4}$$

The proofs of these three facts are all easy. For (2), expand p in powers of $(z - \alpha)$; this yields

$$p(z) = \sum_0^n c_j (z - \alpha)^j, \ c_j = \frac{p^{(j)}(\alpha)}{j!}, \ j = 0, \ldots, n.$$

However, $c_0 = p(\alpha) = 0$ so that

$$\begin{aligned}
p(z) &= c_1(z - \alpha) + \cdots + c_n(z - \alpha)^n \\
&= (z - \alpha)[c_1 + \cdots + c_n(z - \alpha)^{n-1}] \\
&= (z - \alpha)q(z).
\end{aligned}$$

Next, (3) follows from (2) since each time a zero is divided out, the degree of the quotient goes down by one. Finally,

$$\lim_{|z| \to \infty} \left| \frac{p(z)}{z^n} \right| = |a_n| > 0.$$

Consequently, for $|z| = R$ large enough

$$\tfrac{1}{2}|a_n| \leqslant \left| \frac{p(z)}{z^n} \right| \leqslant 2|a_n|,$$

which is exactly (4).

The following proposition shows how the residue theorem is applied to rational functions.

Proposition: Suppose P and Q are polynomials where the degree of Q exceeds the degree of P by 2 or more. If $Q(x) \neq 0$ for all real x, then

$$\int_{-\infty}^{\infty} \frac{P(x)}{Q(x)} \, dx = 2\pi i \sum_U \text{Res}(P/Q; z_j),$$

where the sum is taken over all zeros of Q that lie in the upper half-plane $U = \{z : \text{Im } z > 0\}$.

Proof. Let γ_R be the contour in Figure 2.16.

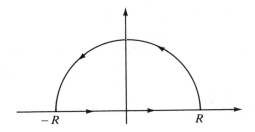

Figure 2.16

There is no loss of generality in assuming that $Q(z) \neq 0$ for $z \in \gamma_R$. On γ_R, when R is big, we have from (4)

$$\left| \frac{P(z)}{Q(z)} \right| \leqslant \frac{2\,|a_n|\,R^n}{\frac{1}{2}|b_m|R^m} \leqslant 4 \left| \frac{a_n}{b_m} \right| R^{-2} = \frac{C}{R^2},$$

since $m \geqslant n + 2$. Hence,

$$\left| \int_{|z|=R} \frac{P(z)}{Q(z)}\,dz \right| \leqslant C\,\frac{2\pi R}{R^2} \to 0 \quad \text{as} \quad R \to \infty.$$

The result now follows from the residue theorem. $\square$

Example 1   Compute

$$\int_{-\infty}^{\infty} \frac{x^2}{(1+x^2)(4+x^2)}\,dx.$$

Solution   The polynomials are $P(z) = z^2$ and $Q(z) = (1+z^2)(4+z^2)$, respectively, and Q has zeros at $z_1 = i$ and $z_2 = 2i$ in U, the upper half-plane. Now

$$\frac{P(z)}{Q(z)} = \frac{z^2}{(z-i)(z+i)(z-2i)(z+2i)}.$$

Example 5 of Section 5 then gives

$$\text{Res}(P/Q; i) = \frac{-1}{(2i)3} = \frac{-1}{6i}$$

and

$$\text{Res}(P/Q; 2i) = \frac{-4}{(-3)(4i)} = \frac{1}{3i}.$$

Hence,

$$\int_{-\infty}^{\infty} \frac{x^2}{(1+x^2)(4+x^2)}\,dx = 2\pi i \left[\frac{-1}{6i} + \frac{1}{3i} \right] = \frac{\pi}{3}.$$ $\square$

Example 2   Compute

$$\int_{-\infty}^{\infty} \frac{dx}{(1 + x^2)^2}.$$

Solution   Here $P(z) = 1$ for all z and $Q(z) = (1 + z^2)^2 = (z + i)^2(z - i)^2$. Hence, Q has just one zero in the upper half-plane U but the zero has order 2. We must find the series expansion of

$$g(z) = \frac{1}{(z + i)^2} = b_0 + b_1(z - i) + b_2(z - i)^2 + \cdots$$

near $z_0 = i$;

$$b_0 = g(i) = \frac{1}{-4} \text{ and } b_1 = g'(i) = \frac{-2}{(i + i)^3} = \frac{1}{4i}.$$

Hence,

$$\frac{1}{Q(z)} = \frac{-1/4}{(z - i)^2} + \frac{1/4i}{(z - i)} + \cdots$$

and so $\text{Res}(1/Q; i) = 1/4i$. Consequently,

$$\int_{-\infty}^{\infty} \frac{dx}{(1 + x^2)^2} = 2\pi i(1/4i) = \pi/2. \qquad \square$$

Integrals over the real axis involving trigonometric functions

For integrals of the form $\int_{-\infty}^{\infty} R(x)\sin x \, dx$ or $\int_{-\infty}^{\infty} R(x)\cos x \, dx$, where R is a rational function, we apply the residue theorem to the function $f(z) = R(z)e^{iz}$ and, at the end of the computation, take the imaginary or real part of the resulting complex number.

Example 3   Compute

$$\int_{-\infty}^{\infty} \frac{\cos x}{x^2 + \alpha^2} \, dx, \qquad \alpha > 0.$$

Solution   Set $f(z) = e^{iz}/(z^2 + \alpha^2)$; f has a pole at $i\alpha$ in U with residue

$$\text{Res}(f; i\alpha) = \frac{e^{-\alpha}}{2i\alpha}.$$

Let γ_R be the contour in Figure 2.16, so that

$$\pi \frac{e^{-\alpha}}{\alpha} = 2\pi i \left(\frac{e^{-\alpha}}{2i\alpha} \right) = \int_{\gamma_R} f(z) \, dz$$

$$= \int_{-R}^{R} \frac{e^{ix}}{x^2 + \alpha^2} \, dx + i \int_{0}^{\pi} f(Re^{i\theta})Re^{i\theta} \, d\theta.$$

The integral over the semicircle can be estimated by

$$\left| \int_{0}^{\pi} f(Re^{i\theta})Re^{i\theta} \, d\theta \right| \leqslant R \int_{0}^{\pi} \frac{e^{-R \sin \theta}}{R^2 - \alpha^2} \, d\theta$$

$$\leqslant \frac{\pi R}{R^2 - \alpha^2} \to 0 \text{ as } R \to \infty.$$

Thus,

$$\frac{\pi e^{-\alpha}}{\alpha} = \int_{-\infty}^{\infty} \frac{\cos x}{x^2 + \alpha^2} \, dx + i \int_{-\infty}^{\infty} \frac{\sin x}{x^2 + \alpha^2} \, dx.$$

The integral with $\sin x$ is zero since the integrand is odd; hence,

$$\int_{-\infty}^{\infty} \frac{\cos x}{x^2 + \alpha^2} \, dx = \pi \frac{e^{-\alpha}}{\alpha}, \quad \alpha > 0. \qquad \square$$

Example 4 In the foregoing result, replace x by βx and set $\delta = \alpha/\beta$. This results in the formula

$$\int_{-\infty}^{\infty} \frac{\cos \beta x}{x^2 + \delta^2} \, dx = \pi \frac{e^{-\delta\beta}}{\delta}; \quad \beta, \delta > 0. \tag{5}$$

Differentiate both sides of this identity with respect to β; this yields

$$\int_{-\infty}^{\infty} \frac{x \sin \beta x}{x^2 + \delta^2} \, dx = \pi e^{-\delta\beta}; \quad \beta, \delta > 0. \tag{6}$$

Letting $\delta \downarrow 0$ with $\beta = 1$, we obtain

$$\int_{-\infty}^{\infty} \frac{\sin x}{x} \, dx = \pi.$$

In these applications of the residue theorem it often becomes necessary to estimate the magnitude of the integral of e^{iz} as z traverses the semicircle $|z| = Re^{i\theta}, 0 \leqslant \theta \leqslant \pi$. For this estimate we have **Jordan's lemma**:

$$\left| \int_{\substack{|z| = R \\ \text{Im } z \geqslant 0}} e^{iz} \, dz \right| \leqslant \pi.$$

Indeed the integral is certainly no larger than

$$\int_{0}^{\pi} |e^{i(R \cos \theta + iR \sin \theta)}| \, |Re^{i\theta}| \, d\theta = R \int_{0}^{\pi} e^{-R \sin \theta} \, d\theta.$$

However, elementary calculus shows that $\sin \theta \geqslant (2/\pi) \theta$ for $0 \leqslant \theta \leqslant \pi/2$ and so $-R \sin \theta \leqslant -(2R/\pi) \theta$ for $0 \leqslant \theta \leqslant (\pi/2)$. Hence,

$$R \int_0^\pi e^{-R \sin \theta} \, d\theta = 2R \int_0^{\pi/2} e^{-R \sin \theta} \, d\theta$$

$$\leqslant 2R \int_0^{\pi/2} e^{-(2R/\pi) \theta} \, d\theta$$

$$= \frac{\pi R}{R} \left[1 - e^{-R} \right] < \pi.$$

Example 5   Compute the value of

$$\int_{-\infty}^\infty \frac{x^3 \sin x}{x^4 + 16} \, dx.$$

Solution   We set

$$f(z) = e^{iz} \frac{z^3}{z^4 + 16}$$

and integrate f over the contour in Figure 2.16. The function f has poles in U of order 1 at $\sqrt{2}(1 + i)$ and $\sqrt{2}(-1 + i)$. The residues are

$$\mathrm{Res}(f; \sqrt{2}(1 + i)) = \tfrac{1}{4} e^{-\sqrt{2}} e^{i\sqrt{2}}$$

and

$$\mathrm{Res}(f; \sqrt{2}(-1 + i)) = \tfrac{1}{4} e^{-\sqrt{2}} e^{-i\sqrt{2}}.$$

By Jordan's lemma (or, better, its proof) we can make the estimate

$$\left| \int_{\substack{|z|=R \\ \mathrm{Im}\, z \geqslant 0}} f(z) \, dz \right| \leqslant \frac{R^3}{R^4 - 16} \int_0^\pi e^{-R \sin \theta} R \, d\theta \leqslant \frac{\pi R^3}{R^4 - 16} \to 0$$

as $R \to \infty$. Thus,

$$\int_{-\infty}^\infty \frac{x^3 \sin x}{x^4 + 16} \, dx = \mathrm{Im}\{2\pi i [\mathrm{Res}(f; \sqrt{2}(1 + i)) + \mathrm{Res}(f; \sqrt{2}(-1 + i))]\}$$
$$= \pi e^{-\sqrt{2}} \cos(\sqrt{2}). \qquad \square$$

Example 6   Find the value of

$$\int_0^\infty \frac{\sin^2 x}{x^2} \, dx.$$

Solution   We use the identity $2 \sin^2 x = 1 - \cos 2x = \mathrm{Re}(1 - e^{2ix})$. Set $f(z) = (1 - e^{2iz})/z^2$ and take γ to be the contour in Figure 2.17.

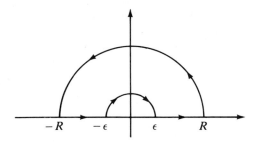

Figure 2.17

The integral of f over γ is zero by Cauchy's Theorem. On the semi-circle $z = Re^{it}$, $0 \leqslant t \leqslant \pi$, we have the estimate

$$|f(Re^{it})| \leqslant (1 + e^{-2R\sin t})/R^2 \leqslant 2/R^2$$

so that the integral over the large semicircle can be estimated by

$$\left| \int_{\substack{|z|=R \\ z \in \gamma}} f(z)\, dz \right| \leqslant \frac{2}{R^2}\, 2\pi R \to 0 \text{ as } R \to \infty.$$

Furthermore, near $z = 0$, we have the expansion

$$f(z) = \frac{1 - \left[1 + 2iz + \dfrac{(2iz)^2}{2!} + \cdots \right]}{z^2}$$

$$= \frac{-2i}{z} + 2 - \frac{4i}{3} z + \cdots .$$

Hence,

$$\int_\pi^0 f(\varepsilon e^{it}) i\varepsilon e^{it}\, dt = -2\pi + (\text{terms that go to zero as } \varepsilon \to 0).$$

Thus,

$$0 = \int_\varepsilon^R \frac{1 - e^{2ix}}{x^2}\, dx + \int_{-R}^{-\varepsilon} \frac{1 - e^{2ix}}{x^2}\, dx - 2\pi + E,$$

where $E \to 0$ as $R \to \infty$ and $\varepsilon \to 0$. Consequently,

$$\int_0^\infty \frac{\sin^2 x}{x^2}\, dx = \pi/2. \qquad \square$$

Integrals of trigonometric functions over $[0, 2\pi]$

Integrals of this type were discussed once already, in Section 3 of this chapter, in connection with Cauchy's Theorem. Here we can approach them more generally by allowing multiple poles or poles of order two or more.

Example 7   Compute the value of

$$\int_0^{2\pi} \frac{d\theta}{2 + \cos^2\theta}$$

Solution   As in Example 1, Section 3, we use the substitution $z = e^{i\theta}$. Thus, $d\theta = dz/iz$ and $\cos\theta = \frac{1}{2}(z + (1/z))$. This gives

$$\int_0^{2\pi} \frac{d\theta}{2 + \cos^2\theta} = \frac{1}{i}\int_{|z|=1} \frac{dz}{z\left(2 + \frac{1}{4}\left(z^2 + 2 + \frac{1}{z^2}\right)\right)}$$

$$= 2\pi\left\{\frac{1}{2\pi i}\int_{|z|=1} \frac{4z}{z^4 + 10z^2 + 1}\, dz\right\}.$$

The rational function

$$R(z) = \frac{4z}{z^4 + 10z^2 + 1}$$

has two poles, each of order 1, within the circle $|z| = 1$, at $z_1 = i\sqrt{5 - 2\sqrt{6}}$ and $z_2 = -i\sqrt{5 - 2\sqrt{6}}$. The residues are, respectively,

$$\text{Res}(R; z_1) = \frac{4z_1}{4z_1^3 + 20z_1} = \frac{1}{z_1^2 + 5}$$

$$= \frac{1}{2\sqrt{6}}$$

and

$$\text{Res}(R; z_2) = \frac{4z_2}{4z_2^3 + 20z_2} = \frac{1}{z_2^2 + 5}$$

$$= \frac{1}{2\sqrt{6}}.$$

Hence,

$$\int_0^{2\pi} \frac{d\theta}{2 + \cos^2\theta} = 2\pi\left\{\frac{1}{2\sqrt{6}} + \frac{1}{2\sqrt{6}}\right\}$$

$$= \frac{2\pi}{\sqrt{6}}. \qquad \Box$$

Integrals involving log x or fractional powers of x

Example 8   Find the value of

$$\int_0^\infty \frac{\log x}{(1 + x^2)^2} \, dx.$$

Solution   We take

$$f(z) = \frac{\log z}{(1 + z^2)^2}$$

on the domain D obtained by deleting the negative imaginary axis; $\log z$ is determined on D by requiring the imaginary part to have values in the interval $(-\pi/2, 3\pi/2)$. We integrate f around the contour in Figure 2.17, for $R > 1$ and $0 < \varepsilon < 1$. f has a single pole at i and it is of order 2. Hence,

$$\text{Res}(f; i) = \text{value of } \left(\frac{\log z}{(z + i)^2} \right)' \text{ at } z = i$$

$$= (\pi + 2i)/8 \text{ (see Example 3, Section 5).}$$

Thus,

$$\int_\gamma \frac{\log z}{(z^2 + 1)^2} \, dz = -\pi/2 + i\pi^2/4.$$

On that part of γ that lies on the circle $|z| = R$, f can be estimated by

$$|f(z)| \leqslant \frac{2 \log R}{(R^2 - 1)^2}, \quad |z| = R$$

so that

$$\left| \int_{\substack{|z| = R \\ z \in \gamma}} f(z) \, dz \right| \leqslant \pi R \frac{2 \log R}{(R^2 - 1)^2} \to 0 \text{ as } R \to \infty.$$

Furthermore, on that part of γ that lies on the circle $|z| = \varepsilon$, we have the estimate

$$|f(z)| \leqslant \frac{-2 \log \varepsilon}{(1 - \varepsilon^2)^2}, \quad |z| = \varepsilon.$$

Thus,

$$\left| \int_{\substack{|z| = \varepsilon \\ z \in \gamma}} f(z) \, dz \right| \leqslant \pi \varepsilon \frac{(-2 \log \varepsilon)}{(1 - \varepsilon^2)^2} \to 0 \text{ as } \varepsilon \to 0.$$

On the segment of γ on the negative real axis

$$f(z) = \frac{\log z}{(z^2 + 1)^2} = \frac{\log|x| + i\pi}{(x^2 + 1)^2}, \quad z = x, \quad -R \leqslant x \leqslant -\varepsilon.$$

All this together yields

$$-\pi/2 + i\pi^2/4 = \int_\varepsilon^R \frac{\log x}{(1 + x^2)^2} \, dx + \int_{-R}^{-\varepsilon} \frac{\log(-x) + i\pi}{(1 + x^2)^2} \, dx + E,$$

where E goes to zero as $\varepsilon\downarrow 0$ and $R\uparrow\infty$. Taking the real part of both sides we find that

$$\int_0^\infty \frac{\log x}{(1+x^2)^2}\, dx = -\pi/4. \tag{7}$$

The imaginary part yields

$$\int_0^\infty \frac{dx}{(1+x^2)^2} = \pi/4,$$

which we knew already from Example 2.   □

Example 9   Let $0 < \alpha < 1$; find the value of

$$\int_0^\infty \frac{x^{\alpha-1}}{1+x}\, dx.$$

Solution   Let D be the simply connected domain obtained by deleting the non-positive real numbers from the complex plane and let

$$f(z) = \frac{z^{\alpha-1}}{1-z}, \quad z \in D, \; z \neq 1$$

The function $z^{\alpha-1}$ is analytic on D when we choose the principal branch of $\log z$ and set $z^{\alpha-1} = \exp\{(\alpha-1)\mathrm{Log}\, z\}$.

Let γ be the "keyhole" contour pictured in Figure 2.18.

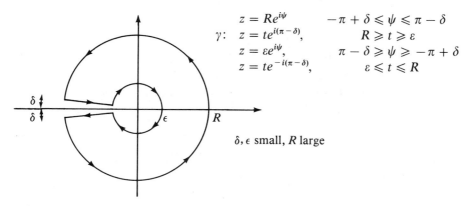

$$\gamma: \quad \begin{array}{ll} z = Re^{i\psi} & -\pi+\delta \leqslant \psi \leqslant \pi-\delta \\ z = te^{i(\pi-\delta)}, & R \geqslant t \geqslant \varepsilon \\ z = \varepsilon e^{i\psi}, & \pi-\delta \geqslant \psi \geqslant -\pi+\delta \\ z = te^{-i(\pi-\delta)}, & \varepsilon \leqslant t \leqslant R \end{array}$$

δ, ε small, R large

Figure 2.18

From Cauchy's formula or the residue theorem we have

$$\frac{1}{2\pi i}\int_\gamma \frac{z^{\alpha-1}}{z-1}\, dz = \{\text{value of } z^{\alpha-1} \text{ at } 1\} = 1$$

Let γ_R be that part of γ on the circle $|z| = R$ and γ_ε that part of γ on the circle $|z| = \varepsilon$. On γ_R, we have the estimate

$$\left| \int_{\gamma_R} \frac{z^{\alpha-1}}{z-1} \, dz \right| \leqslant \frac{R^{\alpha-1}}{R-1} \, 2\pi R \to 0 \text{ as } R \to \infty.$$

On γ_ε we have the estimate

$$\left| \int_{\gamma_\varepsilon} \frac{z^{\alpha-1}}{z-1} \, dz \right| \leqslant \frac{\varepsilon^{\alpha-1}}{1-\varepsilon} \, 2\pi \varepsilon \to 0 \text{ as } \varepsilon \to 0.$$

We parametrize the segment from $\varepsilon e^{-i\theta_1}$ to $Re^{-i\theta_1}$ (recall $\theta_1 = \pi - \delta$) by $z = te^{-i\theta_1}$, $\varepsilon \leqslant t \leqslant R$, and so the integral over this segment is

$$\int_\varepsilon^R \frac{t^{\alpha-1} e^{-i\theta_1(\alpha-1)}}{te^{-i\theta_1} - 1} \, e^{-i\theta_1} \, dt.$$

As $\theta_1 \to \pi$, this approaches

$$-e^{-i\pi\alpha} \int_\varepsilon^R \frac{t^{\alpha-1}}{1+t} \, dt.$$

Likewise, the integral over the remaining segment, from $Re^{i\theta_1}$ to $\varepsilon e^{i\theta_1}$, approaches

$$e^{+i\pi\alpha} \int_\varepsilon^R \frac{t^{\alpha-1}}{1+t} \, dt$$

as $\theta_1 \to \pi$. Now letting $\varepsilon \downarrow 0$ and $R \uparrow \infty$ we find that

$$2\pi i(1) = (e^{i\pi\alpha} - e^{-i\pi\alpha}) \int_0^\infty \frac{t^{\alpha-1}}{1+t} \, dt.$$

Equivalently,

$$\int_0^\infty \frac{t^{\alpha-1}}{1+t} \, dt = \frac{\pi}{\sin(\alpha\pi)}, \quad 0 < \alpha < 1. \qquad \square \quad (8)$$

A number of other formulas can be obtained from this last result. In the integral in (8), replace t by t^β, $0 < \beta < \infty$. We then obtain

$$\int_0^\infty \frac{t^{\alpha\beta-1}}{1+t^\beta} \, dt = \frac{1}{\beta} \frac{\pi}{\sin(\alpha\pi)}. \qquad (9)$$

For instance, by taking $\beta = 1/\alpha$, we find that

$$\int_0^\infty \frac{dt}{1+t^\beta} = \frac{1}{\beta} \frac{\pi}{\sin(\pi/\beta)}, \quad 1 < \beta < \infty. \qquad (10)$$

Exercises for Section 2.6

Use the method of Examples 1 and 2 to compute these integrals.

1. $\displaystyle\int_{-\infty}^{\infty} \frac{x^4}{1 + x^8}\, dx$

2. $\displaystyle\int_{-\infty}^{\infty} \frac{x^2}{x^4 - 4x^2 + 5}\, dx$

3. $\displaystyle\int_{-\infty}^{\infty} \frac{dx}{(x^2 + a^2)(x^2 + b^2)}\,; \; a, b > 0$

Use the method of Examples 3 to 6 to compute the following integrals.

4. $\displaystyle\int_{-\infty}^{\infty} \frac{\cos \alpha x}{(x^2 + 1)(x^2 + 4)}\, dx$

5. $\displaystyle\int_{-\infty}^{\infty} \frac{x \sin x}{x^4 + 1}\, dx$

6. $\displaystyle\int_{-\infty}^{\infty} \frac{\sin x}{x^2 + 6x + 10}\, dx$

7. $\displaystyle\int_{-\infty}^{\infty} \frac{\cos x}{(x + \alpha)^2 + \beta^2}\, dx$

8. $\displaystyle\int_{0}^{\infty} \frac{\cos \gamma x}{(x^2 + \alpha^2)(x^2 + \beta^2)}\, dx$ Answer: $\dfrac{\pi}{2\alpha\beta(\alpha^2 - \beta^2)}\left[\alpha e^{-\gamma\beta} - \beta e^{-\gamma\alpha}\right]$

Use the method of Example 7 to compute these integrals.

9. $\displaystyle\int_{0}^{2\pi} \frac{d\theta}{(2 - \sin \theta)^2}$

10. $\displaystyle\int_{0}^{2\pi} \frac{d\theta}{(1 + \beta \cos \theta)^2}, \quad -1 < \beta < 1$

11. $\displaystyle\int_{0}^{2\pi} \frac{\cos 2\theta}{1 - 2a \cos \theta + a^2}\, d\theta$

12. $\displaystyle\int_{0}^{2\pi} \sin^{2K} \theta\, d\theta$ Answer: $\pi\, \dfrac{(2K)!}{(K!)^2 2^{2K-1}}$

Use the "keyhole" contour in Figure 2.19 below in the manner of Example 9 to compute the following integrals

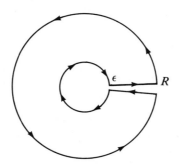

Figure 2.19

13. $\displaystyle\int_{0}^{\infty} \frac{x^\alpha}{x^2 + 3x + 2}\, dx, \; 0 < \alpha < 1$

14. $\displaystyle\int_{0}^{\infty} \frac{\sqrt{x}}{x^2 + 2x + 5}\, dx$ Answer: $\dfrac{\pi}{2\sqrt{2}}\sqrt{1 + \sqrt{5}}$

15. $$\int_0^\infty \frac{x^\lambda}{x^2 + 2x \cos \omega + 1} \, dx$$

Use the method of Example 8 in the following problems. (Note that the integrals are improper since log 0 is not defined.)

16. $$\int_0^\infty \frac{\log x}{(1 + x^2)(4 + x^2)} \, dx$$

17. $$\int_0^\infty \frac{\log x}{1 + x^2} \, dx$$

18. $$\int_0^\infty \frac{[\log x]^2}{1 + x^2} \, dx \quad \textit{Answer: } \pi^3/8$$

19. $$\int_0^\infty \frac{[\log x]^4}{1 + x^2} \, dx$$

20. (a) In the identity (8) differentiate both sides with respect to α to obtain the formula

$$\int_0^\infty \frac{t^{\alpha-1} \log t}{1 + t} \, dt = -\pi^2 \cot(\alpha\pi)\csc(\alpha\pi), \ 0 < \alpha < 1.$$

(b) Use the result of (a) to derive the formula

$$\int_0^\infty \frac{\log s}{1 + s^\gamma} \, ds = -\left(\frac{\pi}{\gamma}\right)^2 \cot\left(\frac{\pi}{\gamma}\right)\csc\left(\frac{\pi}{\gamma}\right), \ 1 < \gamma < \infty.$$

(**Hint:** Let $s = t^\alpha$.)

21. Show that $$\int_{-\infty}^\infty \frac{\cos x}{e^x + e^{-x}} \, dx = \frac{\pi}{e^{\pi/2} + e^{-\pi/2}}.$$

(**Hint:** Integrate $f(z) = e^{iz}/(e^z + e^{-z})$ over the rectangle with vertices at $\pm R$ and $\pm R + i\pi$.)

22. Use (9) or (10) to compute these integrals

(a) $$\int_0^\infty \frac{dx}{8 + x^3}$$ (b) $$\int_0^\infty \frac{x}{x^4 + 16} \, dx$$

(c) $$\int_0^\infty \frac{x^\gamma}{1 + x^\beta} \, dx, \ 0 \leqslant \gamma < \beta - 1.$$

23. Show that (a) $$\int_{-\infty}^\infty \frac{dx}{(x^2 + \alpha^2)^2} = \frac{\pi}{2\alpha^3}$$ (b) $$\int_{-\infty}^\infty \frac{x^2}{(x^2 + \alpha^2)^3} \, dx = \frac{\pi}{8\alpha^3}$$

24. Let p be a polynomial of degree n with distinct zeros at $z_1, \ldots, z_n$ and let q be a polynomial of degree $n - 2$ or less. Show that

$$\sum_{k=1}^n \operatorname{Res}\left(\frac{q}{p}; z_k\right) = 0.$$

(**Hint:** Integrate q/p over the circle $|z| = R$ and let $R \to \infty$.)

The residue theorem for functions with essential singularities★

25. Suppose that D is a star-shaped or a simply-connected domain and

that g is analytic in D except for a finite number of isolated singularities at points $z_1, \ldots, z_N$ of D. Let γ be a piecewise smooth positively oriented simple closed curve in D that does not pass through
any of the points $z_1, \ldots, z_N$. Show that

$$\frac{1}{2\pi i} \int_\gamma g(z)\, dz = \sum_{z_j \text{ inside } \gamma} \text{Res}(g; z_j)$$

where the summation is over those singularities of g that lie inside γ.

26.   Use Exercise 25 to calculate the following integrals

(a) $\dfrac{1}{2\pi} \displaystyle\int_0^{2\pi} e^{e^{-i\theta}} e^{in\theta}\, d\theta$

(b) $\dfrac{1}{2\pi} \displaystyle\int_0^{2\pi} e^{\cos\theta} \cos n\theta\, d\theta$ (**Hint:** $g(z) = z^{n-1} \exp(\tfrac{1}{2}(z + 1/z))$)

(c) $\dfrac{1}{2\pi} \displaystyle\int_0^{2\pi} e^{\sin\theta} \sin n\theta\, d\theta$

The summation of certain series by means of residues⋆

27.   Show that the residue of $C(z) = \pi \cot \pi z$ at each of its poles $z_k = k$,
$k = 0, \pm 1, \pm 2, \ldots$, is precisely 1.

28.   Show that $C(z) = \pi \cot \pi z$ is bounded on the square γ_N with vertices
at $\pm(N + \tfrac{1}{2}) \pm i(N + \tfrac{1}{2})$, N a positive integer, by a bound that does *not*
depend on N.

29.   Let $f(z) = 1/z^2$; show that

$$\int_{\gamma_N} f(z)C(z)\, dz \to 0 \text{ as } N \to \infty.$$

use the Residue Theorem to prove that

$$\sum_1^\infty \frac{1}{n^2} = \pi^2/6.$$

30.   Suppose that f is analytic on the plane except for poles at $\zeta_1, \ldots, \zeta_s$
and suppose that $\lim_{|z| \to \infty} |zf(z)| = 0$. Show that

$$\sum_{-\infty}^{\infty} f(n) = -\sum_{j=1}^{s} \text{Res}(f(z)\pi \cot \pi z; \zeta_j) \qquad \textbf{(11)}$$

31.   Use (11) to find the sum of the following series

(a) $\displaystyle\sum_{n=1}^{\infty} \frac{1}{n^2 + a^2}$

(b) $\displaystyle\sum_{-\infty}^{\infty} \frac{1}{(n + a)^2}$; $a > 0$, a not an integer   *Answer:* $\left(\dfrac{\pi}{\sin \pi a}\right)^2$

(c) $\displaystyle\sum_{n=1}^{\infty} \frac{1}{(2n-1)^2}$ (d) $\displaystyle\sum_{n=1}^{\infty} \frac{1}{n^4}$.

Further use of the residue theorm on $\pi \cot \pi z\star$

32.   Let γ_N be the curve from Exercise 28. Show that

$$\frac{1}{2\pi i} \int_{\gamma_N} \frac{C(z)}{z(z-w)}\, dz = \frac{\pi \cot \pi w}{w}$$

$$+ \sum_{\substack{k=-N \\ k\neq 0}}^{N} \frac{1}{k(k-w)} - \frac{1}{w^2}, \; w \neq 0, \pm 1, \pm 2, \ldots, \pm N$$

33.   Let $N \rightarrow \infty$; show that

$$C(w) = \pi \cot \pi w = \frac{1}{w} + \sum_{k=1}^{\infty} \frac{2w}{w^2 - k^2}.$$

34.   Integrate $\pi \cot \pi w - (1/w)$, which is analytic at $w = 0$, from 0 to z to obtain

$$\text{Log}\left(\frac{\sin \pi z}{z}\right) = \sum_{k=1}^{\infty} \text{Log}\left(1 - \frac{z^2}{k^2}\right) + \log \pi,$$

z not a negative real number.

Conclude that

$$\sin \pi z = \pi z \prod_{k=1}^{\infty} \left(1 - \frac{z^2}{k^2}\right). \tag{12}$$

Further reading

Each of the following books gives a more sophisticated and far-reaching development of the theory of analytic functions than is presented here.

Ahlfors, L. V. *Complex analysis.* 3rd ed. New York: McGraw-Hill, 1979.

Burckel, R. B. *An introduction to classical complex analysis.* New York: Academic Press, 1979.

Caratheodory, C. *Theory of functions.* Vols. 1 and 2. New York: Chelsea, 1964.

Heins, M. *Complex function theory.* New York: Academic Press, 1968.

Hille, E. *Analytic function theory.* Vols. 1 and 2. New York: Blaisdell, 1959.

Knopp, K. *Theory of functions,* parts I and II; and *Problem book in the theory of functions,* parts I and II. New York: Dover, 1945; 1952.

The books by Burckel and Hille contain extensive and informative historical and biographical information on the theorems of complex analysis and their creators. These notes are well worth reading.

3

Analytic functions as mappings

The theme of this chapter is the "geometry" of analytic functions, in particular the nature of their range. This theme manifests itself first in several theorems in Section 1 which pinpoint the number of solutions to an equation of the form $f(z) = w_0$. The answer is formulated in a number of different ways, all connected closely with the behavior of the function f on a closed curve. This ability to count the number of solutions of $f(z) = w_0$ is exploited in Section 2 to prove that the range of a nonconstant analytic function is an open set. This is a profound and elegant property of analytic functions that leads in a natural way to the maximum modulus principle. Even more important we are led directly to the heart of the study of conformal mapping, a tool of enormous beauty and power both in applications and theory.

3.1 The zeros of an analytic function

Suppose that f is analytic and not identically zero in a domain D. We saw in (5) and (6) of Section 4, Chapter 2 that each zero of f has a certain **order**; that is, if $f(z_0) = 0$, then there is an integer m, $m \geqslant 1$, such that

$$f(z) = (z - z_0)^m g(z),$$

where g is analytic in D and $g(z_0) \neq 0$. The integer m is frequently called the **multiplicity** of the zero of f at z_0. In this section we shall collect several results about the zeros of an analytic function. We begin with a fact closely related to (5) of Section 4, Chapter 2.

> Suppose that f is analytic in a domain **D**. If there are
> distinct points $z_1, z_2, \ldots$ in D with f(z_n) = 0, n = 1, 2, … (1)
> and if the sequence $\{z_n\}$ converges to a point z_0 of D,
> then f(z) = 0 for all z ∈ D.

To show why (1) is true we make use of (5) in Section 4, Chapter 2; we shall show that the hypotheses in (1) imply that $f^{(k)}(z_0) = 0$ for $k = 0, 1, 2, \ldots$. We note first that since $z_n \to z_0$ and since f is continuous, we must have $f(z_0) = 0$. Now $f(z)$ has a power series valid for $|z - z_0| < \delta$:

$$f(z) = a_1(z - z_0) + a_2(z - z_0)^2 + \cdots, \quad |z - z_0| < \delta.$$

Suppose we know that $a_0 = \cdots = a_{N-1} = 0$; we shall show that $a_N = 0$, too.

Define a function g by

$$g(z) = \begin{cases} \dfrac{f(z)}{(z - z_0)^N}, & z \neq z_0 \\[2mm] a_N, & z = z_0 \end{cases}$$

Clearly g is analytic on $D - \{z_0\}$. In the disc $|z - z_0| < \delta$, we have

$$g(z) = a_N + a_{N+1}(z - z_0) + \cdots ,$$

where the series converges for $|z - z_0| < \delta$; hence, g is analytic for $|z - z_0| < \delta$, as well. However,

$$g(z_n) = \frac{f(z_n)'}{(z_0 - z_n)^N} = 0, \ n = 1, 2, \dots \ \text{and} \ z_n \neq z_0$$

so that $a_N = g(z_0) = \lim_{n \to \infty} g(z_n) = 0$. Thus, all the coefficients a_k of the power series for $f(z)$ about the point z_0 are zero and hence $f(z) = 0$ for all z in D by (5) in Section 4, Chapter 2.

To reiterate,

> *if an analytic function* f *on a domain* D *vanishes on a sequence of distinct points* $\{z_n\}$ *of* D, *which converge to a point* z_0 *of* D, *then* f *vanishes identically in* D.

Example 1   The hypothesis that z_0 lies in D is not superfluous. For instance, $f(z) = \sin(\pi/z)$ is analytic for all $z \neq 0$ and $f(1/n) = \sin(\pi n) = 0$ for $n = 1, 2, \dots$ and $z_n = 1/n \to 0$ as $n \to \infty$, yet f is not identically zero. (Obviously, f is not analytic on any disc centered at the origin.)

The mathematical way to phrase (1) is to say that the zeros of a nonconstant analytic function are **isolated** (from each other). This allows us to count them, as the next several theorems show.

Suppose that h is a function analytic on a domain D except for a finite number of poles. Let γ be a piecewise smooth simple closed curve in D whose inside lies in D and which passes through no zero or pole of h. Let $z_1, \dots, z_N$ be the distinct zeros of h inside γ and let $w_1, \dots, w_M$ be the distinct poles of h inside γ (Figure 3.1).

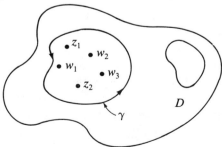

Figure 3.1

Let n_j be the order (or multiplicity) of the zero of h at $z_j, j = 1, \ldots, N$. Then

$$h(z) = (z - z_j)^{n_j} G_j(z),$$

where G_j is analytic near z_j and $G_j(z_j) \neq 0$. Hence,

$$\frac{h'(z)}{h(z)} = \frac{n_j}{z - z_j} + \frac{G_j'(z)}{G_j(z)},$$

so that

$$\operatorname{Res}\left(\frac{h'}{h}; z_j\right) = n_j. \tag{2}$$

Likewise, if m_k is the order of the pole of h at $w_k, k = 1, \ldots, M$, then

$$h(z) = \frac{H_k(z)}{(z - w_k)^{m_k}},$$

where H_k is analytic near w_k and $H_k(w_k) \neq 0$. Thus,

$$\frac{h'(z)}{h(z)} = \frac{-m_k}{z - w_k} + \frac{H_k'(z)}{H_k(z)}$$

for z near w_k so that

$$\operatorname{Res}\left(\frac{h'}{h}; w_k\right) = -m_k. \tag{3}$$

We now apply the residue theorem (Theorem 1 from Section 6, Chapter 2) to the function (h'/h). The result is

$$\frac{1}{2\pi i} \int_\gamma \frac{h'(z)}{h(z)}\, dz = \sum_{j=1}^N n_j - \sum_{k=1}^M m_k. \tag{4}$$

Here is the formal statement of (4).

Theorem 1: Suppose that h is analytic in a domain D except for a finite number of poles. Let γ be a piecewise smooth simple closed curve in D, which does not pass through any pole or zero of h and whose inside lies in D. Then,

$$\frac{1}{2\pi i} \int_\gamma \frac{h'(z)}{h(z)}\, dz = \begin{Bmatrix} \text{number of zeros} \\ \text{of } h \text{ inside } \gamma \end{Bmatrix} - \begin{Bmatrix} \text{number of poles} \\ \text{of } h \text{ inside } \gamma \end{Bmatrix}.$$

All zeros and poles are counted with multiplicities.

The integral in Theorem 1 has another interpretation. Since h is not zero on γ, each point z on γ is the center of a small disc on which h is not zero. Hence, there is a branch of $\log h$ that is analytic on this disc and $(\log h)' = h'/h$. Thus, if γ is the range of the curve $z(t), a \leqslant t \leqslant b$, we have

$$\int_\gamma \frac{h'(z)}{h(z)}\, dz = \int_a^b \frac{d}{dt} \log\{h(z(t))\}\, dt$$

$$= \log\{h(z(b))\} - \log\{h(z(a))\}$$

$$= i\Delta_\gamma(\arg\, h(z)),$$

where Δ_γ stands for the total net change of arg $h(z)$ as z traverses γ once in the positive direction. The last equality is the result of the fact that $\log w = \log |w| + i$ arg w and although $\log |w|$ is single-valued, arg w is not. The net change in arg $h(z)$ may be nonzero, though it is always an integer multiple of 2π. When we combine these observations with Theorem 1, we obtain the following result.

Theorem 2: The Argument principle. Suppose h is analytic on a domain D except for isolated poles. Let γ be a piecewise smooth simple closed curve in D whose inside lies in D and which does not pass through any zeros or poles of h. Then,

$$\frac{1}{2\pi} \left\{\begin{array}{l}\text{change in arg } h(z) \\ \text{as } z \text{ traverses } \gamma\end{array}\right\} = \left\{\begin{array}{l}\text{number of zeros} \\ \text{of } h \text{ inside } \gamma\end{array}\right\} - \left\{\begin{array}{l}\text{number of poles} \\ \text{of } h \text{ inside } \gamma\end{array}\right\}.$$

Example 2   Show that for each complex number w with w *not* in the interval $[-1, 1]$, the equation

$$z + \frac{1}{z} + 2w = 0$$

has precisely one solution z with $|z| < 1$.

Solution   On the circle $|z| = 1$ we have $z = e^{i\theta}$ and so

$$z + \frac{1}{z} + 2w = 2(\cos \theta + w).$$

Now $w = s + it$ where t is *nonzero* if $-1 \leqslant s \leqslant 1$. Hence, if $|s| > 1$, then $\text{Re}(z + (1/z) + 2w) = 2(\cos \theta + s)$ is of constant sign (positive if $s > 0$, negative if $s < 0$) and so $z + (1/z) + 2w$ lies always in the half-plane $\text{Re}\, \zeta > 0$ (if $s > 0$) or always in the half-plane $\text{Re}\, \zeta < 0$ (if $s < 0$). On the other hand, if $-1 \leqslant s \leqslant 1$, then

$$\text{Im}\left(z + \frac{1}{z} + 2w\right) = 2t$$

and so $z + (1/z) + 2w$ lies always in the half-plane $\text{Im}\, \zeta > 0$ (if $t > 0$) or always in the half-plane $\text{Im}\, \zeta < 0$ (if $t < 0$). Thus, for each w the function $f(z) = z + (1/z) + 2w$ has no net change in its argument as z traverses the circle $|z| = 1$ once. Therefore, as a consequence of Theorem 2, f has as many zeros as poles within the circle $|z| = 1$. But f clearly has precisely one pole (at $z = 0$) and so f has exactly one zero has well (Fig. 3.2).

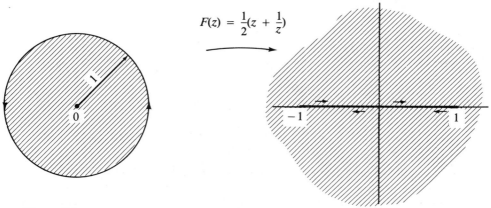

$$F(z) = \frac{1}{2}\left(z + \frac{1}{z}\right)$$

Figure 3.2

Note that replacing z by $1/z$ has no effect on the function $z + (1/z) + 2w$ and so we see as well that the function $F(z) = \frac{1}{2}(z + 1/z)$ is a one-to-one analytic function that maps the region $|z| > 1$ onto the plane with the interval $[-1, 1]$ removed. The curves

$$y - \frac{y}{x^2 + y^2} = 2c, \qquad z = x + iy$$

are mapped onto the horizontal lines, $\mathrm{Im}\, w = c$. When c is very big, these curves are virtually horizontal lines. ◻

Example 3   Find the number of zeros of $z^3 - 2z^2 + 4$ in the first quadrant.

Solution   We examine $f(z) = z^3 - 2z^2 + 4$ on the contour shown in Figure 3.3; R is very big. On the segment $0 \leqslant x \leqslant R$, $f(x) = x^3 - 2x^2 + 4$ is real and

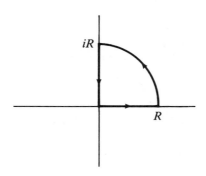

Figure 3.3

greater than 2. On the quarter-circle $z = Re^{it}$, $0 \leqslant t \leqslant \pi/2$, we have

$$f(Re^{it}) = R^3 e^{3it}\left(1 - \frac{2}{Re^{it}} + \frac{4}{R^3 e^{3it}}\right) = R^3 e^{3it}(1 + \zeta),$$

where $|\zeta| \leqslant 6/R < 1/2$ for R large. Thus, $\arg f(Re^{it})$ is approximately $\arg(e^{3it}) = 3t$ for large R and so $\arg f(Re^{it})$ increases from 0 to about $3\pi/2$ as t increases from 0 to $\pi/2$. On the segment $z = iy$, $R \geqslant y \geqslant 0$, we have

$$f(iy) = -iy^3 + 2y^2 + 4.$$

For $R \geqslant y > 0$, this point lies in the fourth quadrant since

$$\mathrm{Re}(f(iy)) = 4 + 2y^2 > 0$$
$$\mathrm{Im}(f(iy)) = -y^3 \quad < 0.$$

Hence as y decreases from R to 0, $f(iy)$ lies in the fourth quadrant and moves toward the point $w = 4$. Consequently, as z traverses the contour in figure 3.3, $\arg f(z)$ increases by exactly 2π and thus $f(z) = z^3 - 2z^2 + 4$ has precisely *one* zero in the first quadrant.   ☐

Here is another theorem, closely related to the argument principle

Theorem 3: Rouché's* Theorem. Suppose f and g are analytic on an open set containing a piecewise smooth simple closed curve γ and its inside. If

$$|f(z) + g(z)| < |f(z)| \quad \text{for all } z \in \gamma, \tag{5}$$

then f and g have an equal number of zeros inside γ, counting multiplicities.

Proof. Note that the hypothesis (5) ensures that neither f nor g is zero on γ. We may assume further that all the common zeros of f and g inside γ have been canceled; this affects neither the hypotheses nor the conclusion. Let $h = g/f$; then we have

$$|h(z) + 1| < 1 \quad \text{for all } z \in \gamma$$

so that the range of h on γ lies in the disc of radius 1 centered at the point -1. In particular, $\arg h(z)$ has no net change as z traverses γ (Fig. 3.4).

By Theorem 2 the number of zeros of h inside γ equals the number of poles of h inside γ. But the number of zeros of h is just the number of zeros of g and the number of poles of h is just the number of zeros of f.   ☐

---

* Eugene Rouché, 1832–1910.

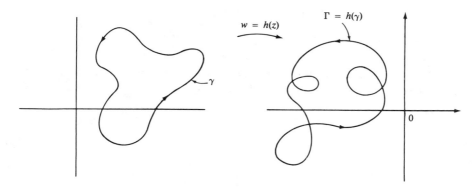

Figure 3.4

Rouché's Theorem has a number of useful applications as the following examples show.

Example 4   Show that all the zeros of

$$p(z) = 3z^3 - 2z^2 + 2iz - 8$$

lie in the annulus $1 < |z| < 2$.

Solution   On the circle $|z| = 1$ we have

$$|p(z) + 8| \leqslant 3 + 2 + 2 = 7 < 8$$

so that $p(z)$ and $f(z) \equiv -8$ have the same number of zeros within $|z| = 1$; thus, $p(z)$ does not vanish in the disc $|z| \leqslant 1$. Furthermore, on $|z| = 2$ we have

$$|p(z) - 3z^3| \leqslant 2(4) + 2(2) + 8 = 20 < 24 = |3z^3|$$

and so $p(z)$ and $f(z) = 3z^3$ have an equal number of zeros within the circle $|z| = 2$; that is, $p(z)$ has all three of its zeros within $|z| = 2$. Consequently, all the zeros of p lie in the annulus $1 < |z| < 2$.   ☐

Example 5   Show that for each $\lambda > 1$, the equation

$$z + e^{-z} = \lambda \tag{6}$$

has exactly one root z_0 with Re $z_0 > 0$.

Solution   Let $f(z) = z - \lambda$ and $g(z) = z + e^{-z} - \lambda$. Then on the imaginary axis

$$|g(iy) - f(iy)| = |e^{-iy}| = 1$$
$$< \sqrt{\lambda^2 + y^2} = |f(iy)|.$$

Furthermore, on the semicircle $z = Re^{i\theta}$, $-\pi/2 \leqslant \theta \leqslant \pi/2$, we have

$$|g(Re^{i\theta}) - f(Re^{i\theta})| = e^{-R\cos\theta} \leqslant 1$$
$$< \sqrt{\lambda^2 + R^2 - 2\lambda R \cos\theta} = |f(Re^{i\theta})|$$

for any $R > \lambda + 1$. Hence, f and g have an equal number of zeros within the contour of Figure 3.5. Since f has exactly one zero there (if $R > \lambda$) we see that g has exactly one zero for all $R > \lambda + 1$. ☐

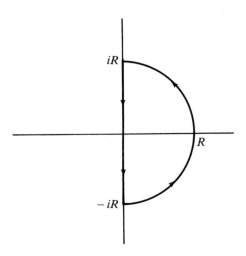

Figure 3.5

Example 6   Find the number of roots of the equation

$$\frac{z^2 - 4}{z^2 + 4} + \frac{2z^2 - 1}{z^2 + 6} = 0 \tag{7}$$

that lie within the unit circle $|z| = 1$.

Solution   When we clear the fractions we see that we must determine the number of roots within the circle $|z| = 1$ of the equation

$$(z^2 - 4)(z^2 + 6) = (z^2 + 4)(1 - 2z^2).$$

Let $p(z) = (z^2 - 4)(z^2 + 6) = z^4 + 2z^2 - 24$ and let $q(z) = (z^2 + 4)(1 - 2z^2) = -2z^4 - 7z^2 + 4$. Then, on $|z| = 1$, we have

$$|p(z)| = |24 - 2z^2 - z^4| \geqslant 24 - 2 - 1 = 21$$

and

$$|q(z)| = |2z^4 + 7z^2 - 4| \leqslant 2 + 7 + 4 = 13.$$

Hence, $|p(z)| > |q(z)| = |\{p(z) - q(z)\} - \{p(z)\}|$, which implies that p and $p - q$ have an equal number of zeros within the circle $|z| = 1$. But clearly p has no zeros within $|z| < 1$ and so (7) has no roots within $|z| < 1$. ☐

The Fundamental Theorem of Algebra

Example 7   A polynomial of degree n has exactly n zeros, counting multiplicities.

Proof.   We do not affect the solutions of $p(z) = 0$ by dividing $p(z)$ by a_n ; thus, we may suppose that

$$p(z) = z^n + a_{n-1}z^{n-1} + \cdots + a_1 z + a_0 .$$

Now for large values of $|z| = R$, we have

$$\left| \frac{p(z) - z^n}{z^n} \right| = |a_{n-1}z^{-1} + \cdots + a_0 z^{-n}|$$

$$\leqslant (|a_0| + \cdots + |a_{n-1}|)/R \leqslant 1/2.$$

Hence, $|p(z) - z^n| \leqslant \frac{1}{2}|z|^n < |z^n|$ for $|z| = R$ and so by Rouché's Theorem (Theorem 3), we learn that $p(z)$ and z^n have an equal number of zeros within the circle $|z| = R$. But z^n has exactly n zeros within $|z| = R$ and hence the same is true for $p(z)$.

Exercises for Section 3.1

Use the technique of Example 3 to determine the number of zeros of f in the first quadrant.

1. $f(z) = z^2 - z + 1$ 2. $f(z) = z^4 - 3z^2 + 3$

3. $f(z) = z^3 - 3z + 6$ 4. $f(z) = z^2 + iz + 2 + i$

5. $f(z) = z^9 + 5z^2 + 3$ 6. $f(z) = z^7 + 6z^3 + 7$

In Exercises 7 and 8 use the technique of Example 3 to determine the number of zeros of the given function in the upper half-plane

7. $z^4 + 3iz^2 + z - 2 + i$ 8. $2z^4 - 2iz^3 + z^2 - 2iz - 1$

9.   Show that the equation $z^4 - 5z^2 + 3 = e^{-z}$ has no solutions on the imaginary axis and precisely two solutions in the half-plane Re $z > 0$.

10.   Suppose that w is not in the interval $[-R, R]$. Show that the equation $z + R^2/z = 2w$ has one solution z with $|z| < R$ and one solution z with $|z| > R$.

11.   Show that there is no entire function F with $F(x) = 1 - \exp[2\pi i/x]$ for $1 \leqslant x \leqslant 2$.

In Exercises 12–15 use the technique of Example 4 to determine how many zeros of the following functions lie in the given annulus.

12. $z^3 - 3z + 1$ in $1 < |z| < 2$ 13. $z^4 - 2z - 2$ in $\frac{1}{2} < |z| < \frac{3}{2}$

14. $ze^z - \frac{1}{4}$ in $0 < |z| < 2$

15. $4z^3 - 12z^2 + 2z + 10$ in $\frac{1}{2} < |z - 1| < 2$

16.   Let f and g be analytic on a domain containing a simple closed curve γ and its inside. Show that if $|f(z)| > |g(z)|$ for all $z \in \gamma$ then the two equations $f(z) = g(z)$ and $f(z) = 0$ have an equal number of solutions inside γ.

17.   Use the result of Exercise 16 to determine how many solutions each of the following equations has in the given region.
       (a) $z^3 - 10 = z^2 + z + 6$ in $|z| < 1$
       (b) $z^3 = z^2 + z + 6$ in $|z| < 1$
       (c) $z + a = e^z$ for $a > 1$, in Re $z < 0$. (**Hint:** let γ be the semicircle of radius R from iR to $-iR$ through $-R$ plus the segment from $-iR$ to iR on the imaginary axis.)

18.   Extend formula (4) to prove the following. Let g be analytic on a domain containing γ and its inside. Then
       $$\frac{1}{2\pi i} \int_\gamma \frac{h'(z)}{h(z)} g(z)\, dz = \sum_{i=1}^N g(z_i) - \sum_{j=1}^M g(w_j),$$
       where $z_1, \ldots, z_N$ are the zeros of h and $w_1, \ldots, w_M$ are the poles of h inside γ, each listed according to its multiplicity.

19.   Suppose that f is analytic on a domain containing $\{z : |z| \leqslant 1\}$ and that $|f(e^{i\theta})| = 1$, $0 \leqslant \theta \leqslant 2\pi$. Use the polar form of the Cauchy–Riemann equations to show that $\arg f(e^{i\theta})$ is an increasing function of θ. Conclude that either f is constant or that f must have at least one zero in the disc $|z| < 1$. You will need to know that $|f(z)| \leqslant 1$ if $|z| < 1$ (see Exercise 15, Section 3, Chapter 2).

20.   Suppose that f is analytic on a domain containing $\{z : |z| \leqslant 1\}$ and that $|f(e^{i\theta})| < 1$, $0 \leqslant \theta \leqslant 2\pi$. Show that f has exactly one **fixed point** in the disc $|z| < 1$; that is, the equation $f(z) = z$ has precisely one solution in the disc $|z| < 1$.

21.   Let p and q be polynomials of degree n. If $p(z) = q(z)$ at $n + 1$ distinct points of the plane, then $p(z) = q(z)$ for all z.

22.   Let p be a polynomial with real coefficients. Show that $p(z) = p_1(z) \cdots p_m(z)$, where each p_j is a polynomial of degree 1 or 2 with real coefficients.

23.   Let P be a polynomial with $|P(e^{it})| = 1$ for all t, $0 \leqslant t \leqslant 2\pi$. Show that $P(z) = \lambda z^N$, $|\lambda| = 1$. (**Hint:** Let N be the degree of P and set $Q(z) = z^N \overline{P(1/\bar{z})}$. Show that Q is a polynomial and that $P(e^{it})Q(e^{it}) = e^{iNt}$, $0 \leqslant t \leqslant 2\pi$. Conclude that Q is constant.)

24.   Let p be a polynomial of degree n all of whose zeros are real. Let
        $x_1 < x_2 < \cdots < x_r$ be the distinct zeros of p with respective multiplic-
        ities $m_1, m_2, \ldots, m_r$.
        (a) Show that p' is not zero in $(-\infty, x_1)$ or in (x_r, ∞)
        (b) Show that p' has precisely one zero in $(x_j, x_{j+1}), j = 1, \ldots, r - 1$.

Rational functions on the extended plane★

Suppose that f is analytic on a plane except for a finite number of poles;
suppose, in addition, that f is either analytic at ∞ or has a pole at ∞. Complete the following exercises to show that f is, in fact, a rational function.
        Let m_0 be the order of the pole of f at ∞ with $m_0 = 0$ if f is analytic at
∞. Let $w_1, \ldots, w_r$ be the distinct poles of f in the plane with respective orders
$m_1, \ldots, m_r$. Set

$$g(z) = (z - w_1)^{m_1} \cdots (z - w_r)^{m_r} f(z).$$

25.   Show that g is an entire function.

26.   Show that

$$|g(z)| \leqslant C|z|^m, \quad |z| > R,$$

        where C is a constant and $m = m_0 + m_1 + \cdots + m_r$.

27.   Conclude that g is a polynomial of degree m or less and thus f is a
        rational function. (**Hint:** use Exercise 21, Section 4, Chapter 2.)

Mapping properties of rational functions★

Let P and Q be polynomials of respective degrees N and M and assume that P
and Q have no common zero. Let $R = P/Q$ and define $d = $ **degree of** R to be
the maximum of N and M.

28.   Suppose that $M \geqslant N$ (a) Show that R has a removable singularity at
        ∞. (b) If $M > N$, then R has a zero of order $M - N$ at ∞. (c) If
        $M = N$, then $R(\infty) \neq 0$.

29.   Suppose that $M < N$. Show that R has a pole at ∞ of order $N - M$.

30.   Counting multiplicities and counting the zeros or poles at ∞, if any,
        show that R has d zeros and d poles.

31.   Let α be a complex number. Show that $R - \alpha$ is a rational function
        with the same degree as R.

32.   Show that the equation $R(z) = \alpha$ has exactly d solutions, counting
        multiplicities, for each complex number α. Conclude that R maps the
        complex plane plus ∞ onto itself and each point is "covered" exactly
        d times.

Legendre polynomials★

We define the nth Legendre polynomial, P_n, by

$$P_n(t) = \frac{1}{2^n n!} \frac{d^n}{dt^n}(t^2 - 1)^n. \tag{8}$$

33.   Let γ be a smooth simple closed curve around t. Show that

$$P_n(t) = \frac{1}{2\pi i} \frac{1}{2^n} \int_\gamma \frac{(z^2 - 1)^n}{(z - t)^{n+1}} \, dz. \tag{9}$$

34.   In Exercise 33 take γ to be a circle of radius $\sqrt{|t^2 - 1|}$ centered at t; show that

$$P_n(t) = \frac{1}{2\pi} \int_0^{2\pi} (t + \sqrt{t^2 - 1} \cos \theta)^n \, d\theta. \tag{10}$$

35.   Show that P_n is a solution of **Legendre's differential equation:**

$$(1 - z^2)f''(z) - 2zf'(z) + n(n + 1)f(z) = 0, \qquad f(1) = 1.$$

36.   Prove the **generating formula:** $\dfrac{1}{\sqrt{1 - 2zt + t^2}} = \displaystyle\sum_{n=0}^\infty P_n(z)t^n$

(**Hint:** In the series on the right, substitute for P_n the formula (10).)

37.   Show that $\{P_n\}$ satisfies both of the following **recursion relations:**
(a) $(n + 1)P_{n+1}(z) - (2n + 1)zP_n(z) + nP_{n-1}(z) = 0$
(b) $(2n + 1)P_n = P'_{n+1} - P'_{n-1}$
(**Hint:** For (a) let $F(z, t) = (1 - 2zt + t^2)^{-1/2}$; show that
$(1 - 2zt + t^2) \, \partial F/\partial t - (z - t)F = 0$. Now use the formula in Exercise
36 for $F(z, t)$ in terms of $P_n(z)t^n$ and collect equal powers of t.)

3.1.1   The stability of solutions of a system of linear differential equations★

The operation of many physical processes is described by a system of ordinary differential equations, together with a set of initial conditions. It is a matter of great relevance whether small perturbations in the initial conditions will over the long run produce substantial or even unbounded changes in the corresponding solution. As an elementary example, imagine placing a ball on the curve $y = x^2$. If placed at the origin $(x_0, y_0) = (0, 0)$, the ball remains stationary; if placed at some point (x_0, y_0), $x_0 \neq 0$, the ball rolls back and forth about $(0, 0)$ indefinitely [or, if there is friction, tends to $(0, 0)$ as time goes to infinity]. On the other hand if the ball is placed on the curve $y = -x^2$ then it moves further and further from $(0, 0)$ as time goes to infinity *unless* it was placed (with a steady hand!) exactly at $(0, 0)$ to begin with. Obviously the first situation is what we can call "stable" while the second is "unstable." In this section we shall show that if the zeros of a certain polynomial associated with the system of differential equations all have negative real parts, then the solutions of the differential equations are stable, in the sense that small perturbations in the initial conditions produce only small perturbations in the solution. A general knowledge of the solution of a linear first-order system of differential equations would be helpful but is not strictly necessary in reading this section.

The system we shall investigate is

$$\dot{x} = Ax + b, \tag{1}$$

where A is an $n \times n$ matrix with constant entries, b is an $n \times 1$ column vector with constant entries and

$$x = x(t) = \begin{pmatrix} x_1(t) \\ \vdots \\ x_n(t) \end{pmatrix}.$$

The solutions of this system involve the zeros of the polynomial p given by

$$p(z) = \det(zI - A), \tag{2}$$

where I is the $n \times n$ identity matrix and 'det' stands for determinant. The polynomial p has degree n and so has n zeros in the plane, counting multiplicities. Let $\lambda_1, \ldots, \lambda_r$ be the *distinct* zeros and let $m_1, \ldots, m_r$ be their respective multiplicities so that

$$p(z) = \prod_1^r (z - \lambda_j)^{m_j}.$$

The study of linear first-order systems of which (1) is representative tells us that each solution of the **homogeneous** system $Ax = \dot{x}$ has the form

$$x_l(t) = \sum_{j=1}^r p_{jl}(t)e^{\lambda_j t}, \qquad l = 1, \ldots, n, \tag{3}$$

where each p_{jl} is a polynomial of degree at most $m_j - 1$. Furthermore, given a set of initial conditions $x(0) = x_0$ there is a unique solution $x(t)$ of (1) with $x(0) = x_0$. We wish to examine the difference between the solution $x_1(t)$ corresponding to initial conditions $x_1(0) = x_1$ and the solution $x_0(t)$ and watch how this difference behaves as $t \to \infty$. Note that the linearity of the system implies that

$$A(x_0 - x_1) = (\dot{x}_0 - b) - (\dot{x}_1 - b) = \dot{x}_0 - \dot{x}_1$$

and

$$x_0(0) - x_1(0) = x_0 - x_1.$$

Hence, if we write $w = x_0 - x_1$, then w solves the homogeneous equation

$$Aw = \dot{w} \qquad \text{and} \qquad w(0) = w_0 = x_0 - x_1.$$

We shall show now that if all the zeros of the polynomial p in (2) lie in the left half-plane, then any solution $x(t)$ of the homogeneous linear system $Ax = \dot{x}$

tends to zero as $t \to \infty$. Indeed, the proof is elementary. Let $x_l(t)$ be the lth component of the solution. From (3) we have

$$|x_l(t)| \leqslant \sum_{k=1}^{r} |p_{kl}(t)| e^{t \, \mathrm{Re} \, \lambda_k}.$$

Now $p_{kl}(t)$ has degree at most $m_k - 1$ and so we know that

$$|p_{kl}(t)| \leqslant A_{kl} t^{m_k - 1}, \ t \geqslant 1,$$

where A_{kl} is a positive constant. Furthermore, our assumption about p implies that there is a positive number σ with

$$\mathrm{Re} \, \lambda_j \leqslant -\sigma, \ j = 1, \dots, r,$$

because $\mathrm{Re} \, \lambda_j < 0$ for $j = 1, \dots, r$. Assuming $t \geqslant 1$, as we can do, we find that

$$|x_l(t)| \leqslant C_l t^{n-1} e^{-\sigma t}, \ t \geqslant 1,$$

where C_l is some positive constant. However, it is elementary that

$$\lim_{t \to \infty} (t^{n-1} e^{-\sigma t}) = 0,$$

and so $x_l(t) \to 0$ as $t \to \infty$, as we wished to show.

Since no eigenvalue of A is zero we know that A is invertible and so there is a (constant) vector v_0 with $Av_0 = -b$. Set $x_0(t) = v_0$ for all t. Then $x_0(t)$ solves (1) and so *any* other solution of (1) tends to v_0 as $t \to \infty$. The precise nature of how $x(t)$ converges to v_0 depends, of course, on the properties of the matrix A.

It is interesting to note that if the polynomial p given in (2) has all its zeros in the half-plane $\mathrm{Re} \, z < 0$, then *any* set of initial conditions produces a solution that, over the long run, converges to the constant solution $x(t) = v_0$. That is, if the operation of a mechanical system is determined by (1), then no matter how the system starts, it will (eventually) stabilize at the same "steady state."

Stable polynomials

A polynomial p whose zeros all lie in the half-plane $\mathrm{Re} \, z < 0$ is called **stable**. The choice of terminology and the importance of such polynomials have been made clear above. Here we shall look for conditions that ensure that p is stable. We give a full solution if the degree of p is 2 or 3. We also give the result (but not the justification) for general polynomials. These results go under the general heading of the **Routh–Hurwitz criteria**.

> **Proposition 1:** Suppose that A and B are real numbers. The real parts of both the roots of
>
> $$z^2 + Az + B = 0$$
>
> are negative if and only if both A and B are positive.

Proof. The roots of the equation are

$$z_1 = \tfrac{1}{2}(-A + \sqrt{A^2 - 4B})$$

and

$$z_2 = \tfrac{1}{2}(-A - \sqrt{A^2 - 4B}).$$

If $\operatorname{Re} z_1$ and $\operatorname{Re} z_2$ are negative, then

$$-A = \operatorname{Re}(z_1 + z_2) < 0.$$

Furthermore, if $A^2 - 4B \leqslant 0$, then $4B \geqslant A^2 > 0$ and therefore $B > 0$. Otherwise, $A^2 - 4B > 0$ and

$$0 > 2 \operatorname{Re} z_1 = -A + \sqrt{A^2 - 4B}$$

so that $A > \sqrt{A^2 - 4B}$, which implies that $B > 0$.

Conversely, if both A and B are positive then virtually the same arguments show $\operatorname{Re} z_1$ and $\operatorname{Re} z_2$ are both negative. $\square$

Example 1 The polynomial $p(z) = z^2 + z + 6$ is stable but the polynomial

$$q(z) = z^2 + 2z - \varepsilon$$

is not stable no matter how small the positive number ε is made. (One of its zeros lies at $z = -1 + \sqrt{1 + \varepsilon}$, which is positive.)

The situation for cubic polynomials is not quite as simple as it is for quadratic polynomials but it is not terribly difficult either.

Proposition 2: Let A, B, and C be real numbers. The real part of each of the roots of the cubic polynomial

$$z^3 + Az^2 + Bz + C = 0$$

is negative if and only if A, B, and C are all positive and $AB > C$.

Proof. Let $p(z) = z^3 + Az^2 + Bz + C = (z - z_1)(z - z_2)(z - z_3)$. Then

$$A = -z_1 - z_2 - z_3$$
$$B = z_1 z_2 + z_1 z_3 + z_2 z_3$$
$$C = -z_1 z_2 z_3.$$

We first examine the case when the root z_1 is real and the other two roots z_2 and z_3 are complex conjugates of each other. We write

$$z_1 = a, \qquad z_2 = b + ic, \qquad z_3 = b - ic$$

where a, b, and c are all real. Thus, we find that

$$A = -a - 2b, \qquad B = 2ab + b^2 + c^2, \qquad C = -a(b^2 + c^2).$$

Suppose first that a and b are negative; it clearly follows from the formulas above that A, B, and C are all positive. Furthermore,

$$C - AB = 2b(a^2 + 2ab + b^2 + c^2) < 0$$

so that $AB > C$. Conversely, suppose that all of A, B, C and $AB - C$ are positive; we must show that a and b are negative. Since

$$0 < C = -a(b^2 + c^2)$$

we clearly have $a < 0$. Furthermore, from above we have

$$0 < AB - C = -2b(a^2 + 2ab + b^2 + c^2)$$

so that b must also be negative.

The other case to be examined is when all the roots z_1, z_2, z_3 are real; for simplicity we denote them by a, b, and c, respectively. Clearly we have

$$A = -a - b - c$$
$$B = ab + bc + ac$$
$$C = -abc,$$

so that if a, b, and c are all negative, then A, B, and C are all positive. Furthermore,

$$AB - C = -b(a^2 + c^2) - c(a^2 + b^2) - a(b^2 + c^2) - 2abc > 0$$

if a, b, and c are all negative.

Conversely, suppose all four numbers, A, B, C, and $AB - C$ are positive. First of all, since $0 < C = -abc$ either a, b, and c are all negative or one of these numbers is negative and the other two are positive. Hence, in order to reach a contradiction, let us assume that $a < 0$ and that b and c are positive. But then

$$-a = A + b + c$$

and so

$$\begin{aligned}0 < B &= a(b + c) + bc \\ &= -(A + b + c)(b + c) + bc \\ &= -A(b + c) - b^2 - c^2 - bc < 0,\end{aligned}$$

a contradiction. This completes the proof.   □

Example 2   The polynomial

$$p(z) = 6 + 4z + 2z^2 + z^3$$

is stable since $A = 2$, $B = 4$, and $C = 6$ are all positive and $AB = 8 > 6 = C$.

Example 3   The polynomial

$$p(z) = 12 + 5z + 2z^2 + z^3$$

is not stable since $C = 12 > 10 = (5)(2) = AB$.

The general case is set forth in the following theorem.

Theorem 4: Routh*–Hurwitz† condition. Suppose that $a_1, a_2, \ldots, a_n$ are real numbers; set $a_j = 0$ if $j > n$. All the roots of the polynomial $p(z) = a_n + a_{n-1}z + a_{n-2}z^2 + \cdots + a_1 z^{n-1} + z^n$ have negative real parts if and only if for *each* $k = 1, \ldots, n$ the determinant of the $k \times k$ matrix

$$M_k = \begin{bmatrix} a_1 & a_3 & a_5 & \cdots & a_{2k-1} \\ 1 & a_2 & a_4 & \cdots & a_{2k-2} \\ 0 & a_1 & a_3 & \cdots & a_{2k-3} \\ 0 & 1 & a_2 & & \\ \vdots & \vdots & \vdots & & \\ 0 & 0 & 0 & \cdots & a_k \end{bmatrix}$$

is positive.

Example 4   Let the polynomial in question be

$$p(z) = z^4 + 4z^3 + z^2 + 2z + \tfrac{1}{4}.$$

According to Theorem 4, p is stable if and only if the determinant of each of these four matrices is positive:

$$M_1 = [4], \qquad M_2 = \begin{bmatrix} 4 & 2 \\ 1 & 1 \end{bmatrix}, \qquad M_3 = \begin{bmatrix} 4 & 2 & 0 \\ 1 & 1 & \tfrac{1}{4} \\ 0 & 4 & 2 \end{bmatrix}$$

and

$$M_4 = \begin{bmatrix} 4 & 2 & 0 & 0 \\ 1 & 1 & \tfrac{1}{4} & 0 \\ 0 & 4 & 2 & 0 \\ 0 & 1 & 1 & \tfrac{1}{4} \end{bmatrix}$$

A computation shows that the determinants of M_1 and M_2 are positive but those of M_3 and M_4 are zero. Hence, p is not stable.

---

* Edward John Routh, 1831–1907.
† Adolf Hurwitz, 1859–1919.

Exercises for Section 3.1.1

1. Use the Routh–Hurwitz criteria to test the following polynomials for stability.

    (a) $3 - 4z + 5z^2$

    (b) $2 + z + \frac{1}{2}z^2$

    (c) $-\frac{1}{4} + 2z - \frac{1}{4}z^2$

    (d) $1 + z - 3z^2 + z^3$

    (e) $1 + z + z^3$

    (f) $3 + 4z + z^2 + z^3$

    (g) $3 + 2z + 2z^2 + 5z^3$

    (h) $1 + 4z^2 + 4z^4$

    (i) $z^4 + z^3 + 2z^2 + z + \frac{1}{2}$

    (j) $z^4 + 2z^3 + 3z^2 + 3z + 4$

    (k) $z^5 + z^4 + 5z^3 + 7z^2 + 4z + 8$

    (l) $z^4 + 4z^3 + 9z^2 + 8z + 5.$

2. Let γ_R be the curve consisting of the semicircle Re^{it}, $-\pi/2 \leqslant t \leqslant \pi/2$, followed by the segment iy, $R \geqslant y \geqslant -R$. Show that a polynomial p, which does not vanish on the imaginary axis, is stable if and only if

    $$\int_{\gamma_R} \frac{p'(z)}{p(z)} \, dz = 0$$

    for all large R.

3. Determine whether or not the systems determined by the following matrices are stable.

    (a) $A = \begin{bmatrix} -1 & 2 \\ 1 & -3 \end{bmatrix}$

    (b) $A = \begin{bmatrix} 0 & -3 \\ 1 & -2 \end{bmatrix}$

    (c) $A = \begin{bmatrix} 1 & 2 \\ 2 & -6 \end{bmatrix}$

    (d) $A = \begin{bmatrix} 1 & w \\ -w & 1 \end{bmatrix}$, w complex

    (e) $A = \begin{bmatrix} 1 & -1 & 1 \\ 1 & -2 & 1 \\ -2 & -4 & 0 \end{bmatrix}$

    (f) $A = \begin{bmatrix} 1 & 3 & 1 \\ 1 & 2 & 3 \\ 2 & 0 & -1 \end{bmatrix}$.

4. Suppose the polynomial p given in (2) has all its zeros in the closed half-plane Re $w \leqslant 0$, and any zeros that lie on the imaginary axis are of order 1. Show that any solution of the system (1) remains bounded as $t \to \infty$.

    Work out in detail the solution of the system with

    $$A = \begin{bmatrix} 0 & \omega \\ -\omega & 0 \end{bmatrix}, \quad \omega > 0,$$

    to verify what happens in one special case.

5. Show by means of an *example* that if p given in (2) has a zero in the half-plane Re $z > 0$, then the homogeneous system $Ax = \dot{x}$ has unbounded solutions, and hence small perturbations in the initial conditions can result in unboundedly large variations in the resulting solution.

6.   Examine the behavior as $t \to \infty$ of the solutions of the two equations

$$y'' = -y; \qquad y(0) = x_0, \ y'(0) = 0$$

and

$$y'' = y; \qquad y(0) = x_0, \ y'(0) = 0.$$

The first corresponds to the system

$$\begin{bmatrix} 0 & 1 \\ -1 & 0 \end{bmatrix} x = \dot{x}$$

and the second to the system

$$\begin{bmatrix} 0 & 1 \\ 1 & 0 \end{bmatrix} x = \dot{x}.$$

The matrix $\begin{bmatrix} 0 & 1 \\ -1 & 0 \end{bmatrix}$ has eigenvalues of $\pm i$ while the matrix $\begin{bmatrix} 0 & 1 \\ 1 & 0 \end{bmatrix}$

has eigenvalues of ± 1.

7.   In the theory of **feedback-control systems** the question of the stability of the system takes the form of determining whether a function of the form $1 + (1/c)F$, has any zeros in the right half-plane; here F is a rational function and c is a nonzero constant. One answer to this question is provided by the Routh–Hurwitz condition. Write $F = P/Q$, where P and Q are polynomials with no common zero. Then $1 + (1/c)F$ has no zeros in the right half-plane if and only if the polynomial $cQ + P$ is stable. Another answer is provided by the **Nyquist stability criterion**, which is a variation on the argument principle. Let γ_R be the curve shown in Figure 3.6a and let Γ_R be its image under F shown in Figure 3.6b.

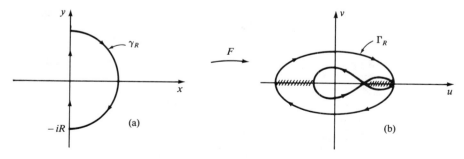

Figure 3.6

(Note that γ_R is oriented negatively; this is done because it is traditional to evaluate $F(iy)$ as y increases.) Show that if Γ_R winds around

$-c$ a net positive number of times in the clockwise direction, then $1 + (1/c)F$ has a zero in the right half-plane (and so the system is unstable.) In Figure 3.6b those areas marked by slashes are values of c for which $1 + (1/c)F$ has a zero. More on the derivation of the stability criterion for feedback-control systems can be found in the book by A. Kyrala cited in the references at the end of Chapter 4.

3.2   Maximum modulus and mean value

Suppose that f is an analytic function on a domain D. We saw in the beginning of Chapter 2 that if the range of f lies in a circle or on a straight line then f is necessarily constant. We shall see now that a great deal more is true; either f is constant on D or the range of f is an open set.

Suppose that f is not identically constant and let $w_0 = f(z_0)$ be an arbitrary point in the range of f. The function $f(z) - w_0$ is not identically zero since f is not identically constant and so $f(z) - w_0$ has a zero of order $m \geqslant 1$ at z_0. Choose $r > 0$ so small that $f(z) - w_0$ has no zero in the region $0 < |z - z_0| \leqslant r$; this is possible because the zeros of a nonconstant analytic function are isolated (see Section 1). Let δ be the minimum value of $|f(z) - w_0|$ for all z with $|z - z_0| = r$ and let w be any point with $|w - w_0| < \delta$. Then on the circle $|z - z_0| = r$ we have

$$|(f(z) - w) - (f(z) - w_0)| = |w - w_0| < \delta \leqslant |f(z) - w_0|. \tag{1}$$

Therefore, Rouché's Theorem, Theorem 3 of Section 1, implies that the two functions $f - w$ and $f - w_0$ have an equal number of zeros within the circle $|z - z_0| = r$. However, $f - w_0$ has exactly m zeros and hence so does $f - w$. This shows that each point w_0 in the range of f lies at the center of a small disc, which is also within the range of f. Hence, we have established this theorem.

> **Theorem 1:** Suppose that f is a nonconstant analytic function on a domain D. Then the range of $f(z)$, as z varies over D, is an open set.

We can draw an immediate conclusion from the work that preceded the statement of Theorem 1.

> *Suppose that f is a nonconstant analytic function on a domain D and that f − f(z_0) has a zero of order m at z_0. Then f is m-to-1 near z_0; in particular, if f'(z_0) = 0, then f is not one-to-one in any disc containing z_0.* (2)

Another very important consequence of Theorem 1 follows.

> **Corollary 1: The maximum-modulus principle.** If f is a nonconstant analytic function on a domain D, then $|f|$ can have no local

maximum on D.

Proof. If $|f(z_0)| \geqslant |f(z)|$ for all z with $|z - z_0| < r$, then $f(z_0)$ lies on the boundary of the open set $W = \{f(z): |z - z_0| < r\}$. This is in contradiction to the fact that W is an open set containing $f(z_0)$. □

The maximum-modulus principle expressed in Corollary 1 implies, for example, that if we observe a nonconstant, sourceless, irrotational flow on a domain D, then at no point of D is the speed of the flow largest, or even locally largest. No matter where we look, the speed is always strictly greater at some nearby point.

The maximum-modulus principle (Corollary 1) has ramifications for the real and imaginary parts of an analytic function. Suppose that f is a nonconstant analytic function on a domain D; let $u = \text{Re}(f)$ and $g = e^f$. Then both g and $1/g$ are analytic on D so that neither of the functions

$$|g| = e^u \qquad \text{and} \qquad |1/g| = e^{-u}$$

has a local maximum in D; consequently, u has no local maxima and no local minima in D. This is stated formally in this way:

> If f is analytic and nonconstant on a domain D, then (3)
> Re f has no local maxima and no local minima on D.

Let us continue this line of thought a bit further. Let D be a bounded domain and let B be the boundary of D. The set $D \cup B$ is both closed and bounded. For such sets, there is a result from real variables that asserts that a continuous real-valued function on a closed and bounded set actually *attains* its maximum value. Let us apply this to each of the three functions $|f|$, Re f, and $-\text{Re } f$, where f itself is analytic on D and continuous on $D \cup B$. We know that each of these three functions is continuous on $D \cup B$ and so must attain its maximum somewhere on $D \cup B$. If f is nonconstant, then the maximum can not be attained on D itself. Hence, the maximum must be attained on B, the boundary of D. Quite obviously the same conclusion holds if, in fact, f is constant on D. Thus, in all cases, $|f|$, Re f, and $-\text{Re } f$ attain their maximum values on B, the boundary of D.

> If f is analytic on a bounded domain D and continuous
> on D ∪ B, where B is the boundary of D, then each of |f|, (4)
> Re f, and − Re f attain its maximum value on B.

In particular, one consequence of (4) is that if Re f is zero everywhere on B, then Re f is zero throughout D and so f itself is constant on all of D. It should be stressed that D's being bounded is not a superfluous hypothesis here. For instance, $f(z) = iz$ is analytic and nonconstant on the whole plane, Re $f = 0$ on the real axis, and yet f is not constant on the half-plane $\{\text{Im } z > 0\}$, whose boundary is the real axis.

One of the best-known applications of the maximum-modulus principle is the following result, known for historical reasons as Schwarz's lemma.

Theorem 2: Schwarz's Lemma. Suppose that f is analytic in the disc $|z| < 1$, that $f(0) = 0$, and that $|f(z)| \leq 1$, for all z in the disc. Then

$$|f(z)| \leq |z|, \quad |z| < 1.$$

Equality can hold for some $z \neq 0$ only if $f(z) = \lambda z$, where λ is a constant of absolute value one.

Proof. Since $f(0) = 0$, we know $g(z) = f(z)/z$ is also analytic on $|z| < 1$. For $|z| = r$ we have

$$|g(z)| = \frac{|f(z)|}{r} \leq 1/r.$$

By the maximum-modulus principle, the inequality $|g(z)| \leq 1/r$ is true for $|z| < r$ as well. Since r can be made as near 1 as we like, we must have $|g(z)| \leq 1$ if $|z| < 1$; thus, $|f(z)| \leq |z|$ for $|z| < 1$. Furthermore, if $|f(z_0)| = |z_0|$ for some $z_0 \neq 0$, then $1 = |g(z_0)|$, and consequently, $|g(z)|$ has an interior maximum. This implies that $g(z)$ is a constant λ, $|\lambda| = 1$, and gives the conclusion that $f(z) = \lambda z$. $\square$

The reader should contrast the result in Theorem 2 to the case of real-valued functions of a real variable. For instance, the function

$$u(x) = \frac{2x}{x^2 + 1}, \quad -1 \leq x \leq 1,$$

satisfies $-1 \leq u(x) \leq 1$, $u'(x)$ is continuous, $u(0) = 0$, yet $|u(x)| > |x|$ if $1 > |x| > 0$.

Mean Value

Theorem 3 of Section 3, Chapter 2 (Cauchy's formula) gives us

$$f(z_0) = \frac{1}{2\pi i} \int_\gamma \frac{f(\zeta)}{\zeta - z_0} \, d\zeta$$

when γ is a circle and z_0 is inside γ. If we take z_0 to be the center of the circle, then $\zeta = z_0 + re^{it}, 0 \leq t \leq 2\pi, d\zeta = ire^{it} \, dt$, and we obtain

$$f(z_0) = \frac{1}{2\pi} \int_0^{2\pi} f(z_0 + re^{it}) \, dt. \tag{5}$$

The result expressed in (5) is the **mean-value theorem** for analytic functions.

This result can also be obtained by replacing $f(z_0 + re^{it})$ by the power-series expansion for f about the point z_0 and integrating term by term.

In formula (5) take the real part of both sides and write $u = \operatorname{Re} f$. We find that

$$u(z_0) = \frac{1}{2\pi} \int_0^{2\pi} u(z_0 + re^{it})\, dt. \tag{6}$$

As we will see in Chapter 4, those functions u, which are the real parts of some analytic function f, are of great importance in applications. Formula (6), which shows us how to "recover" $u(z_0)$ from the values of u on the circle $|z - z_0| = r$, is but a special case of a more general formula that allows us to recover $u(\zeta)$ for all ζ inside $|z - z_0| = r$ from the values of u on the circle $|z - z_0| = r$.

Exercises for Section 3.2

1. Let $f(z) = z^2/(z + 2)$; find the maximum value of $|f(z)|$ as z varies over the disc $|z| \leqslant 1$.

2. Let $f(z) = ze^z$; find the minimum value of $|f(z)|$ as z varies over the region $D = \{x + iy: x^2 + y^2 \leqslant 4, x \geqslant 0, y \geqslant 0\}$.

3. Find the maximum value of $|g(z)|$, $g(z) = z/(4z^2 - 1)$, as z varies over the region $\{z: |z| \geqslant 1\}$.

4. (a) If $|z| = r < 1$, show that $|e^z - 1| \geqslant r - r^2/(2 - r)$. (**Hint:** $e^z - 1 = z + z^2/2 + z^3/3! + \cdots$ so that $|e^z - 1| = |z| \,|1 + z/2 + z^2/3! + \cdots| \geqslant r\{1 - \sum_{k=1}^{\infty} r^k/2^k\}$.)
    (b) Use (a) to show that the equation $z(e^z - 1) = w$ has exactly two solutions z with $|z| < \frac{1}{2}$ whenever $|w| < \frac{1}{6}$. (**Hint:** Look at the work around (1), preceding the statement of Theorem 1.)

5. The function $S(z) = \exp[(1 + z)/(1 - z)]$ is analytic on the disc $\{z: |z| < 1\}$ and $|S(z)| = 1$ if $|z| = 1$, $z \neq 1$. However, $S(0) = e > 1$. Why does this not contradict the maximum-modulus principle?

6. Let D be a bounded domain with boundary B. Suppose that f and g are both analytic on D and continuous on $D \cup B$ and suppose further that $\operatorname{Re} f(z) = \operatorname{Re} g(z)$ for all $z \in B$. Show that $f = g + i\alpha$ in D, where α is a real constant.

7. Let F be analytic and nonconstant on the disc $|z - z_0| < R$ and suppose that $\operatorname{Re}(F(z_0)) = 0$. Show that on every circle $|z - z_0| = r$, $0 < r < R$, $\operatorname{Re} F$ assumes both positive and negative values.

8. Let f be analytic on a bounded domain D and continuous on $D \cup B$, where B is the boundary of D. Show that if f is never zero on D, then the minimum of $|f|$ is assumed on B. You will need to use the fact that $|f|$ does, indeed, assume a minimum somewhere on $D \cup B$.

9.   Suppose that f is analytic on a domain D, which contains a simple closed curve γ and the inside of γ. If $|f|$ is constant on γ, then either f is constant or f has a zero inside γ.

10.   Let f be a nonconstant entire function and U be an open set in the plane. Show that there is a z_0 such that $f(z_0) \in U$.

11.   Suppose that p is a polynomial of degree n and that $|p(z)| \leq M$ if $|z| = 1$. Show that $|p(z)| \leq M|z|^n$ if $|z| \geq 1$. (**Hint:** Apply the maximum-modulus principle to $f(z) = p(z)/z^n$ on the domain $|z| > 1$, including the point " ∞.")

In Exercises 12 to 14, f is an analytic function on the disc $|z| < 1$ that satisfies $|f(z)| \leq M$ if $|z| < 1$.

12.   Suppose that $f(\alpha) = 0$ for some α, $|\alpha| < 1$. Show that

$$|f(z)| \leq M \left| \frac{z - \alpha}{1 - \bar{\alpha}z} \right|, \quad |z| < 1.$$

13.   Show that

$$M \left| \frac{f(z) - f(z_0)}{M^2 - \overline{f(z_0)}f(z)} \right| \leq \left| \frac{z - z_0}{1 - \bar{z}_0 z} \right| \quad \text{for all } z, z_0.$$

14.   Suppose that $f^{(k)}(0) = 0$ for $k = 0, \ldots, N$. Show that $|f(z)| \leq M|z|^{N+1}$ for all z, $|z| < 1$.

15.   Let g be analytic and nonconstant on a simply connected domain D; for instance, D could be a disc. Show that the set $\{z: \operatorname{Re} g(z) = c\}$ contains no simple closed curve. (**Hint:** Use Exercise 9 with $f = \exp(g)$.)

16.   Use (6) to show that the real part of a nonconstant analytic function has no strict local maxima or local minima.

17.   Let f and g be analytic on a domain that contains a simple closed curve γ and the inside of γ. Assume further that f is never zero inside γ. Show that if $|f(z)| \geq |g(z)|$ for all z in γ, then $|f(z)| \geq |g(z)|$ for all z inside γ as well. Give an example to show that the hypothesis "$f \neq 0$ inside γ" is essential for the validity of the conclusion.

18.   Suppose that f is analytic in $\{z: 0 < |z - z_0| < R\}$ and that f has a pole of order m, $m \geq 1$, at z_0. Show that there is a large number M with this property: for each w_0 with $|w_0| > M$ the equation $f(z) = w_0$ has exactly m solutions z, z near z_0.

Positive rational functions and stable polynomials★

If p is a polynomial, define $\tilde{p}(z) = \overline{p(-\bar{z})}$.

19.   Show that $\tilde{\tilde{p}} = p$. For which polynomials Q is $\tilde{Q} = Q$?

20. Suppose that p is stable. Show that p and $\tilde{p}$ have no common zero.

21. Suppose that p is stable. Show that the rational function $r = \tilde{p}/p$ maps the right half-plane, Re $z > 0$, into the disc $\{w: |w| < 1\}$. Conversely, show that if $r = \tilde{p}/p$ maps Re $z > 0$ into $|w| < 1$ and if p and $\tilde{p}$ have no common zero, then p is stable.

22. A rational function R is **positive** if Re$\{R(z)\} > 0$ whenever Re $z > 0$. Show that
    (a) R is positive if and only if $1/R$ is positive.
    (b) if R is positive, then R has no zero or pole in the half-plane Re $z > 0$ and any zero or pole on the imaginary axis has order one. (**Hint:** By (a) it suffices to consider just zeros.)

23. Suppose that p is stable. Show that $R = (p - \tilde{p})/(p + \tilde{p})$ is positive. Conversely, show that if p and $\tilde{p}$ have no common zero and if $R = (p - \tilde{p})/(p + \tilde{p})$ is positive, then p is stable.

3.3 Linear fractional transformations

A **linear fractional transformation** T is a rational function of the special form

$$T(z) = \frac{az + b}{cz + d},$$

where a, b, c, and d are complex numbers and $ad - bc \neq 0$. The restriction $ad - bc \neq 0$ is essential for otherwise

$$T'(z) = \frac{ad - bc}{(cz + d)^2} = 0 \text{ for all } z,$$

and so T is identically constant.

A linear fractional transformation is a one-to-one function. For suppose that

$$\frac{az_1 + b}{cz_1 + d} = T(z_1) = T(z_2) = \frac{az_2 + b}{cz_2 + d}.$$

Then

$$(az_1 + b)(cz_2 + d) = (az_2 + b)(cz_1 + d)$$

and, after some cancellation, we find that

$$bcz_2 + adz_1 = adz_2 + bcz_1$$

or

$$(ad - bc)z_1 = (ad - bc)z_2.$$

Consequently, $z_1 = z_2$. Hence, the function T maps distinct points onto distinct images. Note as well that T has a pole of order 1 at $-d/c$ and

$\lim_{|z| \to \infty} T(z) = a/c$.

Since T is one-to-one there is a function T^{-1} that is the inverse of T in the sense of the composition of functions:

$$T^{-1}(T(z)) = z$$

for all z. T^{-1} is also a linear fractional composition as this computation shows.

$$T(z) = w = \frac{az + b}{cz + d}$$

implies that $czw - az = b - dw$, and thus,

$$T^{-1}(w) = z = \frac{-dw + b}{cw - a}.$$

Hence, a linear fractional transformation is a one-to-one mapping of the complex plane plus the point at ∞ onto itself. Conversely, a one-to-one (analytic) mapping of the complex plane plus "∞" onto itself is a linear fractional transformation (see the exercises).

Fixed Points and Triples to Triples

A linear fractional transformation that is not identically equal to z has at most two distinct **fixed points**; that is, points z for which

$$T(z) = z.$$

This is evident from the fact that z is a solution of the equation $T(z) = z$ exactly when z is a root of the quadratic equation

$$0 = cz^2 + (d - a)z - b,$$

and, of course, a quadratic equation has at most two distinct roots. It is a consequence of this that if T and S are two linear fractional transformations that are equal at three distinct points, say $T(z_j) = S(z_j)$ for $j = 1, 2, 3$, then $T(z) = S(z)$ for all z. This is because the linear fractional transformation $S^{-1}(T(z))$ has three distinct fixed points

$$S^{-1}(T(z_j)) = z_j, \ j = 1, 2, 3.$$

Hence, $S^{-1}(T(z)) = z$ for all z, and thus, $T(z) = S(z)$ for all z.

On the other hand, if three distinct complex numbers z_1, z_2, and z_3 are given and if any other three distinct complex numbers w_1, w_2, and w_3 are chosen, then there is a necessarily unique linear fractional transformation L with $L(z_j) = w_j, j = 1, 2, 3$. We derive its form in this way. Set

$$T(z) = \left(\frac{z - z_1}{z - z_3} \right) \left(\frac{z_2 - z_3}{z_2 - z_1} \right).$$

Then $T(z_1) = 0$, $T(z_2) = 1$, and $T(z_3) = \infty$. Let

$$S(w) = \left(\frac{w - w_1}{w - w_3}\right)\left(\frac{w_2 - w_3}{w_2 - w_1}\right),$$

so that $S(w_1) = 0$, $S(w_2) = 1$, and $S(w_3) = \infty$. L is given by

$$L(z) = S^{-1}(T(z)).$$

We can carry out a similar sort of computation if one of z_1, z_2, and z_3 or w_1, w_2, and w_3 is ∞; this is left for the exercises at the end of this section.

Example 1 Find the linear fractional transformation that sends 0, 1, 2 to −1, 0, 4, respectively.

Solution We have

$$T(z) = \left(\frac{z - 0}{z - 2}\right)\left(\frac{1 - 2}{1 - 0}\right) = \frac{z}{2 - z}$$

and

$$S(w) = \left(\frac{w - (-1)}{w - 4}\right)\left(\frac{0 - 4}{0 - (-1)}\right) = -4\,\frac{w + 1}{w - 4}.$$

Then

$$S^{-1}(z) = 4\,\frac{z - 1}{z + 4}$$

and so

$$L(z) = S^{-1}(T(z)) = \frac{8z - 8}{-3z + 8}. \qquad \square$$

Lines and Circles

It is an important property of a linear fractional transformation that it maps each circle onto another circle or onto a straight line and each straight line onto another straight line or a circle. We show this below.

First, if $T(z) = az + b$, $a \neq 0$, then the assertion that T maps circles and straight lines to the same type of figure is quite clear. The circle

$$C: \{z: |z - z_0| = r\}$$

is transformed to the circle

$$C': \{w: |w - (az_0 + b)| = |a|r\}$$

and the straight line

$$L: \{z: \operatorname{Re}(Az + B) = 0\}$$

is transformed to the straight line

$$L': \{w: \text{Re}[(A/a)w + B - b(A/a)] = 0\}.$$

Thus, we henceforth assume that

$$T(z) = \frac{az + b}{cz + d}, \quad c \neq 0.$$

Now

$$T(z) = \frac{az + b}{cz + d} = \frac{1}{c}\left\{\frac{bc - ad}{cz + d} + a\right\}, \quad c \neq 0$$

and we see from this that T is actually the composition of several simpler linear fractional transformations. Precisely, set

$$U(z) = cz + d, \qquad V(w) = 1/w, \qquad \text{and } W(\zeta) = \frac{1}{c}[(bc - ad)\zeta + a].$$

Then

$$T(z) = W(V(U(z))).$$

From the foregoing discussion we know that both U and W send circles to circles and lines to lines. Consequently, the desired conclusion will be reached if we can show that V maps circles and lines to circles and lines. The equation

$$\alpha(x^2 + y^2) + \beta x + \gamma y = \delta,$$

where α, β, γ, and δ are real and not all of α, β, γ are zero, represents either a circle (iff $\alpha \neq 0$ and $\beta^2 + \gamma^2 + 4\alpha\delta > 0$) or a straight line (iff $\alpha = 0$). Now

$$\frac{1}{z} = \frac{x}{x^2 + y^2} + i\frac{-y}{x^2 + y^2} = u + iv,$$

so that replacing z by $1/z$ yields the equation

$$\delta(u^2 + v^2) - \beta u + \gamma v = \alpha.$$

This is again the equation (in u, v coordinates) of either a line or a circle. You will have no difficulty verifying that a circle or line through the origin is transformed by inversion to a line, and that a circle or a line *not* through the origin is transformed by inversion to a circle (which goes through the origin when and only when a line was inverted) (Fig. 3.7).

The circulation produced by a linear fractional transformation⋆

Let $T(z) = (z + 2)/(2z + 1)$; this linear fractional transformation fixes the two points 1 and -1. We shall show that if C is a circle centered on the imaginary axis that passes through both 1 and -1, then T maps C onto itself. (Of course, 1 and -1 are the only fixed *points* of T.) Let $i\alpha$ be the center of C, α real. Then

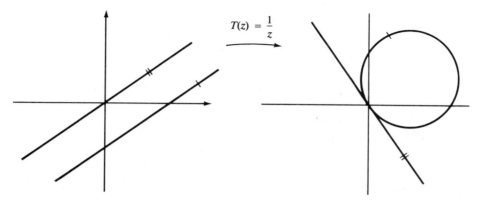

$$T(z) = \frac{1}{z}$$

Figure 3.7

the radius of C is $r = \sqrt{\alpha^2 + 1}$ and so a point z lies on C precisely when $|z - i\alpha|^2 = 1 + \alpha^2$. Hence,

$$x^2 + (y - \alpha)^2 = 1 + \alpha^2$$

or

$$x^2 + y^2 - 2\alpha y = 1.$$

Now

$$|T(z) - i\alpha|^2 = \left| \frac{z + 2}{2z + 1} - i\alpha \right|^2$$

$$= \frac{(x + 2 + 2\alpha y)^2 + (y - \alpha - 2\alpha x)^2}{(2x + 1)^2 + (2y)^2}$$

and a modest amount of computation, using the fact that $x^2 + y^2 - 2\alpha y = 1$, shows that

$$|T(z) - i\alpha|^2 = 1 + \alpha^2,$$

which, of course, is exactly the statement that $T(z)$ lies in the circle C if z is in C. Set

$$T_1(z) = T(z) = \frac{z + 2}{2z + 1},$$

and then

$$T_2(z) = T_1(T_1(z)) = \frac{5z + 4}{4z + 5},$$

and, in general,

$$T_{n+1}(z) = T_n(T_1(z)), \quad n = 1, 2, 3, \ldots.$$

For instance,

$$T_3(z) = \frac{13z + 14}{14z + 13}, \qquad T_4(z) = \frac{41z + 40}{40z + 41}.$$

In general, we see that $T_n(z) \to 1$ as $n \to \infty$ so long as $z \neq -1$. Thus, T sets up a **circulation** in the plane with a **source** at -1 and a **sink** at $+1$: each point z in the plane is on precisely one circle C through z, 1, and -1 (the real axis counts as a circle for this purpose); T carries C onto itself and pushes z toward 1 along the circle. Thus, the points in the plane move toward 1 and away from -1 under the action of T. We note that $T'(1) = -1/3$ and $T'(-1) = -3$. Thus, the sink is located at that fixed point at which the derivative is less than 1 (Fig. 3.8).

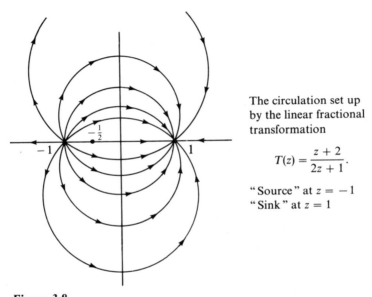

The circulation set up by the linear fractional transformation

$$T(z) = \frac{z + 2}{2z + 1}.$$

"Source" at $z = -1$
"Sink" at $z = 1$

Figure 3.8

You should convince yourself that the critical issue in the foregoing discussion is that the linear fractional transformation fixes the two points 1 and -1 and not its special form. A similar phenomenon occurs for a linear fractional transformation that fixes the two distinct points p and q (see Exercise 9 at the end of this section).

You should also look back at the Circles of Appolonius (Fig. 1.10), with the foregoing in mind.

Example 2   This example continues the preceding discussion. We shall investigate what happens when the two fixed points move toward each other. Set

$$T_\varepsilon(z) = \frac{z + \varepsilon^2}{z + 1}, \quad 0 < \varepsilon < 1.$$

T_ε fixes ε and $-\varepsilon$ and so, exactly as above, T_ε sets up a circulation using the family of circles centered on the imaginary axis and passing through ε and $-\varepsilon$. As $\varepsilon \to 0$, $T_\varepsilon(z) \to T_0(z) = z/(z + 1)$ and we see that this T_0 sets up a circulation for the family of circles centered on the imaginary axis and passing through the origin. This circulation is termed a **dipole.** Note that $T_0(z) = z/(z + 1)$ has only one fixed point and that it is a double root of $T_0(z) = z$ at $z = 0$. A similar dipole will appear as the circulation of any linear fractional transformation with a single (distinct) fixed point p, when p is a double root of $T(z) = z$ (Fig. 3.9).

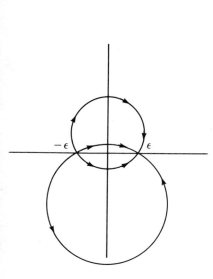

Circulation of $T_\varepsilon(z) = \dfrac{z + \varepsilon^2}{z + 1}$

with fixed points $\pm \varepsilon$

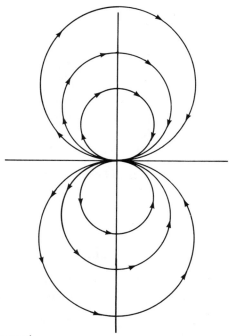

A dipole located at the origin; the circulation is given by

$$T(z) = \frac{z}{z + 1}$$

Figure 3.9

Exercises for Section 3.3

1. Let z_1, z_2, z_3 and w_1, w_2, w_3 be two triples of distinct complex numbers. Set $\alpha = (z_2 - z_3)/(z_2 - z_1)$ and $\beta = (w_2 - w_3)/(w_2 - w_1)$.

Show that the linear fractional transformation that maps z_j to w_j for $j = 1, 2, 3$ is given explicitly by

$$L(z) = \frac{z(\alpha w_3 - \beta w_1) + (\beta w_1 z_3 - \alpha z_1 w_3)}{z(\alpha - \beta) + (\beta z_3 - \alpha z_1)}.$$

2.   Verify that the linear fractional transformation

$$S(z) = (z - z_1)/(z_2 - z_1)$$

maps z_1 to 0, z_2 to 1, and ∞ to ∞.

3.   Verify that the linear fractional transformation

$$T(z) = (z_2 - z_1)/(z - z_1)$$

maps z_1 to ∞, z_2 to 1, and ∞ to 0.

4.   In each case find the linear fractional transformation that maps the first triple (z_1, z_2, z_3) onto the second triple (w_1, w_2, w_3).
   (a) $(1, i, -1)$ onto $(-1, i, 1)$
   (b) $(1, 4, \infty)$ onto $(0, 1 - i, 1 + i)$
   (c) $(1, 0, i)$ onto $(1, 0, 1 + i)$
   (d) $(\infty, -1, i)$ onto $(1, 0, 1 - i)$
   (e) $(i, 2, -i)$ onto $(-i, 3, i)$

5.   Find a linear fractional transformation that carries
   (a) the circle $|z| = 1$ onto the line $\operatorname{Re}((1 + i)w) = 0$
   (b) the circle $|z| = 1$ onto the circle $|w - 1| = 1$
   (c) the real axis onto the line $\operatorname{Re} w = 1/2$
   (d) the circle $|z - z_0| = r$ onto the circle $|w| = 1$
   (e) the line $\operatorname{Re} z = 1/2$ onto the circle $|w - 4i| = 4$

6.   Show that a linear fractional transformation T that maps the circle $|z| = 1$ onto itself has the form

$$T(z) = \lambda \frac{z - \gamma}{1 - \bar{\gamma}z}, \quad |\lambda| = 1, \ |\gamma| \neq 1, \qquad \text{or} \qquad Tz = \frac{\lambda}{z}, \ |\lambda| = 1.$$

7.   In each of the following find a linear fractional transformation T with the given property
   (a) T fixes the points 0 and 1 and $T(i) = \infty$.
   (b) T maps the real axis onto itself and the imaginary axis onto the circle $|w - \frac{1}{2}| = \frac{1}{2}$.
   (c) T maps the real axis onto itself and the imaginary axis onto the circle $|w - \frac{5}{4}| = \frac{3}{4}$.
   (d) T maps the real axis onto itself and the line $y = x$ onto the circle

$$|w + i| = \sqrt{2}.$$

8.   (a) In Exercise 7(c) determine what happens to the first quadrant

under the action of T.

(b) In Exercise 7(d) determine what happens to the sector $\{z: 0 \leqslant y \leqslant x\}$ under the action of T.

9. Discuss the circulation set up by each of the following linear fractional transformations; find the fixed points (if any) and which way the circulation rotates.

(a) $T(z) = z/(z + 2)$

(b) $T(z) = (z + 4)/(z + 1)$

(c) $T(z) = 3z - 2$

(d) $T(z) = -\frac{1}{2} z$ (spiral behavior)

(e) $T(z) = z + 1$ (no fixed points)

10. (a) If T is a linear fractional transformation that maps the circle C_1 onto the circle C_2, show that T carries the inside of C_1 onto either the inside of C_2 or onto the outside of C_2. (**Hint:** Suppose a point z_0 inside C_1 goes to a point w_0 inside C_2. Let z_1 be any other point inside C_1 and L the line segment joining z_0 and z_1. T carries L to an arc of a circle (or a straight line segment) joining w_0 and $w_1 = T(z_1)$ and this arc (or segment) does not meet C_2. [Why?])

(b) T maps the point $z_0 = -d/c$ to ∞; tell how this decides whether T maps the inside of C_1 onto the inside of C_2 or onto the outside of C_2.

Fixed points

11. Show that a linear fractional transformation which fixes the two points 1 and -1 has the form

$$T(z) = \frac{z + \xi}{\xi z + 1}$$

if $\xi = T(0) \neq \infty$. (If $Tz = 1/z$, then $T(1) = 1$, $T(-1) = -1$.)

12. Suppose that T fixes the two points p and q, $p \neq q$. Let S be any linear fractional transformation with $S(1) = p$, $S(-1) = q$, and let $U = S^{-1} \circ T \circ S$. Show that U fixes 1 and -1.

13. Suppose that U is a linear fractional transformation that fixes 1 and -1. Show that $U'(1) = \{U'(-1)\}^{-1}$.

14. Let T be a linear fractional transformation fixing p and q, $p \neq q$. Use Exercise 13 to show that $T'(p) = \{T'(q)\}^{-1}$.

15. Show that the linear fractional transformation $T(z) = (1 - z)/(1 + z)$ satisfies $T(T(z)) = z$ for all z. Find the general form of those linear fractional transformations S with $S(S(z)) = z$. (Equivalently, $S(z) = S^{-1}(z)$.)

Cross-Ratio

The **cross-ratio** of four distinct complex numbers z_0, z_1, z_2, and z_3 is defined to be the complex number

$$(z_0, z_1, z_2, z_3) = \frac{z_0 - z_1}{z_0 - z_2} \frac{z_3 - z_2}{z_3 - z_1}.$$

16.   Show that if T is a linear fractional transformation with $Tz_j = w_j$ for $j = 1, 2, 3$, then

$$(z, z_1, z_2, z_3) = (T(z), w_1, w_2, w_3) \text{ for all } z.$$

17.   Let z_1, z_2, z_3 lie in a circle or on a straight line C. Show the point z_0 is also on C if and only if the cross-ratio (z_0, z_1, z_2, z_3) is real. (**Hint:** By Exercise 16 it is enough to do this for $C =$ real axis and $z_1 = 1$, $z_2 = 0$, $z_3 = -1$. The remainder is a computation: the cross-ratio of $(z, 1, 0, -1)$ is real if and only if $\zeta = (z - 1)/z$ is real. Show ζ is real if and only if z is real.)

18.   Use Exercises 16 and 17 to give another demonstration of the fact that a linear fractional transformation T carries circles and lines to circles and lines.

Reflection*

Let L be a straight line and z be a point in the complex plane. The **reflection** of z over L is the unique point z^* located on the line L' through z and perpendicular to L, lying on the other side of L from z, and at a distance from L equal to that of z from L (Fig. 3.10a). Let C be the circle $|\zeta - \zeta_0| = r$ and z a point of the complex plane, $z \neq \zeta_0$. The **reflection** of z over C is the unique point z^* located on the ray from z_0 to ∞, which passes through z, lying on the other side of C from z and satisfying $|z - \zeta_0| |z^* - \zeta_0| = r^2$.

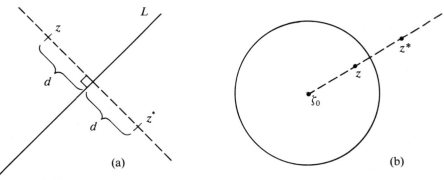

(a)   (b)

Figure 3.10

19.   (a) If z is a point in L or C, show that $z^* = z$.

(b) If L is the real axis, show $z^* = \bar{z}$, the complex conjugate of z.
(c) Show $(z^*)^* = z$ for any L, C, and z.

20.   If C is the circle $|\zeta - \zeta_0| = r$, show that

$$z^* = \frac{r^2}{\bar{z} - \bar{\zeta}_0} + \zeta_0.$$

21.   Let z_1, z_2, and z_3 be points on the circle C of Exercise 20; fill in the reasons for each of the following equalities (all but one use the invariance of the cross-ratio under linear fractional transformations.)

$$(z^*, z_1, z_2, z_3) = \left(\frac{r^2}{\bar{z} - \bar{\zeta}_0} + \zeta_0, z_1, z_2, z_3 \right)$$

$$= \left(\frac{r^2}{\bar{z} - \bar{\zeta}_0}, z_1 - \zeta_0, z_2 - \zeta_0, z_3 - \zeta_0 \right)$$

$$= \left(\bar{z} - \bar{\zeta}_0, \frac{r^2}{z_1 - \zeta_0}, \frac{r^2}{z_2 - \zeta_0}, \frac{r^2}{z_3 - \zeta_0} \right)$$

$$= \overline{(z - \zeta_0, z_1 - \zeta_0, z_2 - \zeta_0, z_3 - \zeta_0)}$$

$$= \overline{(z, z_1, z_2, z_3)}.$$

22.   (a) Conclude from Exercise 21, that z^* is the reflection of z over the circle C if and only if

$$(z, z_1, z_2, z_3) = \overline{(z^*, z_1, z_2, z_3)} \qquad\qquad (*)$$

for any triple z_1, z_2, z_3 of distinct points in C.

(b) Let L be a straight line through z_1 and z_2 ; choose z_3 to be ∞ and modify Exercises 20 and 21 to conclude that $(*)$ also characterizes z^* for reflection over L.

23.   Use Exercise 22 to show that the net result of two successive reflections (over lines/circles) is a linear fractional transformation.

3.4   Conformal mapping

Conformal maps are functions that preserve angles between curves in a sense that we will make precise below. They are indispensable tools in studying flows, fields, and in solving boundary-value problems, as we shall see in Section 4.1 and Chapter 4. In this section and the next we shall study some of the basic facts about conformal mapping.

Suppose that γ is the range of a smooth curve $z(t)$, $a \leqslant t \leqslant b$, and γ passes through the point $z_0 = z(t_0)$, $a < t_0 < b$. The curve γ has a tangent vector $z'(t_0)$ at z_0 and we suppose that $z'(t_0)$ is not zero. We now investigate what happens to the curve γ when we apply an analytic function f to it. The curve γ is transformed into a new curve Γ in the w-plane; Γ is given by

$$\Gamma: w(t) = f(z(t)), \quad a \leqslant t \leqslant b.$$

Furthermore, by the familiar chain rule, we find that the tangent vector to Γ at $w_0 = f(z_0)$ is

$$w'(t_0) = f'(z_0)z'(t_0).$$

See Figure 3.11. In particular, we obtain the two relations

$$|w'(t_0)| = |f'(z_0)|\,|z'(t_0)| \tag{1}$$

and

$$\arg w'(t_0) = \arg(f'(z_0)) + \arg(z'(t_0)). \tag{2}$$

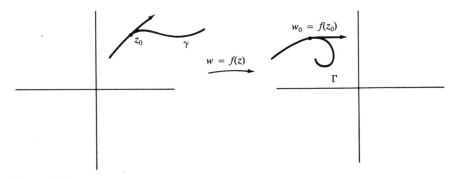

Figure 3.11

We make the natural assumption that $f'(z_0) \neq 0$ and then note that the transformation of γ into Γ by f has these two characteristics: the tangent vector is scaled in length by a factor $|f'(z_0)|$ and the tangent vector is rotated through an angle $\psi_0 = \arg(f'(z_0))$.

Suppose now that γ_1 and γ_2 are two smooth curves that intersect at the point $z_0 = z_1(t_0) = z_2(s_0)$. We define the **angle between** γ_1 **and** γ_2 **at** z_0 to be the angle θ measured counterclockwise from the tangent vector $z_1'(t_0)$ to the tangent vector $z_2'(s_0)$, if neither of these tangent vectors is zero. In this case, we have

$$\theta = \arg(z_2'(s_0)) - \arg(z_1'(t_0))$$

for appropriate determinations of the argument of the tangent vectors $z_2'(s_0)$ and $z_1'(t_0)$. See Figure 3.12. (This definition of the angle between two curves breaks down if one of the vectors $z_1'(t_0)$ or $z_2'(s_0)$ is zero; this will not be a matter of great concern to us.)

Suppose now that ϕ is a function, perhaps not analytic, defined in the disc $|z - z_0| < r$ and satisfying $\phi(z) \neq \phi(z_0)$ if $0 < |z - z_0| < r$. We say that ϕ is **conformal** at z_0 if whenever two curves γ_1 and γ_2 meet at z_0 then the angle from Γ_1 to Γ_2 (their images under ϕ) is equal to the angle from γ_1 to γ_2. That

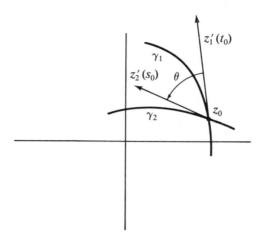

Figure 3.12

is, both the magnitude and the sense of the angle is preserved by ϕ. We shall now show that if f is analytic in the disc $|z - z_0| < r$ and if $f'(z_0) \neq 0$, then f is conformal at z_0.

If γ_1 and γ_2 intersect at z_0, then Γ_1 and Γ_2 intersect at $w_0 = f(z_0)$ and equation (2) shows that the angle from Γ_1 to Γ_2 given by

$$\arg(w_2'(s_0)) - \arg(w_1'(t_0))$$

is equal to

$$\arg(z_2'(s_0)) - \arg(z_1'(t_0)).$$

But the latter is, of course, just the angle from γ_1 to γ_2. Thus, we have established the following result.

> *If f is analytic in the disc $|z - z_0| < r$*
> *and if f'(z_0) $\neq$ 0, then f is conformal at z_0.*   **(3)**

The conclusion expressed in (3) is false whenever $f'(z_0) = 0$, but all is not lost. If $f - f(z_0)$ has a zero of order $m \geq 2$ at z_0, then it can be shown that f increases angles between intersecting curves by a factor of m; the details are presented in the exercises at the end of this section.

In particular, since we have shown in (2) of Section 2 that the derivative of a one-to-one analytic function is never zero we obtain this result.

> *If f is both analytic and one-to-one on a domain D,*
> *then f is conformal at all points of D.*   **(4)**

Many times the word "conformal" is used in complex variables to mean a one-to-one analytic function, although technically speaking it should be used to refer to an analytic function that is just locally one-to-one.

Example 1   The function $f(z) = e^z$ has a nonzero derivative at all points of the plane so it is a conformal mapping at all points. Note that e^z is one-to-one on any strip of the form $\{x + iy: a \leqslant y \leqslant b\}$, $0 < b - a < 2\pi$ but is not one-to-one if $b - a \geqslant 2\pi$.

Example 2   The function $h(z) = \sin z$ is conformal at all points except $z = \pi/2 + n\pi$, $n = 0, \pm 1, \ldots$, which are the zeros of $h'(z) = \cos z$.

Example 3   The function $g(z) = 1/z$ is conformal at all points z except $z = 0$, where it is not defined. In fact, a nonconstant linear fractional transformation $T(z) = (az + b)/(cz + d)$, $ad \neq bc$, is conformal at all points z except $z = -d/c$, where it is not defined. This follows immediately from (4) since we showed in Section 3 that T is one-to-one.

Example 4   Show that each one-to-one analytic function ϕ that maps the disc $\Delta = \{z: |z| < 1\}$ onto itself has the form

$$\phi(z) = \bar{\lambda} \frac{a - z}{1 - \bar{a}z}, \quad |\lambda| = 1, \, a \in \Delta. \tag{5}$$

Solution   We first check that a function ϕ of the form in (5) is a one-to-one mapping of Δ onto Δ. Let

$$\psi(w) = \bar{\lambda} \frac{a\lambda - w}{1 - \bar{a}\bar{\lambda}w}.$$

Then a simple computation yields $\psi(\phi(z)) = z$ and $\phi(\psi(w)) = w$ for all z and w. This shows that both ϕ and ψ are one-to-one and, in fact, are inverse functions of each other in the sense of composition. Furthermore,

$$|\phi(z)|^2 = \frac{|a - z|^2}{|1 - \bar{a}z|^2} = \frac{|a|^2 - 2\,\mathrm{Re}(\bar{a}z) + |z|^2}{1 - 2\,\mathrm{Re}(\bar{a}z) + |a|^2|z|^2}$$

so that $1 > |\phi(z)|^2$ exactly when

$$1 + |a|^2|z|^2 > |a|^2 + |z|^2.$$

This last inequality is equivalent to

$$(1 - |a|^2)(1 - |z|^2) > 0,$$

which is true exactly when $z \in \Delta$. A similar computation shows that $|\psi(w)| < 1$ exactly when $|w| < 1$. Hence, both ϕ and ψ map Δ onto Δ.

Now let f be any one-to-one analytic function mapping Δ onto Δ. Let $f(0) = a$ and set

$$g(z) = \frac{a - f(z)}{1 - \bar{a}f(z)} = \phi(f(z)),$$

where $\phi(z) = (a - z)/(1 - \bar{a}z)$. Then g is a one-to-one analytic function mapping Δ onto Δ and $g(0) = \phi(f(0)) = \phi(a) = 0$. Therefore, $|g(z)| \leqslant |z|$ by Schwarz's lemma (Theorem 2, Section 2). The inverse of g, g^{-1}, is also a one-to-one analytic function of Δ onto Δ, which sends 0 to 0 and so again by Schwarz's lemma we find that $|g^{-1}(z)| \leqslant |z|$. Equivalently, $|\zeta| \leqslant |g(\zeta)|$ for all $\zeta \in \Delta$. Hence, $|z| = |g(z)|$ for all $z \in \Delta$ and this implies that $g(z) = \lambda z$, where λ is a constant of absolute value 1. Therefore,

$$\lambda z = \frac{a - f(z)}{1 - \bar{a}f(z)}, \quad |\lambda| = 1$$

and hence

$$f(z) = \lambda \frac{\beta - z}{1 - \bar{\beta}z}, \quad \beta = \bar{\lambda}a \in \Delta. \qquad \square$$

Level curves

Let p be a real-valued function on a domain D. The set of z, which are solutions of the equation $p(z) = c$, where c is a constant, is called a **level set** or **level curve** of the function p. Level curves are important in describing flows, and thus fields, because the path followed by a particle in the flow is just a level curve of a certain analytic function, as we shall see in Section 4.1. We shall now show that for an analytic function $f = u + iv$, the level curves of u are "almost always" perpendicular to the level curves of v.

Suppose that $f = u + iv$ and $f'(z_0) \neq 0$. We know that f is one-to-one in some small disc $D = \{z : |z - z_0| < r\}$; let $g = \sigma + i\tau$ be the inverse function to f on $\Omega = f(D)$ so that $g(f(z)) = z$ if $|z - z_0| < r$. Let γ_1 consist of those points z with $|z - z_0| < r$ and $u(z) = u(z_0)$. Then

$$\gamma_1 = \{z : u(z) = u(z_0)\} = \{z : \operatorname{Re} f(z) = u(z_0)\}$$
$$= \{g(w) : \operatorname{Re} w = u(z_0), w \in \Omega\},$$

so γ_1 is precisely the range of the function $g(w)$ on the set $\{w \in \Omega : \operatorname{Re} w = u(z_0)\}$ and hence is a smooth arc. Likewise, the set γ_2 consisting of those points z with $|z - z_0| < r$ and $v(z) = v(z_0)$ can be described by

$$\gamma_2 = \{z : v(z) = v(z_0)\} = \{z : \operatorname{Im} f(z) = v(z_0)\}$$
$$= \{g(w) : \operatorname{Im} w = v(z_0), w \in \Omega\}.$$

Hence, γ_2 is precisely the range of $g(w)$ on the set $\{w \in \Omega : \operatorname{Im} w = v(z_0)\}$ and it, too, is a smooth arc. We wish to show that γ_1 and γ_2 meet at a right angle at z_0. But this is now immediate since the function g is conformal and the lines $\operatorname{Re} w = u(z_0)$ and $\operatorname{Im} w = v(z_0)$ meet at a right angle at the point $u(z_0) + iv(z_0) = f(z_0)$ (Fig. 3.13).

In this manner we see that each analytic function f whose derivative does not vanish gives an orthogonal "coordinate" system by means of the curves $u(z) = \text{constant}$, $v(z) = \text{constant}$. The simple case $f(z) = z$ gives the usual

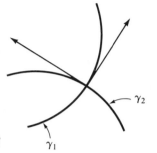

The level curves γ_1, where $u(x, y)$ is constant, and γ_2, where $v(x, y)$ is constant, meet at a right angle.

Figure 3.13

coordinate system. A few others are given in the following examples. It's worth mentioning again that not only do we obtain lovely pictures but the curves Im $f = $ constant represent the paths followed by a particle in the flow given by $\bar{f}'$ and hence these curves are of great importance in applications; more on this is in Section 3.4.1 of this chapter and in Chapter 4.

Example 5 $f(z) = \text{Log } z$ on the plane minus the ray $(-\infty, 0]$. Here the level curves are $\log|z| = $ constant and Arg $z = $ constant, which are, respectively, circles centered at the origin and rays emanating from the origin (Fig. 3.14).

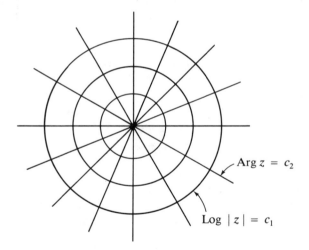

$\text{Arg } z = c_2$

$\text{Log }|z| = c_1$

Figure 3.14

Example 6 $f(z) = z^2 = x^2 - y^2 + i\, 2xy$; the level curves sketched in Figure 3.15 are two families of mutually perpendicular hyperbolas.

Example 7 $f(z) = \log((z - 1)/(z + 1))$; here the level curves of the real part of f are exactly the circles $|z - 1| = \rho|z + 1|$ discussed in Section 2, Chapter 1, while the level curves of the imaginary part of f are (see Figure 3.16) the circles $\text{Arg}((z - 1)/(z + 1)) = $ constant.

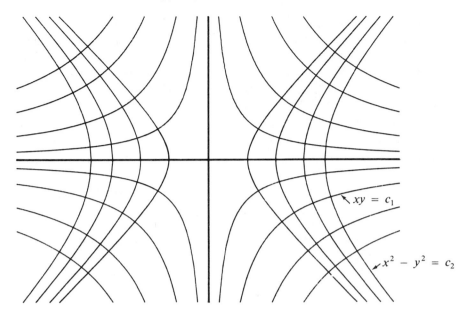

$xy = c_1$

$x^2 - y^2 = c_2$

Figure 3.15

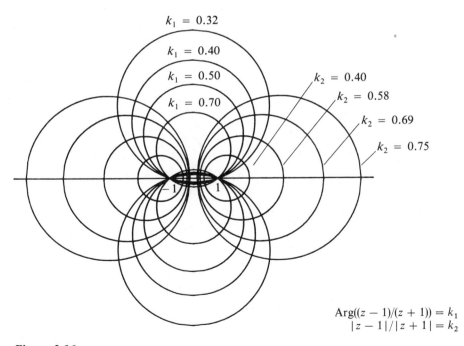

$k_1 = 0.32$

$k_1 = 0.40$

$k_1 = 0.50$

$k_1 = 0.70$

$k_2 = 0.40$

$k_2 = 0.58$

$k_2 = 0.69$

$k_2 = 0.75$

$\operatorname{Arg}((z - 1)/(z + 1)) = k_1$
$|z - 1|/|z + 1| = k_2$

Figure 3.16

Another way to find the level curves is as follows. The function $w = (z - 1)/(z + 1)$ maps circles through 1 and -1 into straight lines through the origin and maps the circles $|(z - 1)/(z + 1)| = c$ to the circles $|w| = c$. The function $\zeta = \log w$ maps the rays $\arg z = $ constant onto the straight lines Im $\zeta = $ constant and maps the circles $|w| = c$ onto the circles $|\zeta| = \log c$. Hence, $\zeta = \log[(z - 1)/(z + 1)] = f(z)$ has as its level curves Im $f(z) = $ constant exactly the circles through 1 and -1.

Example 8 $f(z) = \log(z + \sqrt{z^2 - R^2})$, on the plane minus the segment $[-R, R]$.

The function $g(z) = z + \sqrt{z^2 - R^2} = w$ has as its inverse function

$$z = \tfrac{1}{2}(w + R^2/w) = G(w).$$

A simple computation shows that $G(w)$ carries rays Arg $w = c$ onto the hyperbolas

$$\left(\frac{x}{\cos c}\right)^2 - \left(\frac{y}{\sin c}\right)^2 = R^2, \quad z = x + iy$$

and the circles $|w| = s$, $s > R$, onto the ellipses

$$\frac{4x^2}{(s + R^2/s)^2} + \frac{4y^2}{(s - R^2/s)^2} = 1, \quad z = x + iy.$$

Hence, $g(z)$ carries these hyperbolas and ellipses onto the rays Arg $w = c$ and the circles $|w| = s$, respectively. Thus, $f(z) = \log g(z)$ has these ellipses and hyperbolas as its level curves; see Figure 3.17 for the case $R = 1$.

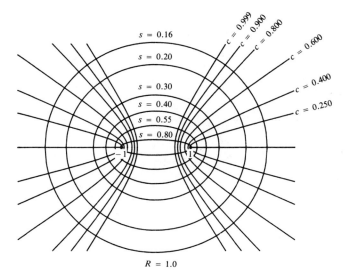

Figure 3.17   The curves $\left(\dfrac{x}{\cos c}\right)^2 - \left(\dfrac{y}{\sin c}\right)^2 = 1$ $\dfrac{4x^2}{(s + 1/s)^2} + \dfrac{4y^2}{(s - 1/s)^2} = 1.$

Example 9 Fix a, $|a| < 1$, and consider

$$f(z) = \log\left(\frac{z - a}{1 - \bar{a}z}\right), \quad |z| < 1.$$

The function $g(z) = (z - a)/(1 - \bar{a}z)$ has constant modulus on the family of circles determined by

$$|z - a| = \rho|z - 1/\bar{a}|.$$

The value $\rho = |a|$ gives the circle $|z| = 1$; the argument of $g(z)$ is constant on the family of circles passing through a and $1/\bar{a}$. Thus, the level curves of $f(z)$ are exactly these two families of mutually perpendicular circles, one family lying entirely within the disc $|z| < 1$ (Fig. 3.18).

Example 10 $f(z) = \sqrt{z}$ on the plane minus the ray $(-\infty, 0)$.
Here

$$u(re^{i\theta}) = \operatorname{Re} f(re^{i\theta}) = \sqrt{r}\,\cos(\theta/2)$$
$$v(re^{i\theta}) = \operatorname{Im} f(re^{i\theta}) = \sqrt{r}\,\sin(\theta/2)$$

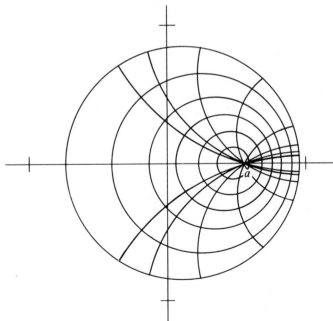

Figure 3.18 The curves

$$|z - a| = \rho|1 - \bar{a}z| \qquad \operatorname{Arg}\left(\frac{z - a}{1 - \bar{a}z}\right) = c, \quad -\pi < c < \pi.$$

for $0 < r < \infty$ and $-\pi < \theta < \pi$. Using the half-angle formula from trigonometry, we find the level curves of $u(re^{i\theta})$ and $v(re^{i\theta})$ are given by the polar equations

$$r(1 + \cos \theta) = \text{constant}$$
$$r(1 - \cos \theta) = \text{constant},$$

respectively. A few of these curves are sketched in Figure 3.19.

You might review the preceding six examples to note the following two phenomena (1) the level curves $\text{Re } f(z) = c_1$ never touch the boundary of D, whereas (2) the level curves $\text{Im } f(z) = c_2$ always meet the boundary of D.

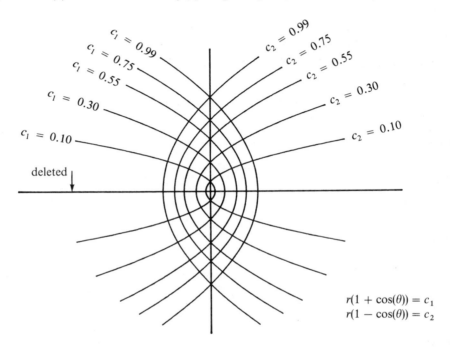

$$r(1 + \cos(\theta)) = c_1$$
$$r(1 - \cos(\theta)) = c_2$$

Figure 3.19

Exercises for Section 3.4

1.   Show that $g(z) = z^2$ is one-to-one on the half-plane $\text{Re } z > 0$ but not on any larger open set.

2.   Find all one-to-one analytic functions that map the upper half-plane U onto itself. (**Hint:** $\phi(z) = i(1 + z)/(1 - z)$ maps the unit disc onto U and ϕ is one-to-one.)

3.   Show that each one-to-one entire function has the form $Az + B$, $A \neq 0$. (**Hint:** Check the behavior at " ∞.")

4.   Let f be a one-to-one analytic function on a domain D and suppose

that there is an analytic function h on D with $h^2 = f$. Show that h is also one-to-one.

5. Let $K(z) = z/(1 - z)^2$. (a) Show that K is one to one on the unit disc $|z| < 1$. (b) Find the range of K. (**Hint:** Find what K does to the circle $|z| = 1$.)

6. Use the basic results on conformality to show that each of the circles $|z - a| = \rho|z - b|$, $0 < \rho < \infty$, is perpendicular to each of the circles $\arg((z - a)/(z - b)) = c$, $0 \leqslant c < 2\pi$. (**Hint:** Consider $f(z) = \log((z - a)/(z - b))$.)

7. Sketch some of the level curves of the real and imaginary parts of the following functions (a) $f(z) = \log((z - 2)/(1 - 2z))$ (b) $g(z) = \sqrt{z - 1}$ on the plane with the interval $(-\infty, 1]$ deleted. (**Hint:** Set $z = 1 + w^2$ with $w = s + it$. Then solve for z when $s = $ constant or $t = $ constant.) (c) $h(z) = (1 + z)/(1 - z)$ (d) $f(z) = z^2 - 2z$.

8. Let f and g be conformal analytic functions with the range of f a subset of the domain of g. Show that the composition $g(f(z))$ is also conformal.

9. Let p be a polynomial of degree 1 or more with r distinct zeros. Let $\gamma_\varepsilon = \{z : |p(z)| = \varepsilon\}$; show that γ_ε consists of r disjoint closed curves if ε is small and one closed curve if ε is very large. These curves are called **lemniscates.** For the specific polynomial $p(z) = z^2 - 1$ show that the lemniscate is a pair of disjoint simple closed curves if $0 < \varepsilon < 1$, a single simple closed curve if $\varepsilon > 1$, and a "figure eight" if $\varepsilon = 1$.

10. Let f be a one-to-one analytic mapping of the unit disc $|z| < 1$ onto itself with 2 fixed points in $|z| < 1$. Show that $f(z) = z$.

11. Let f be analytic on a domain D, let z_0 be a point of D and set $w_0 = f(z_0)$. Suppose that $f(z) - w_0$ has a zero of order m, $m \geqslant 2$, at z_0. Follow the outlined steps to show that f "multiplies angles by m" at z_0.
   (a) The angle from the ray $\gamma_1 = \{z : \arg(z - z_0) = \psi\}$ to the ray $\gamma_2 = \{z : \arg(z - z_0) = \psi + \theta\}$ is θ.
   (b) In a disc $|z - z_0| < r$, we have $f(z) - w_0 = (z - z_0)^m g(z)$, where $g(z_0) \neq 0$.
   (c) If $z_1 \in \gamma_1$ and $z_2 \in \gamma_2$ then $\arg(f(z_1) - w_0) = m\psi + \arg g(z_1)$ and $\arg(f(z_2) - w_0) = m(\psi + \theta) + \arg g(z_2)$.
   (d) Let $\Gamma_1 = f(\gamma_1)$ and $\Gamma_2 = f(\gamma_2)$. The angle from Γ_1 to Γ_2 is

$$\lim_{\substack{z_1 \to z_0 \\ z_2 \to z_0}} [\arg(f(z_1) - w_0) - \arg(f(z_2) - w_0)] = m\theta.$$

12. Suppose that f is analytic on a convex domain D and that $\operatorname{Re}(f'(z)) > 0$ for all $z \in D$. Show that f is one-to-one on D. (**Hint:** $f(z_2) - f(z_1) = \int_\gamma f'(w)\, dw$, where γ is the line segment joining z_1 to z_2.)

13.   Use Exercise 12 to show that the function $g(z) = z + e^z$ is one-to-one on the strip $\{x + iy: -\infty < x < \infty, 0 < y < \pi\}$. (**Hint:** Show that the conclusion of Exercise 12 remains valid if $\operatorname{Im}(f')$ is positive on D.)

14.   Let F be a one-to-one analytic function from a domain D_1 onto a domain D_2. Suppose that $\{z_k\}$ is a sequence of points in D_1 with $z_k \to p$, where p is in the boundary of D_1. Show that if $\{f(z_k)\}$ converges to a point q, then q must necessarily lie in the boundary of D_2. (**Hint:** If $q \in D_2$, then $q = f(z_0)$ for some $z_0 \in D_1$. Show that f carries a small disc centered at z_0 into a small disc centered at q and so derive a contradiction.)

3.4.1   Conformal mapping and flows⋆

One of the most significant applications of conformal mapping is to flows. This section is devoted to the first steps in this application by making the connection between level curves, conformal mapping, and flows. This will not only provide us with an important application of complex variables but as well will make it apparent why we want to search for specific conformal mappings and so will motivate the material of the next section.

We noted already in Section 1.1 and Section 4, Application 3, Chapter 2 that if f is a sourceless and irrotational flow on some domain D, then there must be an analytic function G defined on D, called the **complex potential** of the flow, with

$$G'(z) = \overline{f(z)}, \quad z \in D.$$

We shall show now that the level curves $\operatorname{Im} G = $ constant represent the paths followed by particles in the flow; that is, these level curves are the **streamlines** for the flow.

Let z_0 be a point of D at which $f(z_0) \neq 0$. Then $G'(z_0) = \overline{f(z_0)} \neq 0$ and thus $G(z)$ is one-to-one in some small disc $\{z: |z - z_0| < \delta\}$, centered at z_0. This implies that there is an analytic function $H(w)$ defined at least for $w \in \Omega = \{G(z): |z - z_0| < \delta\}$ with $H(G(z)) = z$ if $|z - z_0| < \delta$. The level curve

$$\Gamma_0 = \{z: \operatorname{Im} G(z) = \operatorname{Im} G(z_0), |z - z_0| < \delta\}$$

can also be described, therefore, as

$$\Gamma_0 = \{H(w): \operatorname{Im} w = \operatorname{Im} G(z_0), w \in \Omega\}$$
$$= \{H(\tau + ic_0): \tau + ic_0 \in \Omega, c_0 = \operatorname{Im} G(z_0)\}.$$

The tangent vector to the curve Γ_0 is just the derivative of $H(\tau + ic_0)$ with respect to τ and this in turn is equal to the derivative of H with respect to w, since H is analytic. By the chain rule, we have

$$1 = H'(G(z))G'(z) = H'(w)\overline{f(z)}, \quad w = G(z).$$

Consequently,

$$\frac{H'(w)}{|H'(w)|} = \frac{|f(z)|}{\overline{f(z)}} = \frac{f(z)}{|f(z)|}.$$

Thus, the unit tangent vector to Γ_0 is parallel to $f(z)$ at each point z of Γ_0. This says that the level curve Γ_0 is the path followed by a particle in the flow (see also Exercise 11 at the end of this section).

Streamlining

Flow problems are typically of two types. The first, illustrated in Figures 3.20a and 3.20b is when an impermeable object is placed in a flow of uniform velocity across the plane. The resulting flow *around* the object is to be found, assuming a [virtually] uniform velocity far from the object.

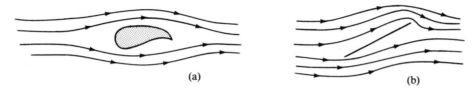

(a)   (b)

Figure 3.20

The second type of flow, illustrated in Figures 3.21a and 3.21b is, for instance, the flow through a channel or around a corner, or, generally in a simply connected region whose boundary consists of straight lines except for a few irregularities or corners. The boundary is impermeable and the flow is assumed to be [virtually] uniform "at ∞," that is, far from the irregular part of the boundary. Both types of flow problem can be handled by consideration of appropriate conformal mappings, as we will now show.

Our objective is to find the path described by a particle as it moves through the domain D when the motion is governed by a sourceless, irrotational flow f that remains in D; that is, the particle does not cross the boundary of D. We let G be an analytic function on D with $G'(z) = \overline{f(z)}$, $z \in D$; we

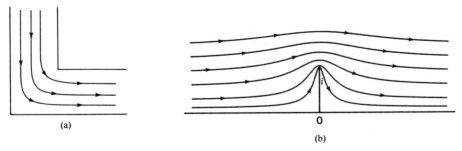

(a)   (b)

Figure 3.21

then know from the foregoing discussion that the path followed by a particle in the flow is just one of the level curves, Im G = constant. The particles never cross the boundary of D and so the boundary of D is one of the streamlines of the flow; that is, Im G is constant on the boundary D. Consequently, to find the path taken by a particle in the flow we need to find an analytic function G on D whose imaginary part is constant on the boundary of D and which has the correct behavior at "∞." This is where conformal mapping enters the picture.

Suppose that G is a one-to-one analytic mapping of the simply connected domain D onto the upper half-plane $U = \{x + iy: y > 0\}$ and G maps the boundary of D onto the real axis. Then Im $G(z)$ is zero if z is in the boundary of D and thus G is the complex potential that we want. That is, the level curves Im G = constant are precisely the paths followed by the particles in the flow and the function $f(z) = \overline{G'(z)}$ is the function that produces the flow. We write $w = G(z)$ and $H(w) = z$; that is, $H(w)$ is the inverse function to $G(z)$; $H(w)$ maps the upper half-plane U onto D. The paths followed by particles in the flow are given by

$$\Gamma_c(t) = \{H(t + ic): -\infty < t < \infty\}, \ 0 < c < \infty.$$

For example, the domain D pictured in Figure 3.20b is mapped onto U by the function $w = G(z)$, where

$$z = (w^2 - 1)^{1/2}$$

as we will see in Example 10 of Section 5. Hence, the curves Γ_c, given by

$$\Gamma_c(t) = [(t + ic)^2 - 1]^{1/2}, \ -\infty < t < \infty,$$

are exactly the paths followed by the particles in this flow. For c very large these are virtually the same as the horizontal lines Im $z = c$. For $c = 0$, the curve is just the boundary of the domain D (the square root must be correctly interpreted, however).

The search for the function G is called "**streamlining**" the domain D and the function $V(x, y) = $ Im $G(x + iy)$ is called the **stream function** of D.

Example 1   The function $G(z) = z^2$ maps the first quadrant conformally onto the upper half-plane U. Its inverse is the function $H(w) = \sqrt{w}$. The stream function is Im$(z^2) = 2xy$.

Example 2   The function $G(z) = \sin z$ maps the strip $\{x + iy: |x| < \pi/2, \ 0 < y < \infty\}$ conformally onto U. Its inverse is $H(w) = \arcsin w$ (see Section 5, Chapter 1). The stream function is $\cos x \cosh y$.

Example 3   Fix a point $a, 0 \leqslant a < 1$. The function

$$G(z) = i \frac{(1 - a)(1 + z)}{(1 - z)(1 + a)}$$

maps the disc $\{|z| < 1\}$ conformally onto U, sending a to i. Its inverse is

$$H(w) = \frac{w(1 + a) - i(1 - a)}{w(1 + a) + i(1 - a)}.$$

The stream function is

$$\frac{(1 - a)}{(1 + a)} \frac{(1 - |z|^2)}{|1 - z|^2}.$$

Exercises for Section 3.4.1

1. Show that the function $G(z) = \frac{1}{2}(z^2 + 1/z^2)$ maps the region D, which is in the first quadrant and also exterior to the circle $|z| > 1$ conformally onto the upper half-plane (Fig. 3.22). Use the function $H = G^{-1}$ to streamline D.

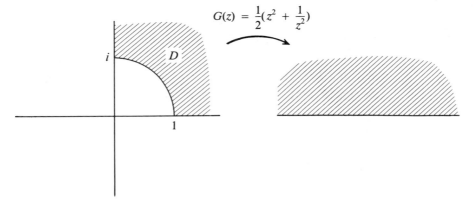

Figure 3.22

2. Show that the function $H(z) = z^\alpha$, $0 < \alpha < 2$, maps the upper half-plane U onto the plane minus a "wedge" of angle $\pi(2 - \alpha)$ (Fig. 3.23). Use H to streamline this region.

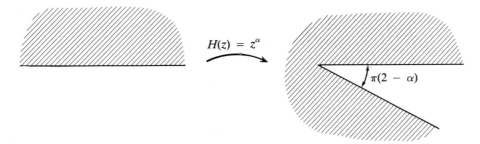

Figure 3.23

3.   Show that the function $K(z) = z/(1-z)^2$ is a conformal mapping of the disc $\{z: |z| < 1\}$ onto the domain D obtained by deleting from the plane the interval $(-\infty, -1/4]$ (Fig. 3.24). Show that $z = (\zeta - i)/(\zeta + i)$ is a conformal mapping of U onto the disc $\{z: |z| < 1\}$. Use these two functions to streamline D (see Exercise 5, Section 4).

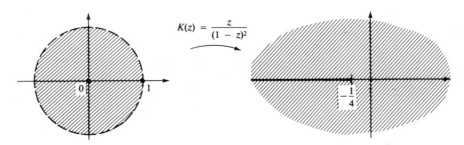

Figure 3.24

4.   Use the function H from Example 3 to streamline the disc $\{z: |z| < 1\}$.

5.   Show that the function $G(z) = -(1-z^2)^2$ maps the region D shown in Figure 3.25 conformally onto U. Use this to streamline D.

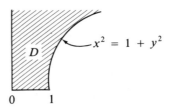

Figure 3.25

6.   Show that the function $-\text{Log}(z - a)$ represents the complex potential of a flow in the plane with a single source at the point a. What are the streamlines for this flow?

7.   Show that the function $\text{Log}((z - 1)/(z + 1))$ represents the complex potential of a flow in the plane with a source at -1 and a sink ($=$ negative source) at $+1$. Find the streamlines for this flow.

8.   Show that the function

$$\text{Log}\left(\frac{z - a}{1 - \bar{a}z}\right), \quad |a| < 1,$$

represents a complex potential of a flow *within* the disc $|z| < 1$, which has a sink at $z = a$ (see Example 9, Section 4).

9.   Sketch some of the streamlines for the flow

$$f(z) = (\bar{p} - \bar{q})/(\bar{z} - \bar{p})(\bar{z} - \bar{q}).$$

(**Hint:** An indefinite integral of $\bar{f}$ is $G(z) = \text{Log}((z - p)/(z - q))$.

10.   The function $G(z) = e^z$ is a conformal mapping of the strip $S = \{x + iy: 0 < y < \pi, -\infty < x < \infty\}$ onto U with $\text{Im } G = 0$ on the edges of S. However, the curves $\text{Im } G = \text{constant}$ are *not* the horizontal lines $y = c$, which are the lines of a uniform flow in S. Explain this apparent discrepancy.

11.   Let f be a sourceless, irrotational flow in a domain D and let G be an analytic function on D with $G' = \bar{f}$. Let z_0 be a point of D at which $f(z_0) \neq 0$ and let δ be a small positive number for which $G'(z) \neq 0$ if $|z - z_0| < \delta$. Let $\Omega = \{G(z): |z - z_0| < \delta\}$ and let $H(w)$ be the analytic function on Ω with $H(G(z)) = z, |z - z_0| < \delta$.

    Let $u(t)$, $a < t < b$, be a continuously differentiable function with values in the line $L: \{\tau + ic_0: \tau + ic_0 \in \Omega, c_0 = \text{Im } G(z_0)\}$, which satisfies the differential equation

$$u'(t) = |H'(u(t))|^{-2}, \quad a < t < b,$$

and set $\gamma(t) = H(u(t))$, $a < t < b$. (That such a function exists is a result from the theory of ordinary differential equations.) Show that the range of the curve $\gamma(t)$ is the level curve

$$\Gamma_0 = \{z: \text{Im } G(z) = c_0\}$$

and that $\gamma'(t) = f(\gamma(t))$ for $a < t < b$. (**Hint:** Look back to the discussion at the beginning of this section.)

3.5   The Riemann mapping theorem and Schwarz–Christoffel transformations

The theorem of greatest theoretical importance in the subject of conformal mapping is the famous theorem of Bernhard Riemann.

Theorem 1: Riemann mapping theorem. Suppose D is a simply connected domain with at least two points in its boundary; let p be a point of D. Then there is a one-to-one analytic function ϕ that maps D onto the open unit disc $\Delta = \{w: |w| < 1\}$ and $\phi(p) = 0$. Furthermore, ϕ is uniquely determined by the requirement that $\phi'(p)$ be positive.

It is a consequence of Theorem 1 that *any* two simply connected domains (each with two or more boundary points) can be linked by a one-to-one analytic function. For if D_1 and D_2 are two such domains and if ϕ_2 maps D_2 onto Δ with $\phi_2(p_2) = 0$, then the analytic function $\psi = \phi_2^{-1} \circ \phi_1$ maps D_1

onto D_2, $\psi(p_1) = p_2$, and ψ is one-to-one. In the language of complex variables, D_1 and D_2 are **conformally equivalent**; the mapping is uniquely determined by the requirements that $\psi(p_1) = p_2$ and $\psi'(p_1)$ be positive. See Figure 3.26.

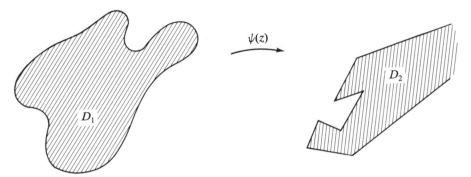

Figure 3.26   Two simply-connected domains that are necessarily conformally equivalent.

The proof of Theorem 1 requires concepts not covered in this book and, in any event, the usual proof provides no clue to how you actually write down the conformal mapping; it is an "existance" proof only (see the notes at the end of the chapter). However, Theorem 1 is of great utility nonetheless since it provides us with the knowledge that a conformal map actually exists and hence might be found explicitly (there is, obviously, little point in searching for an object that does not exist.) We begin our search for useful conformal mappings by reviewing some of the conformal maps that we have already run across. Then we show that a more-or-less explicit rule can be given for constructing conformal maps of the upper half-plane $U = \{w: \operatorname{Im} w > 0\}$ onto a **polygon.** The latter construction produces the **Schwarz–Christoffel** transformations; these transformations are useful tools in analyzing flows and other physical phenomena, as we saw in Section 4.1.

Examples of conformal mappings

Example 1 $f(z) = i\,(1 + z)/(1 - z)$ maps the disc $\Delta = \{z: |z| < 1\}$ onto the half-plane $U = \{w: \operatorname{Im} w > 0\}$ (Fig. 3.27).

Example 2 $g(z) = e^z$ maps the strip $\{-\pi \leqslant y < \pi\}$ onto the punctured plane, those w with $w \neq 0$ (Fig. 3.28) (see Section 5, Chapter 1).

Example 3 $h(z) = \operatorname{Log} z$ maps the region Ω obtained by deleting the ray $(-\infty, 0]$ from the plane onto the strip $\{w: |\operatorname{Im} w| < \pi\}$ (Fig. 3.29) (see Section 5, Chapter 1).

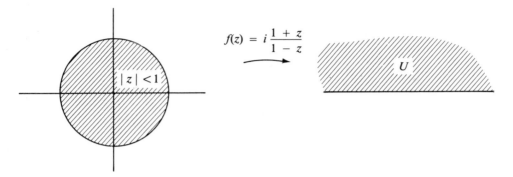

Figure 3.27

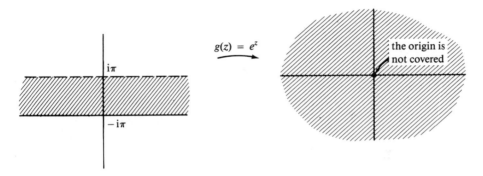

Figure 3.28

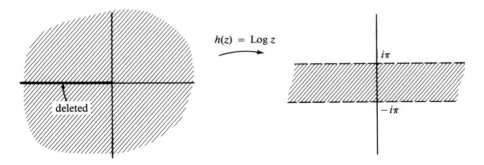

Figure 3.29

Example 4 $F(z) = \sin z$ maps the strip $\{z = x + iy: 0 < x < \pi/2$ and $y > 0\}$ onto the first quadrant, $\{w: \text{Re } w > 0$ and $\text{Im } w > 0\}$ (Fig. 3.30) (see Section 5, Chapter 1).

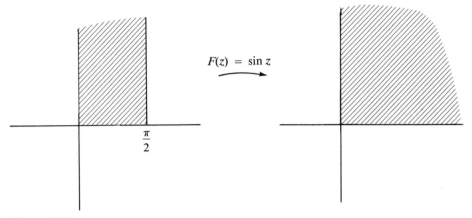

Figure 3.30

Example 5 $f(z) = z^p$, $0 < p < 2$, maps the upper half-plane U onto the region described by $\{w = re^{i\psi} : 0 < \psi < \pi p; \; 0 < r < \infty\}$ (Fig. 3.31).

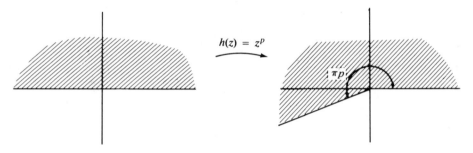

Figure 3.31

Schwarz*–Christoffel† transformations

A Schwarz–Christoffel transformation is an analytic conformal mapping of the upper half-plane onto a polygon. The key to understanding it is the examination of the behavior at the point x_0 of the function f given by

$$f(z) = A(z - x_0)^\beta + B,$$

where x_0 and β are real numbers, $0 < \beta < 2$, and A and B are complex numbers. The root is determined by choosing $\arg(z - x_0)$ to lie in the interval $(-\pi/2, 3\pi/2)$; that is, we delete from the plane the vertical ray from x_0 down.

To begin, suppose $z = x$ is real and $x > x_0$. Then $\arg f'(x) = (\beta - 1)(0) + \arg A$ so that the curve parametrized by f has a tangent

---

* Hermann Amandus Schwarz, 1843–1921.

† Elwin Bruno Christoffel, 1829–1900.

vector of constant slope, arg A; that is, it is a straight line segment. On the other hand, if $x < x_0$ then arg $f'(x) = \pi(\beta - 1) + $ arg A so that $f(x)$ lies on a straight line making an angle $\pi(\beta - 1) + $ arg A with the positive real axis (Fig. 3.32).

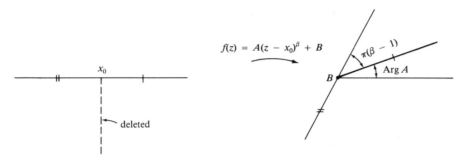

Figure 3.32

Let us set $\alpha = \beta - 1$ so that $-1 < \alpha < 1$; we see that the mapping f carries the real axis into a polygonal curve of only two pieces and the pieces meet at B with angle of $\pi\alpha = \pi(\beta - 1)$. Let us continue this idea, concentrating on f' instead of f since it is the value of arg(f') that determines the slope of the various pieces of the image curve. Let $x_1 < \cdots < x_N$ be points on the real axis and let $\alpha_1, \ldots, \alpha_N$ be real numbers, all in the interval $(-1, 1)$. We shall examine the behavior of the function f, whose derivative is given by

$$f'(z) = A(z - x_1)^{\alpha_1} \cdots (z - x_N)^{\alpha_N}.$$

Once again, the roots are determined by the requirement that arg$(z - x_j)$ lies in the range $(-\pi/2, 3\pi/2)$. If x is real and near x_j, but is slightly more than x_j, then

$$\arg f'(x) = \arg A + \pi\alpha_{j+1} + \cdots + \pi\alpha_N.$$

However, if x is again real and near x_j but is slightly less than x_j, then

$$\arg f'(x) = \arg A + \pi\alpha_j + \pi\alpha_{j+1} + \cdots + \pi\alpha_N.$$

If we add to this the facts that arg $f'(x) = $ arg A whenever $x > x_N$ and that arg $f'(x) = $ arg $A + \pi\alpha_1 + \cdots + \pi\alpha_N$ whenever $x < x_1$, we see that f maps the real axis onto a polygonal curve with $N + 1$ pieces (one or two of which may be infinite in length). See Figure 3.33.

Now we shall apply this knowledge to the mapping of the upper half-plane onto a given polygon. Let us suppose the polygon has $N + 1$ sides with vertices at points $w_0, \ldots, w_N$, arranged in the usual counterclockwise order around the polygon.

Let $\theta_0, \theta_1, \ldots, \theta_N$ be the **exterior** angles at $w_0, \ldots, w_N$, respectively (Fig. 3.34). The angles $\theta_0, \ldots, \theta_N$ lie in the range $(-\pi, \pi)$ and since the polygonal curve $\overrightarrow{w_0 w_1}, \overrightarrow{w_1 w_2}, \ldots, \overrightarrow{w_N w_0}$ is a simple closed positively oriented curve,

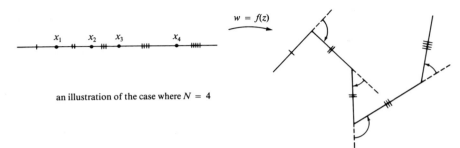

an illustration of the case where $N = 4$

Figure 3.33

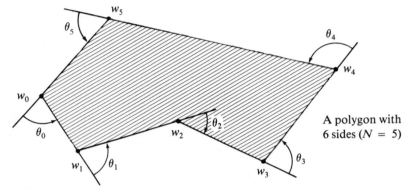

A polygon with 6 sides ($N = 5$)

Figure 3.34

we see that

$$\theta_0 + \theta_1 + \cdots + \theta_N = 2\pi.$$

Set $\alpha_j = -\theta_j/\pi$ so that

$$-1 < \alpha_j < 1 \qquad \text{and} \qquad \sum_{j=0}^{N} \alpha_j = -2.$$

These preliminary comments then prepare the way for the next theorem.

> **Theorem 2: Schwarz–Christoffel.** Let P be a polygon in the plane with vertices $w_0, \ldots, w_N$ and corresponding exterior angles $\theta_0, \ldots, \theta_N$; set $\alpha_j = -\theta_j/\pi$. Then there are real numbers $x_1, \ldots, x_N$ with $x_1 < \cdots < x_N$ and a constant A such that the function $f(x)$ whose derivative is
>
> $$f'(z) = A(z - x_1)^{\alpha_1} \cdots (z - x_N)^{\alpha_N} \tag{1}$$
>
> gives a one-to-one analytic mapping of the upper half-plane Im $z > 0$ onto the polygon P; f maps z_j to w_j for $j = 1, \ldots, N$ and $f(\infty) = \lim_{x \to \pm\infty} f(x) = w_0$.

A full proof of the Theorem 2 result rests on Theorem 1 and on the reflection principle which will be covered in Section 3 of Chapter 4. Here we shall content ourselves with several comments and a number of examples. First, we have arranged the mapping f so that $f(\infty) = w_0$ but this is not necessary and there will be times that we will want to exploit some obvious symmetries in the polygon and *not* require that $f(\infty) = w_0$. In general, we can select any *three* of the vertices w_j and *any* three points x_j on the real line or at ∞ and require that $f(x_j) = w_j$ for these *three* values of j. For example, if we demand that $f(-1) = w_2$, $f(0) = w_5$, and $f(2) = w_6$, then we have $x_2 = -1$, $x_5 = 0$, and $x_6 = 2$ so that $x_1, x_3, x_4, x_7, \ldots$ necessarily satisfy

$$x_1 < -1 < x_3 < x_4 < 0 < 2 < x_7 < \cdots.$$

Second, Theorem 2 is stated for bounded polygons but it holds as well for unbounded polygons and these will be the most useful cases. This can be seen most easily by looking again at the behavior of the function whose derivative is given in (1). Another technique is to obtain the unbounded polygon as a limit of bounded ones. This is sometimes useful to determine the angles $\alpha_1, \ldots, \alpha_N$ and will be illustrated in several of the examples.

Example 6 Find the Schwarz–Christoffel transformation of the upper half-plane U onto the equilateral triangle shown in Figure 3.35.

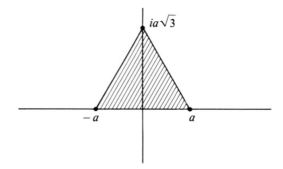

Figure 3.35

Solution The exterior angles of an equilateral triangle are all equal to $\frac{2}{3}\pi$, so that

$$\alpha_j = -\theta_j/\pi = -2/3, \qquad j = 0, 1, 2.$$

We select $x_1 = -1$ and $x_2 = 1$. Then,

$$f'(z) = A(z + 1)^{-2/3}(z - 1)^{-2/3}$$
$$= A(z^2 - 1)^{-2/3}.$$

Hence,

$$f(z) = A \int_1^z \frac{dw}{(w^2 - 1)^{2/3}} + B,$$

where we have selected 1 as the initial point for the integration. (Another choice would produce another B.) To find A and B, we note that

$$a = f(1) = B$$

and

$$i\sqrt{3}a = A \int_1^\infty \frac{dt}{(t^2 - 1)^{2/3}} + B.$$

If we denote by β the value of the integral

$$\beta = \int_1^\infty \frac{dt}{(t^2 - 1)^{2/3}}$$

then we find that

$$A = \frac{a(i\sqrt{3} - 1)}{\beta}$$

and

$$B = a. \qquad\qquad \square$$

Example 7   Find the Schwarz–Christoffel transformation on the upper half-plane U onto the region shown in Figure 3.36.

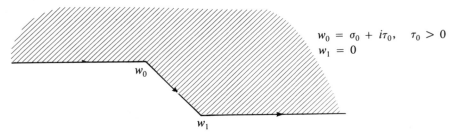

$$w_0 = \sigma_0 + i\tau_0, \qquad \tau_0 > 0$$
$$w_1 = 0$$

Figure 3.36

Solution   The exterior angle at w_0 is θ where

$$\theta = \arctan(\tau_0/\sigma_0) \in (-\pi, 0)$$

and the exterior angle at w_1 is just $-\theta$. If we select $x_0 = -1$ and $x_1 = 1$, then the derivative of the mapping function is

$$f'(z) = A(z + 1)^\alpha (z - 1)^{-\alpha}, \quad \alpha = -\theta/\pi \in (0, 1).$$

The function $f(z)$ cannot usually be given explicitly. However, in the special case when $\sigma_0 = 0$ we have $\theta = -\pi/2$ so that $\alpha = 1/2$ and

$$f'(z) = A\left(\frac{z + 1}{z - 1}\right)^{1/2}.$$

Thus,

$$f(z) = A\{(z^2 - 1)^{1/2} + \text{Log}(z + (z^2 - 1)^{1/2})\} + B.$$

The constants A and B are found from

$$i\tau_0 = f(-1) = A\{i\pi\} + B$$
$$0 = f(1) = A\{0\} + B.$$

Hence, $B = 0$ and $A = \tau_0/\pi$. The mapping is thus

$$f(z) = \frac{\tau_0}{\pi} \{(z^2 - 1)^{1/2} + \text{Log}(z + (z^2 - 1)^{1/2})\}.$$

Some of the streamlines for the flow through the region (when $\sigma_0 = 0$ and $\tau_0 = 1$) are shown in Figure 3.37 below.   □

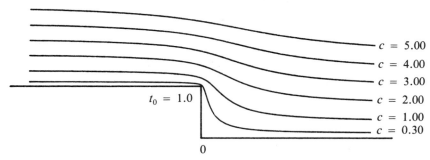

Figure 3.37   The flow over a vertical step. The streamlines are

$$\Gamma_c(t) = \frac{1}{\pi} \{[t + ic)^2 - 1]^{1/2}$$
$$+ \text{Log}(t + ic + [(t + ic)^2 - 1]^{1/2})\}, \quad -\infty < t < \infty; c > 0.$$

Example 8   Find the Schwarz–Christoffel transformation that maps the upper half-plane onto the region shown in Figure 3.38.

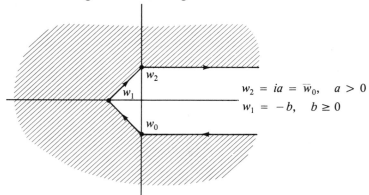

$$w_2 = ia = \overline{w}_0, \quad a > 0$$
$$w_1 = -b, \quad b \geq 0$$

Figure 3.38

Solution   The exterior angles at w_0, w_1, and w_2 are, respectively,

$$\theta_0 = \theta_2 = -\arctan(a/b) \in (-\pi/2, 0)$$

and

$$\theta_1 = -2\theta_0 - \pi \in (-\pi, 0).$$

Hence, $\alpha_0 = \alpha_2 = -\theta_0/\pi \in (0, 1/2)$ and $\alpha_1 = (2\theta_0/\pi) + 1 \in (0, 1)$. Set $\alpha_0 = \alpha_2 = \alpha$ so that $\alpha_1 = -2\alpha + 1$. With the choices $x_0 = -1$, $x_1 = 0$, and $x_2 = 1$ the derivative of the mapping function is

$$f'(z) = A(z + 1)^\alpha z^{-2\alpha + 1}(z - 1)^\alpha$$
$$= A(z^2 - 1)^\alpha z^{-2\alpha + 1}.$$

Once again we cannot find $f(z)$ explicitly, in general. The special case when $b = 0$ gives $\alpha = 1/2$ and

$$f'(z) = A(z^2 - 1)^{1/2}$$

so that

$$f(z) = A\{z\sqrt{1 - z^2} + \arcsin z\} + B.$$

A and B are found by the requirements that

$$-ia = f(-1) = A\{-\pi/2\} + B$$
$$0 = f(0) \quad = B$$

Thus, the mapping is

$$f(z) = i\,\frac{2a}{\pi}\,\{z\sqrt{1 - z^2} + \arcsin z\}. \qquad \square$$

Example 9   Find the Schwarz–Christoffel mapping of the upper half-plane U onto the region in Figure 3.39.

Solution   We find this mapping as the limit of mapping onto the region shown in Figure 3.40.

We select $x_0 = -1$, $x_1 = 0$, and $x_2 = 1$ and think of b as being very large. Thus, the exterior angles at w_0 and w_2 are almost $-\pi/2$ and the exterior angle at w_1 is almost π. Hence, the derivative of the mapping function onto the desired region will be

$$f'(z) = A(z + 1)^{1/2}z^{-1}(z - 1)^{1/2}$$
$$= A\left(\frac{z^2 - 1}{z^2}\right)^{1/2}.$$

Hence,

$$f(z) = A\{\sqrt{z^2 - 1} + \arcsin(1/z)\} + B.$$

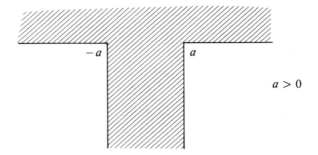

Figure 3.39

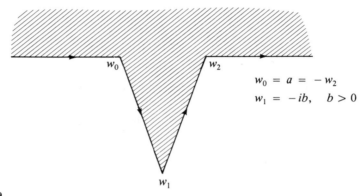

Figure 3.40

We find the constants A and B from the conditions

$$-a = f(-1) = A\{-\pi/2\} + B$$
$$a = f(1) \quad = A(\pi/2) \quad + B.$$

Consequently,

$$f(z) = \frac{2a}{\pi}\{\sqrt{z^2 - 1} + \arcsin(1/z)\}.$$

Figure 3.41 shows some of the streamlines and some of the equipotentials for the flow through this region. This models the flow over a deep hole in a streambed.   □

Example 10   Find the mapping of the upper half-plane U onto the region R obtained by deleting from U the segment $\{it: 0 \leqslant t \leqslant a\}$ (Fig. 3.42).

Solution   We obtain the given region as a limit of the regions shown in Figure 3.43 as $\varepsilon \to 0$. The exterior angle at w_0 is almost $\pi/2$, as is the exterior angle at w_2; the exterior angle at w_1 is almost $-\pi$. Hence, as $\varepsilon \to 0$, we find that the

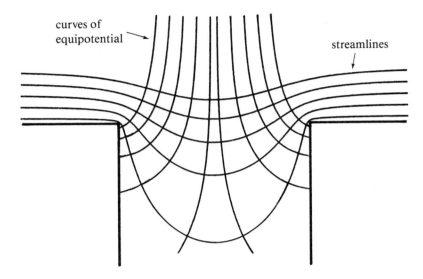

Figure 3.41   The flow over a deep hole in a streambed. The streamlines are

$$\Gamma_c(t) = \frac{2a}{\pi}\left[\sqrt{(t+ic)^2-1} + \arcsin\left(\frac{1}{t+ic}\right)\right], \quad -\infty < t < \infty; \; c > 0.$$

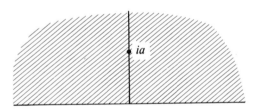

Figure 3.42

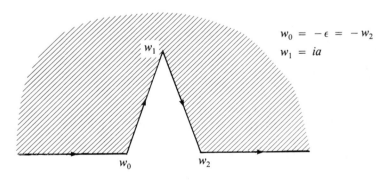

$$w_0 = -\epsilon = -w_2$$
$$w_1 = ia$$

Figure 3.43

derivative of the desired mapping is

$$f'(z) = A(z + 1)^{-1/2}z(z - 1)^{-1/2}$$

$$= A\,\frac{z}{(z^2 - 1)^{1/2}},$$

where, as before, we have selected $x_0 = -1$, $x_1 = 0$, and $x = 1$. Hence,

$$f(z) = A(z^2 - 1)^{1/2} + B.$$

A and B are found by the conditions

$$0 = f(-1) = B, \qquad ia = f(0) = Ai.$$

Hence, the mapping is

$$f(z) = a(z^2 - 1)^{1/2}.$$

Some of the streamlines for the flow through this region are shown in Figure 3.44. The streamlines are given by $z(t) = \sqrt{(t + ic)^2 - 1}$, $-\infty < t < \infty$. This models the flow over an obstacle in a streambed.   □

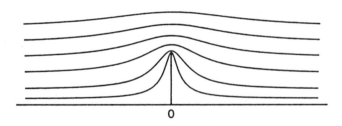

0

Figure 3.44

Example 11 Show that the function $\phi(z) = z + e^z$ maps the strip $\{-\pi < y < \pi\}$ one-to-one and onto the entire complex plane minus the two rays

$$R_1 = \{\sigma + i\pi : \sigma \leqslant -1\}$$
$$R_2 = \{\sigma - i\pi : \sigma \leqslant -1\}.$$

Show further that ϕ maps the two lines $y = \pi$ and $y = -\pi$ onto the two rays R_1 and R_2, respectively, covering each point twice (except the points $-1 \pm i\pi$, which are covered once) (Fig. 3.45).

Solution Since $\phi(\bar{z}) = \overline{\phi(z)}$ we will consider only the region where $0 \leqslant y < \pi$. The function $g(z) = e^z$ maps the strip $\{x + iy : 0 < y < \pi\}$ onto the upper half-plane U so we shall now show by use of the Schwarz–Christoffel transformations that the function $f(z) = z + \text{Log } z$ maps U onto the region Ω

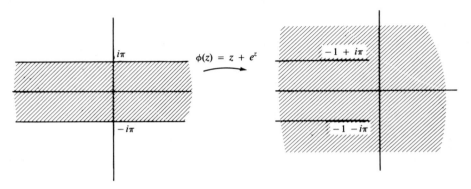

Figure 3.45

obtained by removing from U the ray R_1 (Fig. 3.46). The composition $\phi(z) = f(g(z)) = z + e^z$ will then map the strip $\{x + iy: 0 < y < \pi\}$ onto Ω, as desired.

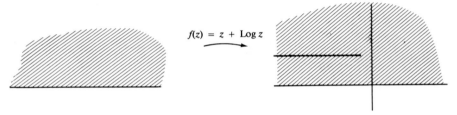

Figure 3.46

Consider the region shown in Figure 3.47. As $R \to \infty$, the region shown approaches the region Ω. We select $x_0 = -1$ and $x_1 = 0$ and note that the exterior angle at w_0 is almost $-\pi$ and the exterior angle at w_1 is almost π, if R is very big. Hence, as $R \to \infty$, we find that

$$f'(z) = A(z + 1)^1 z^{-1}$$
$$= A(1 + 1/z).$$

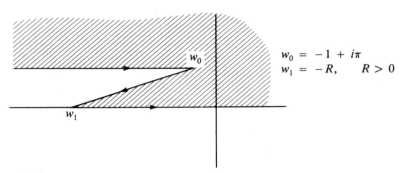

Figure 3.47

Thus, $f(z) = A(z + \text{Log } z) + B$. To determine A and B, we note that

$$\pi = \text{Im } f(x) = \text{Im}(Ax + \log|x| + i\pi + B)$$

if $x \leqslant -1$. Thus,

$$0 = x \text{ Im } A + \text{Im } B, \quad x \leqslant -1,$$

which implies that A and B are real. Furthermore,

$$-1 + i\pi = f(-1) = A(-1 + i\pi) + B,$$

which then gives $A = 1$ and $B = 0$. $\qquad\qquad\square$

Example 12 Find the Schwarz–Christoffel transformation of the upper half-plane U onto the region pictured in Figure 3.48.

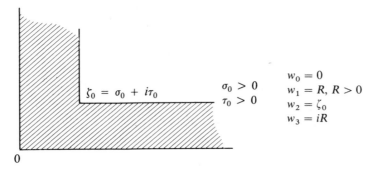

$$\zeta_0 = \sigma_0 + i\tau_0$$

$$\sigma_0 > 0$$
$$\tau_0 > 0$$

$$w_0 = 0$$
$$w_1 = R, R > 0$$
$$w_2 = \zeta_0$$
$$w_3 = iR$$

0

Figure 3.48

Solution Once again we find the desired mapping as a limit of mappings onto regions which "grow" to the desired region. We start by considering the region pictured in Figure 3.49. The exterior angle at w_0 is $\pi/2$; the exterior angles at w_1, w_2, and w_3 are nearly π, $-\pi/2$, and π, respectively. We let w_0

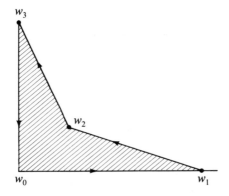

w_3

w_2

w_0 $\qquad\qquad w_1$

Figure 3.49

correspond to -1, w_1 to 0, w_2 to some point a^2, $a > 0$, and w_3 to ∞. The derivative of the desired mapping is then

$$f'(z) = A(z + 1)^{-1/2} z^{-1} (z - a^2)^{1/2},$$

where A is a constant. Our first task is to find a^2 in terms of σ_0 and τ_0, the coordinates of the corner at w_0. Consider Figure 3.50.

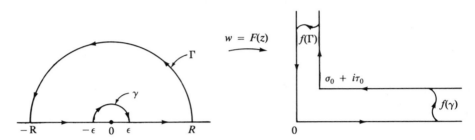

Figure 3.50

Let z lie on the semicircle $\{\varepsilon e^{i\theta}: 0 \leqslant \theta \leqslant \pi\}$, ε small; as z traverses this semicircle clockwise, the imaginary part of $f(z)$ goes from 0 to τ_0. Hence

$$\tau_0 = \text{Im} \int f'(z)\, dz = \text{Im}\left\{ \int_\pi^0 A\left(\frac{\varepsilon e^{i\theta} - a^2}{\varepsilon e^{i\theta} + 1)}\right)^{1/2} i\, d\theta \right\}.$$

Letting $\varepsilon \to 0$ we find that

$$\tau_0 = \text{Im}\{Ai(ia)(-\pi)\} = \pi a\ \text{Im}\ A.$$

Likewise, if z traverses the semicircle $\{Re^{it}: 0 \leqslant t \leqslant \pi\}$ counterclockwise, then the real part of $f(Re^{it})$ decreases from σ_0 to 0. Hence,

$$-\sigma_0 = \text{Re} \int f'(z)\, dz = \text{Re}\left\{ \int_0^\pi A\left(\frac{Re^{it} - a^2}{Re^{it} + 1}\right)^{1/2} i\, dt \right\}.$$

Letting $R \to \infty$, we find that

$$-\sigma_0 = \text{Re}\{iA\pi\} = -\pi\ \text{Im}\ A.$$

Combining these two equations we find that

$$a = \tau_0/\sigma_0, \qquad \text{Im}\ A = \sigma_0/\pi, \qquad A = i\sigma_0/\pi.$$

and

$$f'(z) = i\,\frac{\sigma_0}{\pi}\,\frac{1}{z}\left(\frac{z - a^2}{z + 1}\right)^{1/2}.$$

Thus, the derivative of the Schwarz–Christoffel transformation is completely known and all that remains is to integrate.

To find f we may either consult integral tables or proceed as shown below. Let

$$w = \left(\frac{z+1}{z-a^2}\right)^{1/2}$$

and define g by the relation $g(w) = f(z)$. The chain rule then gives

$$\frac{dg}{dw} = \frac{df}{dz}\frac{dz}{dw} = A\,\frac{1}{z}\left(\frac{z-a^2}{z+1}\right)^{1/2}\frac{dz}{dw}; \quad A = \frac{i\sigma_0}{\pi}.$$

However,

$$z = \frac{1+a^2w^2}{w^2-1}$$

so that

$$\frac{dg}{dw} = A\,\frac{w^2-1}{1+a^2w^2}\frac{1}{w}\frac{2w(1+a^2)(-1)}{(w^2-1)^2}$$

$$= 2A(1+a^2)\,\frac{1}{(1-w^2)(1+a^2w^2)}.$$

This last expression yields immediately to the partial-fraction technique of elementary calculus:

$$\frac{2(1+a^2)}{(1-w^2)(1+a^2w^2)} = \frac{1}{1-w} + \frac{1}{1+w} + \frac{a^2}{1-iaw} + \frac{a^2}{1+iaw}.$$

Integration then gives

$$f(z) = g(w) = A\left[\operatorname{Log}\left(\frac{1+w}{1-w}\right) - ia\,\operatorname{Log}\left(\frac{1+iaw}{1-iaw}\right)\right] + B,$$

where $w = ((z+1)/(z-a^2))^{1/2}$, $a = \tau_0/\sigma_0$, $A = i\sigma_0/\pi$. B is zero since $g(0) = f(1) = 0$. □

Exercises for Section 3.5

In Exercises 1 to 6 use conformal maps or combinations of conformal maps that you already know such as linear fractional transformations, powers, roots, sin z, log z, etc., to find a one-to-one analytic function mapping the given region D onto the upper half-plane U. In each case tell as much as possible about the image of the boundary of D under the mapping.

1. $D = \{z = x + iy\colon x \text{ and } y \text{ are positive}\}$

2. $D = \{z\colon |\operatorname{Arg} z| < \alpha\}, \alpha \leqslant \pi$

3. $D = \{z = x + iy\colon |y - 1| < 2\}$

4. $D = \{z\colon |z - z_0| < r_0\}$

5. $D = \{z: |z| > R \text{ and Im } z > 0\}$

6. $D = \{z: z \notin (-\infty, -1] \cup [1, \infty)\}$

7.   Find a conformal map of the infinite strip $0 < y < \pi$ onto the semi-infinite strip $\{\sigma + i\tau: 0 < \sigma < 1, \quad \tau > 0\}$.

8.   Find a conformal map of the region between the two circles $|z| = 1$ and $|z - 1/2| = 1/2$ onto the disc $|w| < 1$.
      (**Hint:** First apply $z \rightarrow 1/(z - 1)$.)

9.   Find a conformal map of the first quadrant onto the strip $|\text{Im } w| < 1$.

In Exercises 10 to 14 find a Schwarz-Christoffel transformation of the upper half-plane onto the given domain D.

10. $D = \{z: 0 < \text{Arg } z < 4\pi/3\}$

11. $D = \{z = x + iy: -\pi/2 < x < \pi/2 \text{ and } y > 0\}$

12. D is pictured in Figure 3.51a. (**Hint:** Map onto the region in Figure 3.51b and then let $b \rightarrow 0$.)

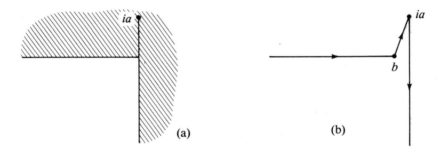

Figure 3.51

13. D is pictured in Figure 3.52

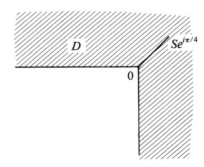

Figure 3.52

14. *D* is pictured in Figure 3.53. (**Hint:** Obtain this region as the limit of the regions in Figure 3.54.) Evaluate the constants as in Example 12.

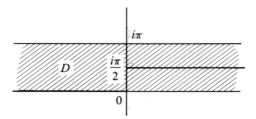

Figure 3.53

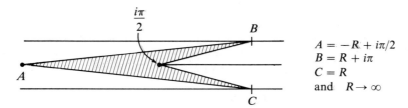

$A = -R + i\pi/2$
$B = R + i\pi$
$C = R$
and $R \to \infty$

Figure 3.54

15. *D* is pictured in Figure 3.55. (**Hint:** Obtain this region as the limit of the regions shown in Figure 3.56.) Let -1 correspond to ia, 0 to $-R$; then let $R \to \infty$ to find $f'(z)$. f may be obtained as shown in Example 12.

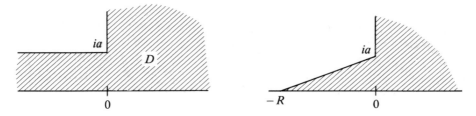

Figure 3.55   **Figure 3.56**

16. Let C_1 and C_2 be two circles with C_2 inside C_1. Show that there is a linear fractional transformation ϕ such that $\phi(C_1)$ and $\phi(C_2)$ are concentric circles. (**Hint:** An initial translation, dilation, and rotation reduce the result to the case when C_1 is the circle $|z| = 1$ and C_2 is centered on the positive real axis. Show that there is an $\alpha \in (0, 1)$ such that $\psi(z) = (z - \alpha)/(1 - \alpha z)$ carries C_2 to a circle centered at the origin. The key is that ψ will carry a diameter of C_2 to a diameter of $\psi(C_2)$.)

17.   Suppose that f is a one-to-one analytic function mapping the disc $|z| < 1$ onto a *bounded* domain D. Show that the **area** of D is given by

$$A(D) = \iint\limits_{|z|<1} |f'(z)|^2 \, dx \, dy, \quad z = x + iy$$

You will need to know the "change-of-variables" formula for double integrals.

18.   Using the formula in Exercise 17, show that the area of D is given by

$$A(D) = \pi \sum_{n=1}^{\infty} n |a_n|^2,$$

where $\sum_{n=0}^{\infty} a_n z^n$ is the power series for f in $|z| < 1$.

19.   Let D be a simply connected domain and ϕ the Riemann mapping of D onto the disc $\{w: |w| < 1\}$. Let $\{z_k\}$ be a sequence of points in D with $z_k \to p$, p in the boundary of D. Show that the sequence $\{|\phi(z_k)|\}$ converges to 1 (see Exercise 14 in Section 4).

Further reading

A lovely treatment of the geometry of complex numbers including an extensive coverage of linear fractional transformations can be found in Schwerdtfeger, H. *Geometry of complex numbers.* N.Y.: Dover, 1979. Linear fractional transformations are frequently called Möbius transformations after August Ferdinand Möbius, 1790–1868, who studied them.

More on conformal mapping, especially of multiply connected domains, and a development of the theory of analytic functions on such domains, is in Nehari, Z. *Conformal mapping.* N.Y.: Dover, 1975. Another possible source is Bieberbach, L. *Conformal mapping.* N.Y.: Chelsea, 1953. The references listed at the end of Chapter 2 also all have more to say on the topics of this chapter and are good places to seek further information. The books by Ahlfors, Burckel, Hille, and Nehari all contain a proof of the Riemann Mapping Theorem. That in Burckel is the somewhat "constructive" proof of the theorem developed by P. Koebe. To understand it, or any of the proofs of this theorem, you must go into certain aspects of function theory not touched on here.

Further information on the stability of solutions to a linear system of ordinary differential equations can be found in Sanchez, D. *Ordinary differential equations.* San Francisco: W.H. Freeman, 1968, or in Lefschetz, S. *Differential equations: geometric theory.* N.Y.: Dover, 1977.

4

Analytic and harmonic functions in applications

4.1   Harmonic functions

A continuous complex-valued function u with continuous first and second partial derivatives on an open set D in the plane is **harmonic** on D if it satisfies **Laplace's equation**

$$\Delta u = \frac{\partial^2 u}{\partial x^2} + \frac{\partial^2 u}{\partial y^2} = 0 \text{ on } D.$$

Harmonic functions are critical components in the solution of numerous physical problems, such as the flow of water through an underground aquifer, steady-state temperature distribution, electrostatic field intensity, elasticity, and diffusion. Many of these physical problems are discussed in Section 2 of this chapter and techniques for their solution are derived in Sections 3 and 4. In this section we review the intimate connection between harmonic and analytic functions and use this connection to establish a number of fundamental properties of harmonic functions.

Suppose first that f is an analytic function on a domain D; we write $f = u + iv$. Both u and v are harmonic, since

$$\Delta u = \frac{\partial}{\partial x}\left(\frac{\partial u}{\partial x}\right) + \frac{\partial}{\partial y}\left(\frac{\partial u}{\partial y}\right)$$

$$= \frac{\partial}{\partial x}\left(\frac{\partial v}{\partial y}\right) + \frac{\partial}{\partial y}\left(\frac{-\partial v}{\partial x}\right)$$

$$= \frac{\partial^2 v}{\partial x\,\partial y} - \frac{\partial^2 v}{\partial y\,\partial x} = 0,$$

and likewise, $\Delta v = 0$. Thus, the real and imaginary parts of an analytic function are harmonic. Moreover, the converse is true, at least locally. If $u(z)$ is a real-valued harmonic function on a disc, $\{z: |z - z_0| < r\}$, then set

$$f(x + iy) = u(x, y) - i \int_{x_0}^{x} \frac{\partial u}{\partial y}(t, y)\, dt + i \int_{y_0}^{y} \frac{\partial u}{\partial x}(x_0, s)\, ds. \qquad (1)$$

We then have

$$\frac{\partial f}{\partial x} = \frac{\partial u}{\partial x} - i \frac{\partial u}{\partial y}.$$

Furthermore,

$$\frac{\partial f}{\partial y} = \frac{\partial u}{\partial y} - i \int_{x_0}^{x} \frac{\partial^2 u}{\partial y^2}(t, y)\, dt + i \frac{\partial u}{\partial x}(x_0, y)$$

$$= \frac{\partial u}{\partial y} + i \int_{x_0}^{x} \frac{\partial^2 u}{\partial x^2}(t, y)\, dt + i \frac{\partial u}{\partial x}(x_0, y)$$

$$= \frac{\partial u}{\partial y} + i \left\{ \frac{\partial u}{\partial x}(x, y) - \frac{\partial u}{\partial x}(x_0, y) \right\} + i \frac{\partial u}{\partial x}(x_0, y)$$

$$= \frac{\partial u}{\partial y} + i \frac{\partial u}{\partial x} = i \frac{\partial f}{\partial x}.$$

Hence, if we write $f = u + iv$ [that is, $v(x, y)$ is the two terms in (1) involving the integrals] then

$$\frac{\partial v}{\partial x} = -\frac{\partial u}{\partial y} \qquad \text{and} \qquad \frac{\partial v}{\partial y} = \frac{\partial u}{\partial x}.$$

These are the Cauchy–Riemann equations, so f is analytic on this disc and $u = \operatorname{Re} f$ there. We summarize this information here.

> A real-valued function u on a domain D is harmonic
> on D if and only if on every disc in D,   (2)
> u is the real part of some analytic function.

The reader is warned that a harmonic function is locally the real part of some analytic function but it need not be the real part of an analytic function on all of D. An example of such a function is

$$u(x, y) = \tfrac{1}{2} \log(x^2 + y^2).$$

u is harmonic on the annulus, $0 < r < x^2 + y^2 < R$, but, because locally $u(x, y)$ is the real part of log z, there is no analytic function on all of the annulus whose real part equals u.

The function v that is the imaginary part of f (that is, $u + iv$ is analytic) is determined completely by u except for adding a purely real constant. For if $u + iv_1$ is also analytic, then $(u + iv) - (u + iv_1)$ is again analytic. But this difference is just $i(v - v_1)$ and so $i(v - v_1)$ must be constant since an analytic function with purely imaginary values is constant. Such a function v is called a **harmonic conjugate of** u.

Example 1   Find a harmonic conjugate of (a) $2xy$ (b) $e^x \cos y$ (c) $(e^x + e^{-x}) \cos y$

Solution   One way to proceed is to use the rule for v given in (1) but it is just as easy to use the Cauchy–Riemann equations:

$$\frac{\partial v}{\partial x} = -\frac{\partial u}{\partial y} \qquad \frac{\partial v}{\partial y} = \frac{\partial u}{\partial x}.$$

For (a):

$$\frac{\partial v}{\partial x} = -\frac{\partial u}{\partial y} = -2x;$$

thus, $v(x, y) = -x^2 + p(y)$. However,

$$\frac{\partial v}{\partial y} = \frac{\partial u}{\partial x} = 2y$$

so that $p'(y) = 2y$. Hence, $v(x, y) = -x^2 + y^2 + c$, where c is a constant.
     For (b):

$$\frac{\partial v}{\partial x} = -\frac{\partial u}{\partial y} = e^x \sin y;$$

thus, $v(x, y) = e^x \sin y + p(y)$. Since

$$\frac{\partial v}{\partial y} = \frac{\partial u}{\partial x} = e^x \cos y,$$

we find that $p'(y) = 0$. Hence, $v(x, y) = e^x \sin y + c$.
     For (c):

$$\frac{\partial v}{\partial x} = -\frac{\partial u}{\partial y} = (e^x + e^{-x}) \sin y$$

and

$$\frac{\partial v}{\partial y} = \frac{\partial u}{\partial x} = (e^x - e^{-x}) \cos y.$$

Thus, $v(x, y) = (e^x - e^{-x}) \sin y$.   □

Maximum modulus

With (2) in mind we can look back at (4) of Section 2, Chapter 3 and conclude

> If u *is a real-valued nonconstant harmonic function*
> *on a domain* D, *then* u *has no local maximum and no*   (3)
> *local minimum in* D.

This is the **maximum principle** for harmonic functions. In fact, a variation of

(3) holds for any harmonic function u whether real- or complex-valued.

> If $u(x, y)$ is harmonic and nonconstant on a domain D,
> then $|u(x, y)|$ has no local maximum in D.   **(4)**

The proof of (4) is left as an exercise.

A theorem from real variables asserts that a real-valued continuous function on a closed and bounded set attains both its maximum and minimum. This, and (3), give the next result.

> Suppose that u is a real-valued harmonic function on a
> bounded domain D and that u is also continuous on the
> union of D and the boundary, B, of D. Then u attains   **(5)**
> both its maximum value and its minimum value over $D \cup B$
> on B. In particular, if $u \equiv 0$ on B, then $u \equiv 0$ on D, as well.

Mean value

Suppose that u is a complex-valued harmonic function on a domain D. Then we can write $u = u_1 + iu_2$, where both u_1 and u_2 are real-valued and harmonic* on D. Hence, in a disc $\{z : |z - z_0| < R\}$, which lies in D, we know that u_1 and u_2 are each the real part of some analytic functions, say f_1 and f_2. Thus, from (5) of Section 2, Chapter 3 we find that for any r, $0 \leqslant r < R$,

$$u(z_0) = u_1(z_0) + iu_2(z_0) = \operatorname{Re} f_1(z_0) + i \operatorname{Re} f_2(z_0)$$

$$= \operatorname{Re}\left\{\frac{1}{2\pi} \int_0^{2\pi} f_1(z_0 + re^{it})\, dt\right\} + i \operatorname{Re}\left\{\frac{1}{2\pi} \int_0^{2\pi} f_2(z_0 + re^{it})\, dt\right\}$$

$$= \frac{1}{2\pi} \int_0^{2\pi} \{\operatorname{Re} f_1(z_0 + re^{it}) + i \operatorname{Re} f_2(z_0 + re^{it})\}\, dt$$

$$= \frac{1}{2\pi} \int_0^{2\pi} u(z_0 + re^{it})\, dt.$$

This is the **mean-value property** of harmonic functions:

$$u(z_0) = \frac{1}{2\pi} \int_0^{2\pi} u(z_0 + re^{it})\, dt. \qquad (6)$$

In the exercises, the reader is asked to show how (6) can be used to derive (4). Property (6) actually characterizes harmonic functions in the sense that any continuous function that satisfies (6) must be harmonic; we shall not prove this here.

---

* The fact that the real part of a harmonic function is again harmonic, but the real part of an analytic function is not analytic, points to a significant property of the Laplacian operator Δ.

Composition

Suppose that u is a harmonic function for z in some domain D. If $\phi(\zeta) = z$ is an *analytic* function of ζ, which maps a domain Ω into D, then

$$w(\zeta) = u(\phi(\zeta))$$

is a harmonic function of ζ, as ζ varies over Ω (Fig. 4.1). This can be shown in several ways; one way is the following. Let ζ_0 be a point of Ω and $z_0 = \phi(\zeta_0)$. Let U be a small disc centered at z_0 in D and let V be a small disc centered at ζ_0 with the property that $\phi(\zeta) \in U$ for all $\zeta \in V$; this is possible because ϕ is continuous at ζ_0. We assume that u is real-valued; thus, there is an analytic function f in the disc U with $u = \operatorname{Re} f$; the function $g(\zeta) = f(\phi(\zeta))$ is then analytic in the disc V and $\operatorname{Re} g = \operatorname{Re}(f \circ \phi) = u \circ \phi = w$ and so $w(\zeta)$ is harmonic on V, being the real part of the analytic function $g(\zeta)$.

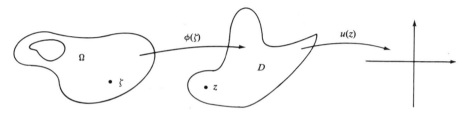

Figure 4.1

A "picture" of the graph of a harmonic function

Suppose D is a bounded domain and $u(x, y)$ is a continuously differentiable function on D whose values on B, the boundary of D, are fixed: $u(x, y) = f(x, y)$ if $(x, y) \in B$ for a given real-valued function $f(x, y)$. The **energy integral** of u is

$$E(u) = \int\int_D \left\{ \left(\frac{\partial u}{\partial x}\right)^2 + \left(\frac{\partial u}{\partial y}\right)^2 \right\} dx\, dy.$$

This integral has a very simple physical interpretation: it is the **potential energy** of a thin membrane in the shape of the surface S in three-dimensional space, which is the graph of the function $u(x, y)$:

$$S = \{(x, y, u(x, y)): x + iy \in D\}.$$

We now pose the problem of minimizing $E(u)$ over all functions $u(x, y)$ with $u = f$ on B; that is, we wish to find that surface with a prescribed edge whose potential energy is minimal. Suppose that u_0 produces this minimum:

$$E(u_0) \leqslant E(u), \text{ all } u \text{ with } u = f \text{ on } B.$$

Let v be any smooth function on D, which is zero on B. Then $u_0 + \varepsilon v$ equals f

on B for any number ε and because u_0 has the minimum energy integral we find that

$$E(u_0) \leqslant E(u_0 + \varepsilon v)$$

$$= E(u_0) + 2\varepsilon \int\int_D \left\{ \frac{\partial u_0}{\partial x} \frac{\partial v}{\partial x} + \frac{\partial u_0}{\partial x} \frac{\partial v}{\partial y} \right\} dx\ dy + \varepsilon^2 E(v).$$

Cancel the $E(u_0)$ term from both sides, divide by ε, and then let $\varepsilon \to 0$. If ε is always positive, we find that

$$\int\int_D \left\{ \frac{\partial u_0}{\partial x} \frac{\partial v}{\partial x} + \frac{\partial u_0}{\partial y} \frac{\partial v}{\partial y} \right\} dx\ dy \geqslant 0$$

and just the reverse inequality if ε is always negative. Thus, we must have

$$\int\int_D \left\{ \frac{\partial u_0}{\partial x} \frac{\partial v}{\partial x} + \frac{\partial u_0}{\partial y} \frac{\partial v}{\partial y} \right\} dx\ dy = 0.$$

We now suppose that D, B, v, and u_0 satisfy the hypotheses of Green's Theorem and we apply Green's identity, from Exercise 13, Section 6, Chapter 1. Thus,

$$0 = \int\int_D \left\{ \frac{\partial u_0}{\partial x} \frac{\partial v}{\partial x} + \frac{\partial u_0}{\partial y} \frac{\partial v}{\partial y} \right\} dx\ dy$$

$$= \int_B v\ \frac{\partial u_0}{\partial n}\ ds - \int\int_D (v)(\Delta u_0)\ dx\ dy.$$

However, $v(x, y) = 0$ if $x + iy$ is in B, so we finally have

$$0 = \int\int_D (v)(\Delta u_0)\ dx\ dy$$

for any (smooth) function $v(x, y)$, which is zero on B. This, of course, implies that $\Delta u_0 = 0$ on D (see Exercise 17). That is, the energy integral is minimized exactly by the function u_0, which is the unique harmonic function on D with values f on B.

Furthermore, we can visualize the graph of the function $u_0(x, y)$ as a thin membrane stretched tightly across a wire frame; the frame is the graph of the function $f(x, y)$—the set of points in three-dimensional space $\{(x, y, f(x, y)): x + iy \in B\}$. This thin membrane can be physically realized as a soap film obtained by dipping an appropriately shaped wire frame into a soap solution (Fig. 4.2).

Exercises for Section 4.1

1.   Decide if each of the following functions u is harmonic; if so, find a harmonic conjugate
       (a) $u(x, y) = x^4 - 6x^2 y^2 + y^4$ (b) $u(x, y) = [\cos(2xy)]\exp(x^2 - y^2)$

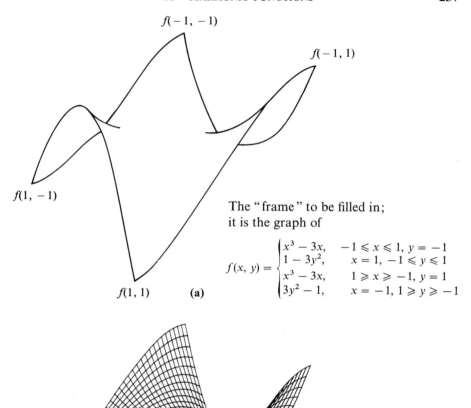

$f(-1, -1)$

$f(-1, 1)$

$f(1, -1)$

The "frame" to be filled in; it is the graph of

$$f(x, y) = \begin{cases} x^3 - 3x, & -1 \leqslant x \leqslant 1,\ y = -1 \\ 1 - 3y^2, & x = 1,\ -1 \leqslant y \leqslant 1 \\ x^3 - 3x, & 1 \geqslant x \geqslant -1,\ y = 1 \\ 3y^2 - 1, & x = -1,\ 1 \geqslant y \geqslant -1 \end{cases}$$

$f(1, 1)$ **(a)**

(b)

The graph of $x^3 - 3xy^2$ which fills in the frame

Figure 4.2

(c) $u(x, y) = \dfrac{x^2}{x^2 + y^2 + 1}$ (d) $u(x, y) = \sin(x^2 - y^2)\cosh(2xy)$

(e) $u(x, y) = \arctan x$

2.   Let f be analytic and never zero in a domain D; show that $\log|f|$ is harmonic in D.

3.   Use Exercise 2 to show that $\log r$ is harmonic if $0 < r < R$.

4.   Use the chain rule to derive this expression for the **Laplace operator in polar coordinates**

$$\Delta u = \frac{\partial^2 u}{\partial x^2} + \frac{\partial^2 u}{\partial y^2} = \frac{1}{r}\frac{\partial u}{\partial r} + \frac{\partial^2 u}{\partial r^2} + \frac{1}{r^2}\frac{\partial^2 u}{\partial \theta^2}.$$

5.   Use the result in Exercise 4 to answer these two questions
      (a) When is a function of θ alone harmonic?
      (b) When is a function of r alone harmonic?

6.   Suppose that u is harmonic on a domain D and $u(x, y) = 0$ for all points $x + iy$ in some open set V in D. Show that $u = 0$ on all of D.

7.   The function $u(x, y) = 1/(x^2 + y^2 + 1)$ is not harmonic since $\Delta u \neq 0$. Explain why the maximum principle also shows that $u(x, y)$ is not harmonic on any open set containing the origin.

8.   Use formula (6) to establish (4). (**Hint:** If $|u(z_0)|$ is greater than $|u(z)|$ for $|z_0 - z| < \delta$, then

$$|u(z_0)| = \left| \frac{1}{2\pi} \int_0^{2\pi} u(z_0 + re^{it})\, dt \right|, \quad 0 < r < \delta$$

$$\leqslant \frac{1}{2\pi} \int_0^{2\pi} |u(z_0 + re^{it})|\, dt \leqslant \frac{1}{2\pi} \int_0^{2\pi} |u(z_0)|\, dt$$

$$= |u(z_0)|.$$

What is wrong with this?)

9.   Show that the function $(1 - r^2)/(1 - 2r\cos\theta + r^2)$ is harmonic for $0 \leqslant r < 1, 0 \leqslant \theta < 2\pi$.

10.   Show that the second-degree polynomial in x and y

$$p(x, y) = ax^2 + bxy + cy^2 + dx + ey + f$$

$(a, b, \ldots, f$ are constants) is harmonic if and only if $a + c = 0$.

11.   Show that the cubic polynomial in x and y

$$q(x, y) = Ax^3 + Bx^2y + Cxy^2 + Dy^3 + Ex^2$$

$$+ Fxy + Gy^2 + Hx + Iy + J$$

is harmonic if and only if

$$3A + C = 0,$$
$$3D + B = 0,$$
$$E + F = 0.$$

12.   Suppose that u is a real-valued harmonic function on a domain D and
        that u^2 is also harmonic on D. Conclude that u is constant. Show that
        the same conclusion holds if u^n is harmonic for some integer n, $n \geqslant 2$.

13.   Let $u(x, y)$ be harmonic and bounded for all $x + iy$ with $y > 0$.
        Suppose further that $u(x, y)$ is continuous for all $x + iy$ with $y \geqslant 0$
        and that $u(x, 0) = 0$ for all x. Show that $u(x, y) = 0$ for all $x + iy$. By
        means of an example show that the hypothesis that $u(x, y)$ is bounded
        cannot be omitted.

14.   Find the harmonic conjugate v on the disc $\{z : |z| < 1\}$ of $u(z) =$
        $\mathrm{Arg}[(1 + z)/(1 - z)]$. Conclude that u is bounded but v is unbounded.

15.   Let $u(x, y)$ be harmonic and real-valued in the disc $|z - x_0| < \delta$, x_0
        real, and suppose that $u(x, 0) = 0$, $x_0 - \delta < x < x_0 + \delta$. Show that
        $u(x, y)$ satisfies the relation $u(x, y) = -u(x, -y)$, $z = x + iy$ and
        $|z - x_0| < \delta$. (**Hint:** Let $f(z)$ be the analytic function on the disc
        $|z - x_0| < \delta$ with $f(x_0) = 0$ and $u = \mathrm{Re}\ f$. Show that $f(z)$ satisfies
        the relation $\overline{f(\bar{z})} = -f(z)$.)

16.   Let u be a real-valued harmonic function that is bounded above on
        the whole plane; that is, $u(z) \leqslant M$ for all z. Show that u is constant.

Vanishing integrals

17.   Suppose that H is a continuous bounded real-valued function on a
        domain D such that $\iint Hv\ dx\ dy = 0$ for every smooth real-valued
        function v on D, which is zero on B, the boundary of D. Conclude that
        $H = 0$ on D. (**Hint:** Suppose that $H(z_0) > 0$ at some point $z_0 \in D$.
        Then there are [small] positive numbers ε and δ such that $H(z) \geqslant \delta$ if
        $|z - z_0| \leqslant \varepsilon$. [Why?] Define

$$v(z) = \begin{cases} (\varepsilon^2 - |z - z_0|^2)^2, & \text{if } |z - z_0| \leqslant \varepsilon \\ 0, & \text{if } |z - z_0| > \varepsilon, z \in D. \end{cases}$$

Show that for this choice of v we have $\iint vH\ dx\ dy > 0$.)

4.2   Harmonic functions
as solutions to physical problems

It is a matter of considerable significance that the theory of functions of a
complex variable can be applied to a great number of "real-world" problems.
That is, the solution to the physical problem can be formulated as a function
that, by means of the physical laws operating within the given context, can be
shown to be harmonic or analytic. We have already seen that this is the case
in the investigation of an irrotational, sourceless flow of an ideal fluid, where

we learned that the complex conjugate of the flow must be an analytic function; indeed it must be the derivative of an analytic function. Once the solution of the given problem is known to be analytic or harmonic, then the arsenal of results we have collected to this point (or that we will develop later) can be applied and the solution can be analyzed or given explicitly.

This section contains an analysis of three types of physical problems: steady-state heat flow; electrostatic fields; and the flow of an ideal fluid, including Bernoulli's Law and a derivation of the lift generated by a flow. There is also a short discussion of diffusion and elasticity.

Steady-state temperature distribution

We imagine a thin plate across which there is a flow of heat; we assume that the temperature at a point $z = x + iy$ in the plate depends on the location but not on time—that is, the temperature is in steady state. Let $T(x, y)$ denote the temperature at a point $z = x + iy$ in the plate; we shall show that T is a harmonic function of x and y.

We envision a (very) small rectangle in the plate with sides of length h and k and lower left corner at $z = x + iy$ (Fig. 4.3). In a unit time period, there is a flux or flow of heat in the amount of $-ck(\partial T/\partial x)(x, y)$ across the left-hand vertical side of the rectangle. Here c is a constant that incorporates the **thermal conductivity** of the material. Likewise, the flux across the right-hand vertical side is $-ck(\partial T/\partial x)(x + h, y)$. The difference

$$-ck\left[\frac{\partial T}{\partial x}(x, y) - \frac{\partial T}{\partial x}(x + h, k)\right]$$

represents the net flux of heat out of the rectangle in the x-direction. Similarly,

$$-ch\left[\frac{\partial T}{\partial y}(x, y) - \frac{\partial T}{\partial y}(x, y + k)\right]$$

represents the net flux of heat out of the rectangle in the y-direction. Of course, both these expressions are only approximately correct but the accuracy of the approximation increases as h and k approach zero. Since there is no net flux of heat out of the rectangle (because there are no heat sources and no heat sinks within the region) we find that

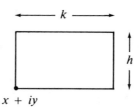

Figure 4.3

$$0 = ckh \left\{ \frac{\frac{\partial T}{\partial x}(x + h, y) - \frac{\partial T}{\partial x}(x, y)}{h} + \frac{\frac{\partial T}{\partial y}(x, y + k) - \frac{\partial T}{\partial y}(x, y)}{k} \right\}$$

and after dividing by kh and letting h and k both approach zero we obtain

$$0 = \frac{\partial^2 T}{\partial x^2} + \frac{\partial^2 T}{\partial y^2},$$

which says, as we wanted, that $T(x, y)$ is a harmonic function of x and y.

Electrostatics

Most electrostatic phenomena involve three independent space variables but certain important situations are basically just two-dimensional. For instance, a very long charged wire, a group of parallel charged wires, and a very long charged cylinder, each produces an electric field in the surrounding space. This electric field, it turns out, is the same in each plane that is perpendicular to the wires or cylinder (we shall derive this fact) and so by considering such a cross-sectional plane we can use complex variables to understand the electric field in this plane and hence in all space.

The basic tool is **Coulomb's Law**, which asserts that two charged particles of charges q and Q, respectively, exert a force on one another whose magnitude is

$$CqQ/r^2,$$

where r is the distance between the particles; the force is directed along the line through the two points and is attractive if q and Q have opposite signs and is repulsive if q and Q have the same sign. C is a constant depending on the choice of units; we shall take $C = 1$.

Imagine now a very long, straight wire carrying a uniform charge of λ coulombs per unit length. We may suppose that the wire is perpendicular to the usual xy-plane (the complex plane) and cuts through the plane at the point z_1. Let r be the distance from z to z_1: $r = |z - z_1|$ and let h be the coordinate along the wire (Fig. 4.4).

The electrostatic force at z, due to the wire, has two components—one is in the xy-plane and is directed toward the wire and the other is perpendicular to this plane. The first we shall temporarily term the planar force and the second the vertical force. We envision the wire as being made up of many small pieces of equal length dh, each therefore carrying a total charge of $\lambda\, dh$. The magnitude of the planar and vertical forces at z due to this short segment are, respectively,

$$dm_p = q \cos \theta \, \frac{\lambda\, dh}{s^2}$$

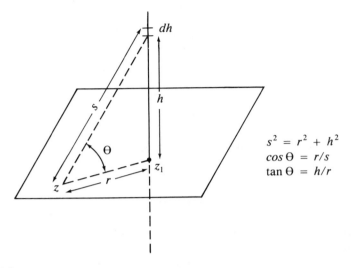

$$s^2 = r^2 + h^2$$
$$\cos \Theta = r/s$$
$$\tan \Theta = h/r$$

Figure 4.4

and

$$dm_v = q \sin \theta \, \frac{\lambda \, dh}{s^2}$$

by Coulomb's Law, where q is the charge at z and s is the distance from z to the small segment on the wire. Some simple trigonometry produces the relations

$$h = r \tan \theta$$
$$s = r \sec \theta.$$

Hence, $dh = r \sec^2 \theta \, d\theta$ and so

$$dm_p = q \cos \theta \, \frac{\lambda r \sec^2 \theta \, d\theta}{r^2 \sec^2 \theta} = q\lambda \cos \theta \, \frac{d\theta}{r}$$

and

$$dm_v = q\lambda \sin \theta \, \frac{d\theta}{r}.$$

If the wire is very long we see that θ varies from $-\pi/2$ to $\pi/2$ and so the total magnitude of the planar force at the point z is

$$m_p = \int_{-\pi/2}^{\pi/2} dm_p = q\lambda \, \frac{1}{r} \int_{-\pi/2}^{\pi/2} \cos \theta \, d\theta = 2q\lambda/r.$$

The vertical force is

$$m_v = \int_{-\pi/2}^{\pi/2} dm_v = \frac{q\lambda}{r} \int_{-\pi/2}^{\pi/2} \sin \theta \, d\theta = 0.$$

That is, the charged wire produces an electric field that is directed perpendicular to the axis of the wire and whose strength is directly proportional to the product of the charge at z and the charge per unit length on the wire and inversely proportional to the distance between the point and the wire. Note how this differs from the situation when just point charges are involved—there the force depends (inversely) on the *square* of the distance between the charges.

The electric force thus has magnitude $2\lambda q/|z - z_1|$ and is directed along the line joining z to z_1. The force is repulsive if q and λ have the same sign and attractive if q and λ have opposite signs. Hence, the force is

$$F(z) = 2\lambda q \frac{z_1 - z}{|z_1 - z|^2} = \frac{2\lambda q}{\bar{z}_1 - \bar{z}}.$$

The real-valued function

$$\phi(x, y) = -q\lambda \log[(x - x_1)^2 + (y - y_1)^2]$$

has the property that its gradient is F; that is,

$$\frac{\partial \phi}{\partial x} + i \frac{\partial \phi}{\partial y} = F.$$

Furthermore, a simple computation shows that $\phi(x, y)$ is a harmonic function of x and y; for

$$\frac{\partial^2 \phi}{\partial x^2} = q\lambda \frac{(x - x_1)^2 - (y - y_1)^2}{[(x - x_1)^2 + (y - y_1)^2]^2}$$

and

$$\frac{\partial^2 \phi}{\partial y^2} = q\lambda \frac{(y - y_1)^2 - (x - x_1)^2}{[(x - x_1)^2 + (y - y_1)^2]^2}.$$

Adding, we find that

$$\frac{\partial^2 \phi}{\partial x^2} + \frac{\partial^2 \phi}{\partial y^2} = 0.$$

The harmonic function ϕ, which carries all the information about the electric field, is called the **electrostatic potential** of the field F.

After understanding the situation of one wire it is quite easy to pass on to other, more complicated, configurations. First, suppose there are many long wires, all perpendicular to the xy-plane. Let $z_1, \ldots, z_n$ be the points at which the wires meet the xy-plane and let $\lambda_1, \ldots, \lambda_n$ be their charge densities, which we assume to be uniform all along the length of each. By superposition of forces we find the force at z to be

$$F(z) = 2q \sum_{j=1}^{n} \lambda_j \frac{z_j - z}{|z_j - z|^2} = 2q \sum_{j=1}^{n} \frac{\lambda_j}{\bar{z}_j - \bar{z}}.$$

Note that $\overline{F(z)} = 2q \sum_{j=1}^{n} (\lambda_j)/(z_j - z)$ is a rational function of z of degree n and has therefore, exactly n zeros. Since one zero is at "infinity," $\bar{F}(z)$ has at most

$n - 1$ zeros in the plane. Each of these zeros is an **equilibrium point** of the field since there the force vanishes. In the case that $\lambda_1, \ldots, \lambda_n$ are all positive integers, we also have

$$\bar{F}(z) = -2qP'(z)/P(z),$$

where $P(z)$ is the polynomial

$$P(z) = (z - z_1)^{\lambda_1} \cdots (z - z_n)^{\lambda_n}.$$

Thus, here we can apply the information studied in Section 1 of Chapter 3 about the location of the zeros of an analytic function. Also there is considerable information on the zeros of a polynomial; see Appendix 1.

The potential function for this field is

$$\phi(x, y) = -q \sum_{j=1}^{n} \lambda_j \log[(x - x_j)^2 + (y - y_j)^2];$$

$\phi(x, y)$ is the sum of harmonic functions and so is harmonic. Furthermore, as before, the gradient of ϕ is F.

Next, imagine a very long charged cylinder with its axis parallel to the vertical axis (see Figure 4.5). We assume the charge is uniform along any line in the cylinder parallel to the axis of the cylinder but not necessarily uniformly distributed in the cross-sectional cuts. Each cross-section is the same and is nothing but a curve γ; γ is not necessarily a closed curve. Suppose that $z(t)$ is a parametrization of γ, $a \leqslant t \leqslant b$. Let $u(z)$ be the strength of the charge at the point z on γ; that is, the line through z and parallel to the axis of the cylinder carries a uniform charge of $u(z)$ coulombs per unit length. To find the electric field produced by the charged curve γ (or rather by the cylinder) we imagine γ broken up into many small segments (see Fig. 4.6). Let ds be the arc length along γ

$$\begin{aligned} ds &= \sqrt{x'(t)^2 + y'(t)^2} \, dt \\ &= |z'(t)| \, dt. \end{aligned}$$

The total charge due to the segment of length ds is (about) $d\lambda = u(z) \, ds$ and so the field it produces is

$$\begin{aligned} dF &= \frac{2q}{\bar{z}(t) - \bar{z}} \, d\lambda \\ &= \frac{2q}{\bar{z}(t) - \bar{z}} \, u(z(t)) \, |z'(t)| \, dt. \end{aligned}$$

Thus, the electric field is just

$$F(z) = 2q \int_a^b \frac{u(z(t))}{\bar{z} - \bar{z}(t)} \, |z'(t)| \, dt.$$

The electrostatic potential of this field is

$$\phi(x, y) = -2q \int_a^b \log\{|z - z(t)|\} u(z(t)) \, |z'(t)| \, dt.$$

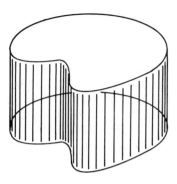

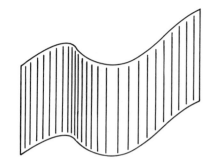

A section of a "closed" cylinder   An "open" cylinder

Figure 4.5

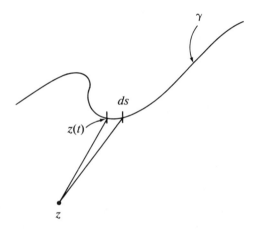

Figure 4.6

Equivalently,

$$\phi(z) = -2q \int_\gamma \{\log |z - w|\} u(w) \, d|w|,$$

where $d|w|$ is short for $|z'(t)| \, dt$; $u(w) \, d|w|$ is the **charge density** along the curve γ. If we write $d\rho(w)$ for the charge density, then (except for the factor of $2q$)

$$\phi(z) = \int_\gamma \log \frac{1}{|z - w|} \, d\rho(w).$$

This is an example of a **potential function**; the study of such functions and the potentials they produce is called **potential theory**.

Flow of an ideal fluid

The analysis of the flow of an ideal fluid provides a setting in which we can apply many of the theorems and formulas that we derived earlier, especially results from conformal mapping.

   We begin by envisioning an incompressible, frictionless fluid flowing through a region in space. If all cross-sections perpendicular to some axis are the same, then the flow can be described by just two space variables, which, as usual, we take to be the regular xy coordinates. Situations such as this arise, for example, in the flow of a fluid through a wide channel with identical cross-sections (Fig. 4.7a) or the flow of a fluid past a long object, each cross-section of which is the same (Fig. 4.7b). The latter model is applicable, for instance, in analyzing the flow past the wing of an aircraft in order to determine the lift (though air is compressible the model is valid at relatively low speeds).

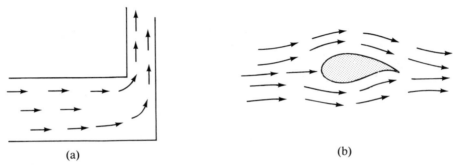

(a)   (b)

Figure 4.7

   In Sections 1.1 and 4 of Chapter 2 and 4.1 of Chapter 3 we have already discussed some aspects of the flow of an ideal fluid. To summarize, we saw that if the velocity of the flow was given by $f = u + iv$ then the twin assumptions that the flow is irrotational and sourceless–sinkless imply that $\bar{f} = u - iv$ is analytic on the domain and, even more, $\bar{f}$ is the derivative of another analytic function G on the domain. We write $G = \phi + i\psi$; ϕ is called the **potential function** of the flow and ψ is the **stream function** of the flow. The terminology results from the fact that the path followed by a particle within the flow is just $\psi(x, y) = $ constant. Furthermore, ϕ is the potential of the flow in just the same sense that this word was used earlier when describing the electrostatic potential: the gradient of ϕ is just $(u, -v)$. Hence, the level lines $\phi(x, y) = $ constant are the curves where the flow has the same potential. Furthermore, since $G' = u - iv$, we see that whenever $f \neq 0$ it follows that the level curves of ϕ and ψ are perpendicular. Finally, both ϕ and ψ are harmonic functions, so we can apply to them any results that we know for harmonic functions in general.

   We begin our discussion of flows by examining sources and sinks; we then derive Bernoulli's Law, which relates the pressure and speed of a flow. Finally, we examine the concept of circulation and show how a circulation produces a lifting force.

Point sources and sinks for flows

We begin our discussion of point sources by noting that the function

$$f(z) = 1/\bar{z}, \; z \neq 0$$

is a locally sourceless and irrotational flow on any domain not containing the origin. However,

$$\int_\Gamma (f \cdot n) \, ds = \mathrm{Im}\left\{ \int_\Gamma \overline{f(z)} \, dz \right\} = \mathrm{Im}\left\{ \int_\Gamma \frac{dz}{z} \right\} = 2\pi$$

for any simple closed curve Γ that surrounds the origin. Thus, flow $f(z) = 1/\bar{z}$ has a source of strength 2π at the origin. Here is how this idea is used.

Example 1   A source of strength Q is located on one of the long sides of a semi-infinite strip. Find the resulting flow.

Solution   We take the strip to be the domain $D = \{x + iy: -\sigma < x < \sigma, y > 0\}$ and locate the source at the point $\sigma + i\tau, \tau \geq 0$ (Fig. 4.8a). The function $\psi(z) = \sin(\pi z/2\sigma)$ maps the strip D onto the upper half-plane, $\mathrm{Im} \, w > 0$, with $\psi(\sigma) = 1$ and $\psi(-\sigma) = -1$ (see Section 5, Chapter 1). The point $\sigma + i\tau$ is carried by ψ to the point

$$A = \tfrac{1}{2}\left(\exp\left\{\frac{\tau\pi}{2\sigma}\right\} + \exp\left\{-\frac{\tau\pi}{2\sigma}\right\} \right) = \cosh\left(\frac{\pi\tau}{2\sigma}\right).$$

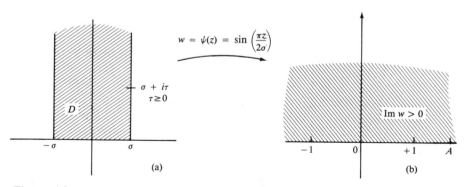

Figure 4.8

The function

$$g(w) = \frac{Q}{\pi} \, \mathrm{Log}(w - A)$$

gives a flow in the domain Im $w > 0$ that has a source of strength Q at the point A. This is actually quite easy to see since the imaginary part of g, which is $(Q/\pi)\text{Arg}(w - A)$, changes from 0, when w is real and greater than A, to Q, when w is real and less than A; indeed, if γ is any curve in Im $w \geq 0$ joining a point on the ray (A, ∞) to a point on the ray $(-\infty, A)$, then Im $g(w)$ changes from 0 to Q as w traverses γ. Hence,

$$G(z) = \frac{Q}{\pi} \text{Log}\left(\sin\left(\frac{\pi z}{2\sigma}\right) - A \right)$$

gives the desired flow in the semi-infinite strip. Note that the velocity of this flow is

$$\overline{G'(z)} = \frac{Q}{2\sigma} \frac{\cos\left(\dfrac{\pi \bar{z}}{2\sigma}\right)}{\sin\left(\dfrac{\pi \bar{z}}{2\sigma}\right) - A}$$

and for any x, $-\pi/2 < x < \pi/2$, and $y > 0$, we have

$$G'\left(\frac{2\sigma z}{\pi}\right) = \frac{Qi}{2\sigma} \frac{e^{ix} + e^{-ix}e^{-2y}}{e^{ix} - e^{-ix}e^{-2y} - 2iAe^{-y}},$$

which approaches

$$v_\infty = i\frac{Q}{2\sigma}$$

as $y \to \infty$. Hence, the velocity of the flow is virtually uniform far from the source. Furthermore, since the width of the strip is 2σ, the total flux out the "end" of the strip (that is, at ∞) is exactly Q, the amount entering at $\sigma + i\tau$. Thus, the amount flowing into the strip per unit time equals the amount flowing out, as it must. Some streamlines of this flow are shown in Figure 4.9. $\square$

Example 2   In a long narrow channel there is source of strength Q at a point on one side. Find the resulting flow.

Solution   This problem is actually a special case of Example 1 despite the different geometric configuration. We take the strip to be $\{x + iy: -\sigma < x < \sigma, -\infty < y < \infty\}$ and locate the source at the point $x_0 = \sigma$, $y_0 = 0$ (Fig. 4.10). The flow has obvious up-down symmetry about the x-axis and hence may be considered as a flow in the semi-infinite strip $\{x + iy: -\sigma < x < \sigma, y > 0\}$ resulting from a source at $x_0 = \sigma$, $y_0 = 0$ of strength $Q/2$. But this flow we already know:

$$G(z) = \frac{Q}{2\pi}\left\{\text{Log}\left(\sin\left(\frac{\pi z}{2\sigma}\right) - 1\right)\right\};$$

here $A = 1$ since $\tau = 0$. Our problem is thus easily solved. The stagnation

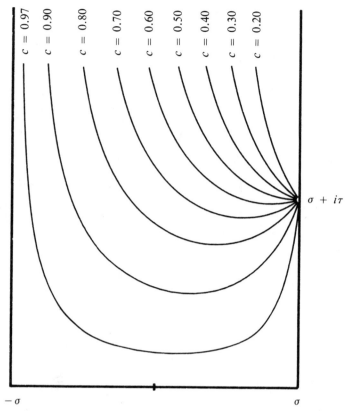

Figure 4.9   The flow in a semi-infinite channel resulting from a source on one of the long sides. The streamlines are

$$\Gamma_c(t) = \frac{1}{\pi} \log\left\{\sin\left[\frac{\pi(t + ic)}{2\sigma} - \cosh\left(\frac{\pi\tau}{2\sigma}\right)\right]\right\}, \quad -\infty < t < \infty; c > 0$$

points of this flow occur where

$$0 = G'(z) = \frac{Q}{4\sigma} \frac{\cos\left(\dfrac{\pi z}{2\sigma}\right)}{\sin\left(\dfrac{\pi z}{2\sigma}\right) - 1}$$

and hence $\cos(\pi z/2\sigma) = 0$; thus

$$\frac{\pi z}{2\sigma} = \frac{\pi}{2} + n\pi, \, n = 0, \, \pm 1, \, \pm 2, \, \ldots$$

which gives $z = \sigma + 2n\sigma$. The only values of z that fall in the strip occur when n is 0 or n is -1. The former is the source and so cannot be counted; the latter is the point $-\sigma$, the point directly opposite the source.   □

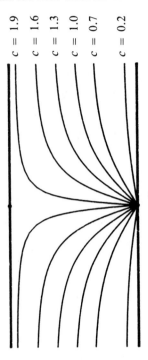

Figure 4.10 The flow in a long channel resulting from a source on one side. The streamlines are

$$\Gamma_c(t) = \frac{1}{\pi} \log\left\{\sin\left[\frac{\pi(t + ic)}{2\sigma} - 1\right]\right\}, \quad -\infty < t < \infty; \, c > 0$$

Example 3 Find the flow out the end of a very long channel.

Solution We take the channel to have sides $\{z = x + iy: y = \pm\pi$ and $-\infty < x < -1\}$. We know from Example 11, Section 5, Chapter 3 that the function $\phi(w) = w + e^w$ is a conformal mapping of the strip $\{w = \sigma + i\tau: -\pi < \tau < \pi, -\infty < \sigma < \infty\}$ onto the complex plane with the two rays $\{x \pm i\pi: -\infty < x < -1\}$ deleted. The uniform flow $(Q/2\pi)w$ of strength Q in the strip $\{\pi > |\tau|\}$ is carried by $\phi^{-1}(z) = w$ to the desired flow out of the channel. Thus, the curves Im $w = $ constant, which are the streamlines in the strip, are carried to the streamlines of the flow out of the channel; the latter, then, are

$$\Gamma_c = \{\sigma + ic + e^\sigma e^{ic}: -\infty < \sigma < \infty\}, \quad c = \text{constant}, \quad -\pi < c < \pi.$$

Some of these are sketched in Figure 4.11. In electrostatics this example gives the "fringing effect" at the end of a parallel plate condenser. □

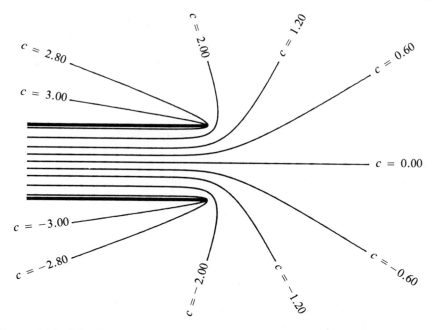

Figure 4.11 The flow out the end of a long channel. The streamlines are

$$\Gamma_c(x) = x + ic + e^{ic}e^x, \quad -\infty < x < \infty; c \in (-\pi, \pi)$$

Bernoulli's* Law

Suppose that $f = u + iv$ represents the velocity of a locally sourceless and locally irrotational flow in a domain D; we shall use the knowledge that $\bar{f} = u - iv$ is analytic on D to derive a fundamental relation between the pressure and the speed of the flow. We begin by imagining a small rectangle in D, as in Figure 4.12. Let $p(x, y)$ be the pressure at the point (x, y). The net force acting on this rectangle in the x-direction is (almost)

$$\{p(x, y) - p(x + h, y)\}k.$$

$p(x, y) = $ pressure

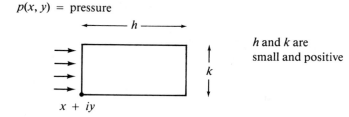

h and k are
small and positive

Figure 4.12

---

* Daniel Bernoulli, 1700–1782.

However, by Newton's Law, force is mass times acceleration. The mass of the little rectangle is its (constant) density ρ, multiplied by its area hk. Furthermore, the acceleration in the x-direction is the time derivative of u, since u is the x-component of the velocity. This yields

$$(\text{mass})(\text{acceleration}) = \rho hk \frac{du}{dt}$$

$$= \rho hk \left\{ \frac{\partial u}{\partial x} \frac{dx}{dt} + \frac{\partial u}{\partial y} \frac{dy}{dt} \right\}.$$

However, $dx/dt = u$ and $dy/dt = v$, which then gives

$$\{p(x, y) - p(x + h, y)\}k = \rho hk \left\{ \frac{\partial u}{\partial x} u + \frac{\partial u}{\partial y} v \right\}.$$

Divide both sides by hk, then let h and k tend to 0. We obtain

$$-\frac{\partial p}{\partial x} = \rho \left\{ \frac{\partial u}{\partial x} u + \frac{\partial u}{\partial y} v \right\}.$$

In an entirely similar manner, working with the y-component of the forces acting on the rectangle we obtain

$$-\frac{\partial p}{\partial y} = \rho \left\{ \frac{\partial v}{\partial x} u + \frac{\partial v}{\partial y} v \right\}.$$

However, $\partial u/\partial x = -\partial v/\partial y$ and $\partial u/\partial y = \partial v/\partial x$ by the Cauchy–Riemann equations, since $u - iv$ is analytic. This gives us two equations:

$$\rho \left(\frac{\partial u}{\partial x} u + \frac{\partial v}{\partial x} v \right) + \frac{\partial p}{\partial x} = 0$$

and

$$\rho \left(\frac{\partial u}{\partial y} u + \frac{\partial v}{\partial y} v \right) + \frac{\partial p}{\partial y} = 0.$$

Integration of the first with respect to x and the second with respect to y yields

$$\tfrac{1}{2}\rho\{u^2 + v^2\} + p = A(y)$$

and

$$\tfrac{1}{2}\rho\{u^2 + v^2\} + p = B(x),$$

where $A(y)$ and $B(x)$ are functions of y alone and of x alone, respectively. Of course, it is clear that $A(y)$ must also equal $B(x)$ and so both are (the same) constant. Thus, we find that

$$\tfrac{1}{2}(u^2 + v^2) + \frac{1}{\rho} p = \text{constant},$$

which says that for an ideal fluid, half the square of the speed plus the pressure divided by the (constant) density adds up to a constant; this is **Bernoulli's Law**.

Circulation and lift

We now investigate the flow around an object, for instance, the situation pictured in Figure 4.13. Let Γ be the boundary of the object, which we take to be a simple closed piecewise smooth curve. We assume that the flow f is both locally irrotational and locally sourceless in the domain exterior to Γ; that is, we assume that $\bar{f} = u - iv$ is analytic in the domain exterior to Γ. Let γ be any smooth simple closed positively oriented curve that contains Γ in its inside. The value of the integral

$$c = \int_\gamma f \cdot \tau \, ds$$

is called the **circulation** of the flow around Γ; here $f \cdot \tau$ is the component of f tangent to the curve γ and ds is arc length. As we showed in Section 1.1, Chapter 2,

$$c = \mathrm{Re}\left\{ \int_\gamma \overline{f(z)} \, dz \right\},$$

so it follows from the fact that $\bar{f}$ is analytic that the integral itself is not dependent on the choice of curve γ. Hence, the value of the circulation depends on the flow f only. Furthermore, because the flow is around Γ (that is, Γ is a streamline) we know that the component of f that is normal to Γ must be zero. (For if $f \cdot n$ is, say, positive at a point p of Γ then there is a net flow over Γ in a small segment of Γ near p, contradicting the fact that Γ is a streamline of the flow.) Hence,

$$0 = \int_\Gamma f \cdot n \, ds = \mathrm{Im}\left\{ \int_\Gamma \overline{f(z)} \, dz \right\}.$$

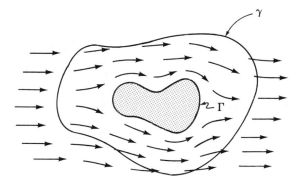

The flow around a solid object with boundary Γ

Figure 4.13

Once again the analyticity of $\bar{f}$ implies that

$$0 = \mathrm{Im}\left\{\int_\gamma \overline{f(z)}\,dz\right\}$$

for any closed curve γ surrounding Γ.

We now calculate the force exerted on Γ by the flow f. For this purpose we make the reasonable assumption that the flow is "uniform at ∞"; that is, the limit

$$\lim_{|z|\to\infty} \overline{f(z)} = a$$

exists. Since $\bar{f}$ is analytic in the region exterior to Γ and bounded at ∞, ∞ is a removable singularity for $\bar{f}$. Hence, for $|z|$ large we may write

$$\overline{f(z)} = a + \frac{c}{2\pi i}\frac{1}{z} + \sum_{k=2}^{\infty} \frac{b_k}{z^k}, \tag{*}$$

where c is the circulation and $b_2, b_3, \ldots$ are constants.

To begin the analysis of the force exerted on Γ by the flow, we let $p(x, y)$ be the pressure at the point (x, y) in Γ (Fig. 4.14). By a simple resolution of forces argument we find that the horizontal component of the force on a little segment in Γ is $-p\,dy$ and so the total *horizontal* component of the force is

$$H = -\int_\Gamma p\,dy.$$

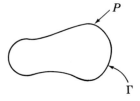
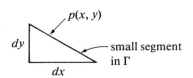

$z(t) = x(t) + iy(t)$ parametrizes Γ
$ds = $ element of arc length along Γ; $(ds)^2 = (dx)^2 + (dy)^2$

Figure 4.14

Likewise, the total *vertical* component of the force on Γ is

$$V = \int_\Gamma p\,dx.$$

From Bernoulli's Law, derived earlier, we have

$$p = -\tfrac{1}{2}\rho(u^2 + v^2) + \text{constant},$$

where ρ is the (constant) density. Hence,

$$V + iH = \int_\Gamma p(dx - i\,dy)$$

$$= -\tfrac{1}{2}\rho \int_\Gamma (u^2 + v^2)(dx - i\,dy)$$

since the integral of the constant term is zero. A bit of manipulation produces the relation

$$(u^2 + v^2)(dx - i\,dy) - (u - iv)^2(dx + i\,dy) = -2i(u - iv)(-v\,dx + u\,dy).$$

However, $-v\,dx + u\,dy$ is exactly the component of the flow f which is normal to Γ. Since Γ is a streamline of the flow, this quantity is zero. Hence

$$V + iH = -\tfrac{1}{2}\rho \int_\Gamma (u - iv)^2(dx + i\,dy)$$

$$= -\tfrac{1}{2}\rho \int_\Gamma [\overline{f(z)}]^2\,dz.$$

The function $\bar{f}^2$ is analytic so that the integral above is identical with the integral of $\bar{f}^2$ over the circle $|z| = R$, where R is so large that (*) is valid. Therefore, after squaring the expression for $\bar{f}$ in (*) and replacing $\bar{f}^2$ by the resulting series we find that

$$V + iH = -\tfrac{1}{2}\rho \int_{|z|=R} [\overline{f(z)}]^2\,dz$$

$$= -\tfrac{1}{2}\rho \int_{|z|=R} \left(a^2 + \frac{ca}{\pi i}\frac{1}{z} + \sum_{k=2}^{\infty}\frac{d_k}{z^k}\right)dz$$

$$= -\rho c a,$$

where $a = \lim_{z \to \infty} \bar{f(z)}$ is the velocity of the flow at ∞.

This result, called the **Kutta–Joukowski Theorem**, is what we have been aiming for. It says that the flow f produces a force on Γ that is directed perpendicular to a, the velocity at ∞, and of magnitude $\rho|c|\,|a|$. In particular, if a is real then the force is vertical and thus is a **lifting force**.

Example 4   We begin our analysis of the lift generated by a flow with the flow around a circle. We make use of the observation in Section 4.1, Chapter 3 that it is enough to find an appropriate function that has the circle as one of the level curves of its imaginary part. Specifically, let the circle be centered at the origin and of radius R and let

$$G_1(z) = s(\lambda z + \bar\lambda R^2/z), \quad s > 0,\ |\lambda| = 1$$

and

$$G_2(z) = -\frac{ic}{2\pi}\log z, \quad c \text{ real}.$$

G_1 is just a scaling of the mapping discussed in Example 2, Section 1, Chapter

3. Here s, λ, and c are all constants. Then on the circle $|z| = R$, we have

$$\text{Im } G_1(Re^{it}) = 0$$

and

$$\text{Im } G_2(Re^{it}) = -\frac{c}{2\pi} \log R.$$

Consequently, the circle $|z| = R$ is a streamline for $G_1 + G_2$. Although $G_2(z)$ is not well-defined (because $\log z$ has multiple values), its derivative is. Set

$$\overline{f(z)} = G_1'(z) + G_2'(z)$$

$$= s(\lambda - \bar{\lambda}R^2/z^2) - \frac{ic}{2\pi}\frac{1}{z}.$$

Then $\bar{f}$ is analytic on the plane minus the origin and f is also analytic at ∞, where

$$\lim_{|z| \to \infty} \overline{f(z)} = \lambda s.$$

Furthermore, the circulation of the flow f is

$$\text{Re}\left\{\int_{|z|=r} \overline{f(z)} \, dz\right\} = c, \; r \geqslant R.$$

Hence, $\bar{f}$ is exactly the flow with circulation c and velocity λs at ∞.

The points z at which $f(z) = 0$ are called **stagnation points** of the flow. For this flow there are stagnation points when

$$s\left(\lambda - \frac{\bar{\lambda}R^2}{z^2}\right) = \frac{ic}{2\pi z}.$$

Multiplying through by z^2 we obtain a quadratic equation in z, which has roots

$$z_1, z_2 = \frac{ic \pm \sqrt{16\pi^2 s^2 R^2 - c^2}}{4\pi s \lambda}.$$

If $0 \leqslant c \leqslant 4\pi sR$, then both roots lie on the circle $|z| = R$; they coincide if $c = 4sR$. Several streamlines of this flow are sketched in Figure 4.15.

Example 5   The Joukowski airfoil.   Let C be a circle with center z_0, which passes through the point $-R$, and contains the point R in its inside, $R > 0$. The function

$$\phi(z) = z + R^2/z$$

maps C onto a simple closed curve Γ (Fig. 4.16). The image curve Γ, which resembles the cross-section of a wing, is called a **Joukowski airfoil** in honor of N.J. Joukowski (1842–1921), who first investigated this particular mapping and the associated flow.

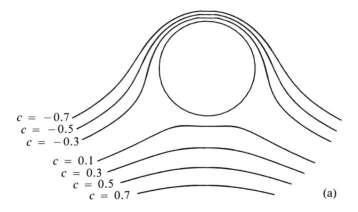

$$\sin \Theta \, (r - 1/r) - a \ln(R) = c$$

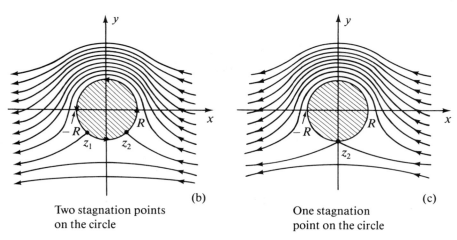

(b)

(c)

Two stagnation points
on the circle

One stagnation
point on the circle

Figure 4.15

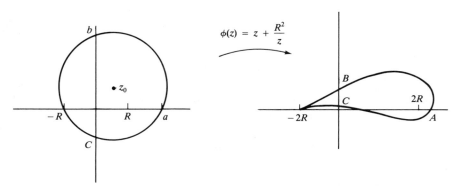

Figure 4.16

Our goal is to find the flow f with Γ as a streamline and with uniform velocity λs at ∞, $s > 0$ and $|\lambda| = 1$; for this we make use of conformal mapping and the results of Example 4.

Our first task is to prove that the mapping ϕ given above is a one-to-one mapping of the exterior of the circle C onto the exterior of the curve Γ. To this end, let Γ_0 be the (unique) circle through both the points R and $-R$, which is tangent to C at $-R$ (Fig. 4.17). The center of Γ_0 lies at the point $i\alpha_0$, α_0 real. It is easy to prove (see Exercise 1 at the end of this section) that a point $z = x + iy$ lies inside, on, or outside Γ_0 if and only if the quantity

$$|z|^2 - 2\alpha_0 y - R^2$$

is negative, zero, or positive, respectively. Thus, z is outside Γ_0 if and only if $w = R^2/z$ is inside Γ_0 (also, see Exercise 2). From this it follows easily that $\phi(z) = z + R^2/z$ is one-to-one on the outside of Γ_0. For if $\phi(z_1) = \phi(z_2)$, then

$$z_1 - z_2 = R^2 \left(\frac{z_1 - z_2}{z_1 z_2} \right).$$

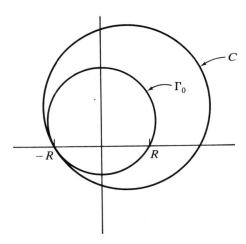

Figure 4.17

If $z_1 \neq z_2$, then $z_2 = R^2/z_1$ and so z_1 and z_2 lie on opposite sides of Γ_0, a contradiction. Since the outside of Γ is within the outside of Γ_0, ϕ is one-to-one outside Γ.

Now we are ready to analyze the flow around the Joukowski airfoil. Let $G(\zeta)$ be the complex potential of this flow and let H be the complex potential of the flow around the circle C; we know H from Example 4. After taking into account the change of scale we find that

$$H(z) = s \left[\lambda(z - z_0) + \bar{\lambda} \frac{|z_0 + R|^2}{z - z_0} \right] - \frac{ic}{2\pi} \log(z - z_0).$$

However, the function $\zeta = \phi(z)$ carries the exterior of C onto the exterior of Γ as we have just shown, so that $H(z)$ and $G(\zeta)$ must be related by $H(z) = G(\phi(z))$.

Consequently,

$$H'(z) = G'(\phi(z))\phi'(z) = G'(\phi(z))(1 - R^2/z^2).$$

Thus, the flow $H'(z)$ has a stagnation point at $z = -R$. Some simple arithmetic then yields

$$c = -4\pi s \; \text{Im}(\lambda(R + z_0))$$

and so the circulation is determined by three factors: (1) the point R; (2) the center, z_0, of the circle C; and (3) λs, the velocity of the flow at ∞. Thus, in turn, the circulation of the flow determines the lift in accordance with the Kutta–Joukowski formula derived earlier. If $\lambda = 1$ (that is, the flow is horizontal at ∞) then the lift is

$$\text{Lift} = 4\pi s^2 \; \text{Im} \; z_0, \quad s = \text{speed of flow at } \infty.$$

Other physical phenomena in which harmonic functions arise

Diffusion

Imagine a level channel connecting a very large body of fresh water to a very large body of salt water, say the ocean (Fig. 4.18). The water in the channel will have various degrees of salinity, depending on location; that is, the salt from the ocean will diffuse through the channel to the fresh water, which being large, will absorb it. If we let $c(x, y)$ be the concentration of salt at the point $x + iy$ in the channel, then an analysis exactly like that for temperature distribution, shows that $c(x, y)$ is a harmonic function of x and y. More generally, if a substance diffuses from a region of high concentration to one of low or zero concentration, then the concentration function $c(x, y)$ is a harmonic function in any domain in which there are no sources or sinks.

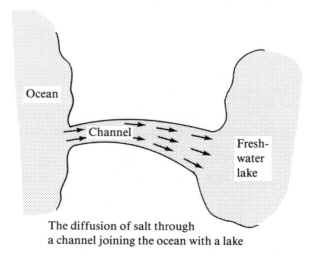

Ocean

Channel

Fresh-
water
lake

The diffusion of salt through
a channel joining the ocean with a lake

Figure 4.18

Planar stress

In an isotropic, homogeneous body there are nine stress components: the normal stress component and two shear stress components in each of the x, y, and z directions. There are compatibility relations among these stress components, which in the case of plane strain with no body forces acting, reduce to

$$\frac{\partial X_x}{\partial x} + \frac{\partial X_y}{\partial y} = 0 \tag{1}$$

$$\frac{\partial X_y}{\partial x} + \frac{\partial Y_y}{\partial y} = 0 \tag{2}$$

$$\left(\frac{\partial^2}{\partial x^2} + \frac{\partial^2}{\partial y^2}\right)(X_x + Y_y) = 0. \tag{3}$$

Here X_x, X_y, and Y_y are the normal and shear stress components in the x-direction and the normal stress component in the y-direction, respectively. We now assume that the cross-section of the body is a simple closed curve together with the domain D inside it. From equation (1) we deduce that there is a function F with

$$\frac{\partial F}{\partial x} = -X_y \quad \text{and} \quad \frac{\partial F}{\partial y} = X_x.$$

From equation (2) we deduce that there is a function G with

$$\frac{\partial G}{\partial x} = Y_y \quad \text{and} \quad \frac{\partial G}{\partial y} = -X_y.$$

From the relations among the partial derivatives of F and G we further deduce that there is a function U with

$$\frac{\partial U}{\partial x} = G \quad \text{and} \quad \frac{\partial U}{\partial y} = F.$$

Thus,

$$\frac{\partial^2 U}{\partial x^2} = Y_y, \quad \frac{\partial^2 U}{\partial y^2} = X_x, \quad \text{and} \quad \frac{\partial^2 U}{\partial x\, \partial y} = -X_y.$$

U is called the **stress function** or the **Airy function**, after G.B. Airy (1801–1892) who first worked with it (circa 1860). Thus, from equation (3) we find that

$$0 = \Delta(X_x + Y_y) = \Delta(\Delta U) = \Delta^2 U.$$

The stress function U is therefore **biharmonic**, that is, it satisfies the equation

$$\frac{\partial^4 U}{\partial x^4} + 2 \frac{\partial^4 U}{\partial x^2\, \partial y^2} + \frac{\partial^4 U}{\partial y^4} = 0.$$

The famous 19th-century physicist J.C. Maxwell (1831-1879) was the first to note that U must be biharmonic.

The representation of biharmonic functions

If U is biharmonic, then $P = \Delta U$ is harmonic. Let φ be an analytic function on D with

$$4\varphi' = P + iQ,$$

where Q is the harmonic conjugate of P on D; such a φ exists because D is simply connected. We write $\varphi = p + iq$. A simple computation shows that

$$U - xp - yq$$

is harmonic on D. Hence, this is the real part of some analytic function f on D, again because D is simply connected. Consequently,

$$U(z) = \operatorname{Re}\{\bar{z}\varphi(z) + f(z)\},$$

where φ and f are analytic and

$$4 \operatorname{Re} \varphi' = \Delta U.$$

This is the most general representation of a biharmonic function U.

Exercises for Section 4.2

1. Let Γ_0 be the circle of radius $\sqrt{R^2 + \alpha_0^2}$ centered at the point $i\alpha_0$, $\alpha_0 > 0$. Show a point z is outside, on, or inside Γ_0 exactly when the quantity $|z|^2 - 2\alpha_0 y - R^2$ is positive, zero, or negative.

2. Let Γ_0 be the circle from Exercise 1. Show that z is inside Γ_0 if and only if R^2/z is outside Γ_0.

3. Let Γ be a simple closed curve. Show that a flow outside Γ (including at ∞) with Γ as a streamline is completely determined by specifying the velocity at ∞ and the circulation. That is, show there is at most *one* analytic function g outside Γ with
    (a) $g(z) = a + \dfrac{c}{2\pi i}\dfrac{1}{z} + \cdots$; a, c fixed, $|z|$ large,
    (b) $\operatorname{Im} G = 0$ on Γ, $G' = g$.

4. Find the stagnation points of the flow in Example 1.

5. Discuss the flow within a disc caused by placing a source and a sink of equal strengths at distinct points on the edge of the disc. Find the streamlines for the flow and sketch a few.

6. Find the flow in the first quadrant created by placing a source of strength Q at a point of the boundary. The case when the source is at the origin is particularly simple. Sketch some of the streamlines.

7. Find the flow in the semi-infinite strip $\{x + iy: \ y > 0$ and $-\pi < x < \pi\}$ established by placing two sources across from each other (on the long sides).

8.   Redo Exercise 7 with one of the sources changed to a sink.

9.   Find the flow in the strip of Exercise 7 if there is a source on the short edge where $x = 0$.

10.   Let D be a region bounded by a smooth simple closed curve Γ and let f be a sourceless flow in D; f is not assumed to be irrotational. The **kinetic energy** of the flow is

$$K = \tfrac{1}{2}\rho \iint_D |f(z)|^2 \, dx \, dy,$$

where ρ is the density (which we assume to be constant). Show that among all sourceless flows in D with given normal velocity at the boundary of D (that is, the normal derivative $\partial f/\partial n$ is specified at Γ), the flow of *smallest* kinetic energy is *irrotational*. This result is known as **Kelvin's* minimum-energy theorem**; you will need to use Green's identity (Exercise 13, Section 6, Chapter 1).

11.   Find the flow in the upper half-plane if there is a source at x_1 of strength Q_1 and a sink at $x_2 \neq x_1$ of strength Q_2, x_1, x_2 real.

12.   A chemical diffuses through a straight channel from a region of high concentration to one of low concentration. Show that the concentration depends *linearly* on the distance from the source (Fig. 4.19).

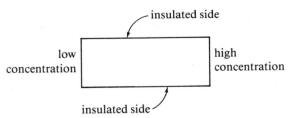

Figure 4.19

13.   Find the electrostatic potential due to a uniformly distributed charge on a straight wire. (**Hint:** Take the wire to be the interval $[-1, 1]$ and the charge density to be $d\rho = c_0 \, dx$.)

14.   Show that the electrostatic potential inside a long hollow uniformly charged tube is constant and that outside the tube the potential is zero.

15.   Find the flow past a long cylinder with an elliptical cross-section (Fig. 4.20).

---

* Sir William Thompson, Baron Kelvin of Largs, 1824–1907.

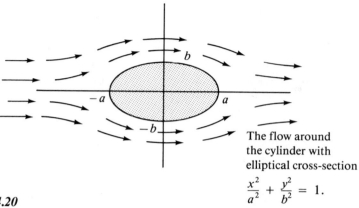

The flow around the cylinder with elliptical cross-section

$$\frac{x^2}{a^2} + \frac{y^2}{b^2} = 1.$$

Figure 4.20

16. Show that the function

$$f(z) = \frac{K}{2\pi i} \frac{1}{\bar{z} - \bar{z}_0}, \; K > 0,$$

produces a flow that rotates clockwise about the point z_0. This flow has a **vortex** at z_0 of strength K. Show that the streamlines are circles centered at z_0.

17. Show that the function

$$g(z) = \frac{K_1}{2\pi i} \frac{1}{\bar{z} - \bar{z}_1} - \frac{K_2}{2\pi i} \frac{1}{\bar{z} - \bar{z}_2},$$

K_1 and K_2 both positive, produces a flow with two vortices, one at z_1 of strength K_1 with clockwise rotation, the other at z_2 of strength K_2 with counterclockwise rotation. What is the circulation of this flow on a curve γ, which is very far from both z_1 and z_2? (A variation on this flow is shown in Figure 4.21).

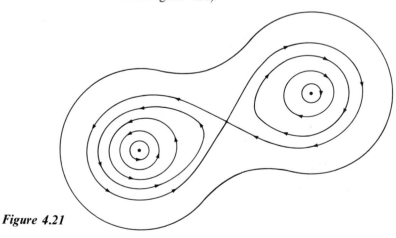

Figure 4.21

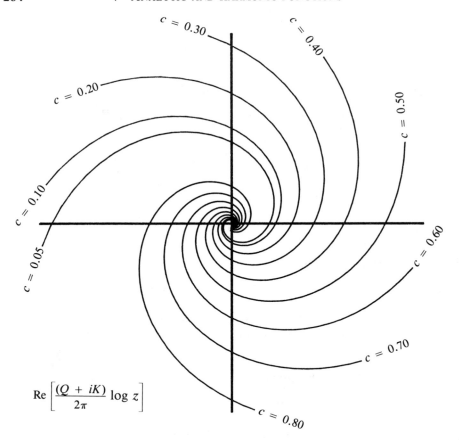

$$\text{Re}\left[\frac{(Q + iK)}{2\pi} \log z\right]$$

Figure 4.22 A spiral vortex produced by the flow $\dfrac{Q + iK}{2\pi i}\dfrac{1}{z}$; $QK \neq 0$

18. The flow

$$h(z) = \frac{Q + iK}{2\pi} \frac{1}{\bar{z} - \bar{z}_0}$$

produces a **spiral vortex** at z_0 (Fig. 4.22). Sketch some of the streamlines and decide how the signs of Q and K affect the streamlines; that is, when will the streamlines spiral into z_0 clockwise and when will they spiral in counterclockwise? See Figure 4.22.

19. Find the flow through a horizontal aperture of length 2σ. (**Hint:** Let the aperture be the segment $[-\sigma, \sigma]$; the function $z = \phi(\zeta) = \sigma \sin \zeta$

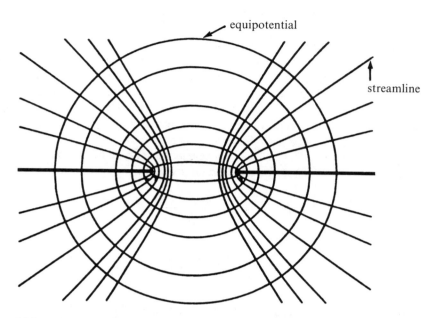

Figure 4.23 The flow from the lower half-plane to the upper half-plane through a horizontal aperture.

maps the strip $\{|\operatorname{Re} \zeta| < \pi/2\}$ onto the complement of the rays $(-\infty, -\sigma)$ and (σ, ∞) with the lines $\operatorname{Re} \zeta = c$ going to the hyperbolas

$$\frac{x^2}{(\sin c)^2} - \frac{y^2}{(\cos c)^2} = \sigma^2$$

(see Section 5, Chapter 1). See Figure 4.23.

20. Use Exercise 19 to find the electrostatic potential produced by two parallel, horizontal charged plates whose edges are separated by a distance 2σ (Fig. 4.24). Sketch some of curves of flux of the field.

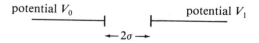

Figure 4.24

4.3   Integral representations of harmonic functions

Integral representation formulas are important tools in solving boundary-value problems. In this section we explore such formulas in two special but important cases: the disc $\{z : |z| < 1\}$ and the upper half-plane $\{z : \operatorname{Im} z > 0\}$.

Suppose first that u is a real-valued function that is harmonic on some open set that includes the entire disc $\{z : |z| \leqslant 1\}$. We shall derive a formula for the values of $u(x, y)$ when $x^2 + y^2 < 1$, which makes use only of the values of $u(x, y)$ when $x^2 + y^2 = 1$. Since u is harmonic on some disc centered at the origin and of radius larger than 1, there is an analytic function f on this same disc with $\operatorname{Re} f = u$. Thus, for a particular z with $|z| < 1$ we have two formulas

$$f(z) = \frac{1}{2\pi i} \int_\Gamma \frac{f(\zeta)}{\zeta - z} \, d\zeta \tag{1}$$

and

$$0 = \frac{1}{2\pi i} \int_\Gamma \frac{f(\zeta)\bar{z}}{1 - \bar{z}\zeta} \, d\zeta. \tag{2}$$

The first, (1), is the Cauchy integral formula with Γ the circle $|\zeta| = 1$ oriented positively. The second (2) is just Cauchy's Theorem applied to the function

$$G(\zeta) = \frac{f(\zeta)\bar{z}}{1 - \bar{z}\zeta},$$

which is analytic on an open set containing $|\zeta| \leqslant 1$. We set $\zeta = e^{it}$, $0 \leqslant t \leqslant 2\pi$, as usual and find that

$$f(z) = \frac{1}{2\pi} \int_0^{2\pi} \frac{f(e^{it})}{1 - ze^{-it}} \, dt \tag{3}$$

and

$$0 = \frac{1}{2\pi} \int_0^{2\pi} \frac{f(e^{it})\bar{z}e^{it}}{1 - \bar{z}e^{it}} \, dt. \tag{4}$$

However, we also have

$$\frac{1}{1 - ze^{-it}} + \frac{\bar{z}e^{it}}{1 - \bar{z}e^{it}} = \frac{1 - |z|^2}{|1 - \bar{z}e^{it}|^2}.$$

When we add (3) and (4) we find that

$$f(z) = \frac{1}{2\pi} \int_0^{2\pi} f(e^{it}) \frac{1 - |z|^2}{|1 - \bar{z}e^{it}|^2} \, dt.$$

The function appearing after $f(e^{it})$ under the integral sign has a special notation. Set

$$P_r(t) = \frac{1 - r^2}{1 - 2r \cos t + r^2} \tag{5}$$

$$= \operatorname{Re}\left\{\frac{1 + z}{1 - z}\right\} \text{ if } z = re^{it}.$$

$P_r(t)$ is called the **Poisson* kernel** for r and t. Thus, we have shown that

$$f(re^{i\theta}) = \frac{1}{2\pi} \int_0^{2\pi} f(e^{it}) P_r(\theta - t) \, dt. \tag{6}$$

Since $P_r(\theta - t)$ is real-valued we may take the real part of both sides in this formula and conclude that

$$u(re^{i\theta}) = \frac{1}{2\pi} \int_0^{2\pi} u(e^{it}) P_r(\theta - t) \, dt. \tag{7}$$

This is the **Poisson integral formula**; it gives the value of the function u in the disc $|z| < 1$ solely in terms of the values of $u(e^{it})$, $0 \leqslant t \leqslant 2\pi$.

Several extensions of this formula are immediate. First, if the disc is not $|z| \leqslant 1$ but rather $|w - w_0| \leqslant R$, then the change of variables $z = (w - w_0)/R$ produces the formula

$$u(w) = \frac{1}{2\pi} \int_0^{2\pi} u(w_0 + Re^{it}) \frac{R^2 - |w - w_0|^2}{|R - (\bar{w} - \bar{w}_0)e^{it}|^2} \, dt. \tag{8}$$

The formula is valid for $|w - w_0| < R$.

Substantially more important, we need not assume that u is harmonic on an open set containing the disc $|z| \leqslant 1$. Suppose, in fact, only that u is a continuous function on the circle $|z| = 1$. Set

$$U(re^{i\theta}) = \frac{1}{2\pi} \int_0^{2\pi} u(e^{it}) P_r(\theta - t) \, dt. \tag{9}$$

It is then true that $U(z)$, $z = re^{i\theta}$, is a harmonic function on the disc $|z| < 1$ and at all points λ on the circle $|\lambda| = 1$ we have

$$\lim_{z \to \lambda} U(z) = u(\lambda). \tag{10}$$

That is, the function defined by

$$\begin{cases} U(z) \text{ if } |z| < 1 \\ u(z) \text{ if } |z| = 1 \end{cases}$$

is continuous on the closed set $\{z: |z| \leqslant 1\}$ and harmonic on the open set $\{z: |z| < 1\}$. Part of this assertion is easy: if u has only real values then

---

* Siméon Denis Poisson, 1781–1840.

$$U(z) = \frac{1}{2\pi} \int_0^{2\pi} u(e^{it}) Re\left(\frac{e^{it} + z}{e^{it} - z}\right) dt$$

$$= Re\left\{ \frac{1}{2\pi} \int_0^{2\pi} u(e^{it}) \frac{e^{it} + z}{e^{it} - z} dt \right\}.$$

The expression within the braces is an analytic function of z and so U is harmonic, being the real part of an analytic function. In general, $u = u_1 + iu_2$, where u_1 and u_2 are real-valued. Hence, $U = U_1 + iU_2$, where both U_1 and U_2 are harmonic; thus, U is also harmonic. The proof that (10) holds is more difficult; it is outlined in the exercises at the end of this section.

There is even a further extension of (7). A function $u(e^{it})$ is **piecewise continuous** if the circle $\{z: |z| = 1\}$ can be broken up into closed arcs $I_1, \ldots, I_N$ that are disjoint except for their endpoints (Fig. 4.25) and, furthermore, on each of the arcs I_j, $u(e^{it})$ is continuous. (This gives two values to u at each of the N endpoints of the arcs; this situation will not bother us because such a set "does not count" in integration). If $u(e^{it})$ is piecewise continuous and U is defined by (9), then we still have that U is harmonic in the disc $|z| < 1$ and, except for the N points that are the endpoints of the arcs $I_1, \ldots, I_N$,

$$\lim_{z \to \lambda} U(z) = u(\lambda), \quad \lambda \text{ is not an endpoint of } I_1, \ldots, I_N.$$

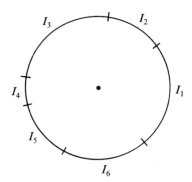

Figure 4.25

Example 1 A long cylinder with circular cross-section is split lengthwise down the middle and the two halves are insulated from one another. One half is given a charge of $+1$ and the other half is given the charge of -1. Find the resulting electrostatic potential inside the cylinder.

Solution We may assume a cross-section of the cylinder is the circle $x^2 + y^2 = 1$. Hence, we must find a function $u(x, y)$ that is harmonic for $x^2 + y^2 < 1$ and that has the value $+1$ on the semicircle $x^2 + y^2 = 1$, $y > 0$, and the value -1 on the semicircle $x^2 + y^2 = 1$, $y < 0$. The answer is given to us immediately by the Poisson integral formula

$$u(re^{i\theta}) = \frac{1}{2\pi} \int_0^{\pi} \frac{1 - r^2}{1 - 2r \cos(\theta - t) + r^2} \, dt - \frac{1}{2\pi} \int_{-\pi}^0 \frac{1 - r^2}{1 - 2r \cos(\theta - t) + r^2} \, dt.$$

We evaluate the integrals in the following way. Since

$$Re\left(\frac{1 + z}{1 - z}\right) = \frac{1 - r^2}{1 - 2r \cos \theta + r^2}, \quad z = re^{i\theta}$$

and since

$$\frac{1 + z}{1 - z} = 1 + \frac{2z}{1 - z}$$

we have

$$\frac{1 + ze^{-it}}{1 - ze^{-it}} = 1 + \frac{2ze^{-it}}{1 - ze^{-it}}.$$

Thus,

$$\int_0^{\pi} \frac{1 + ze^{-it}}{1 - ze^{-it}} \, dt = \pi + 2z\left(\overline{\int_0^{\pi} \frac{e^{it}}{1 - \bar{z}e^{it}} \, dt}\right)$$

$$= \pi + 2z\left(\frac{1}{i} \int_{\gamma} \frac{dw}{1 - \bar{z}w}\right),$$

where γ is the semicircle $\{e^{it}: 0 \leqslant t \leqslant \pi\}$. The antiderivative of $(1 - \bar{z}w)^{-1}$ is $-1/\bar{z} \log(1 - \bar{z}w)$ and so

$$\int_0^{\pi} \frac{1 - r^2}{1 - 2r \cos(\theta - t) + r^2} \, dt = \pi + 2\text{Arg}\left(\frac{1 + z}{1 - z}\right).$$

Likewise

$$\int_{-\pi}^0 \frac{1 - r^2}{1 - 2r \cos(\theta - t) + r^2} \, dt = \pi - 2\text{Arg}\left(\frac{1 + z}{1 - z}\right).$$

Putting these two together, we find that

$$u(z) = \frac{2}{\pi} \text{Arg}\left(\frac{1 + z}{1 - z}\right). \qquad \Box$$

The Poisson integral formula in the upper half-plane

Suppose that $w(t)$ is a bounded, piecewise continuous function on the real axis, $-\infty < t < \infty$. There is an integral formula, much like the Poisson integral formula, that yields a function $W(\zeta)$ that is harmonic in the upper half-plane $\{\zeta = \sigma + i\tau : \tau > 0\}$ and satisfies

$$\lim_{\zeta \to s} W(\zeta) = w(s)$$

for all points s on the real axis at which $w(t)$ is continuous (that is, at all but a finite number of points). The formula is this:

$$W(\sigma + i\tau) = \frac{1}{\pi} \int_{-\infty}^{\infty} w(t) \frac{\tau}{(\sigma - t)^2 + \tau^2} \, dt, \quad \tau > 0. \tag{11}$$

Its derivation is simple. Let

$$\zeta = \psi(z) = i \frac{1 + z}{1 - z}.$$

Then ψ is a linear fractional transformation that maps the disc $\{z: |z| < 1\}$ onto the upper half-plane and ψ maps the circle $|z| = 1$ onto the real axis. Hence,

$$u(e^{i\theta}) = w\left(i \frac{1 + e^{i\theta}}{1 - e^{i\theta}}\right)$$

is bounded and piecewise continuous on the circle $\{e^{i\theta}: 0 \leqslant \theta \leqslant 2\pi\}$. Consequently, the function

$$U(z) = \frac{1}{2\pi} \int_0^{2\pi} u(e^{i\theta}) \frac{1 - r^2}{1 - 2r \cos(s - \theta) + r^2} \, d\theta, \quad z = re^{is}$$

is harmonic in the disc $|z| < 1$ and $U(z) \to u(e^{i\theta})$ as $z \to e^{i\theta}$. Suppose that $w(t)$ is real-valued, then

$$U(z) = \text{Re}\left\{ \frac{1}{2\pi} \int_0^{2\pi} u(e^{i\theta}) \frac{e^{i\theta} + z}{e^{i\theta} - z} \, d\theta \right\}.$$

Now $z = (\zeta - i)/(\zeta + i)$ so that

$$W(\zeta) = U\left(\frac{\zeta - i}{\zeta + i}\right)$$

is harmonic on the upper half-plane. But $e^{i\theta} = (t - i)/(t + i)$ and $e^{i\theta} \, d\theta = 2(t + i)^{-2} \, dt$. With these changes of variables we have

$$W(\zeta) = \text{Re}\left[\frac{1}{2\pi} \int_0^{2\pi} u(e^{i\theta}) \frac{e^{i\theta} + z}{e^{i\theta} - z} \, d\theta \right]$$

$$= \text{Re}\left[\frac{1}{2\pi} \int_{-\infty}^{\infty} w(t) \frac{t\zeta + 1}{i(t - \zeta)} \frac{2}{1 + t^2} \, dt \right]$$

$$= \frac{1}{\pi} \int_{-\infty}^{\infty} w(t) \frac{\tau}{(t - \sigma)^2 + \tau^2} \, dt.$$

This is the desired formula; it is called the **Poisson integral formula for the upper half-plane.**

Example 2   Find the electrostatic potential in the region $y > 0$ if the segment $(-\sigma, \sigma)$ of the real axis is kept at potential $V_0 > 0$ and the remainder of the real axis is kept at potential 0.

Solution The electrostatic potential $V(x, y)$ is a harmonic function for $y > 0$ and must satisfy

$$V(x, 0) = \begin{cases} V_0, & -\sigma < x < \sigma \\ 0, & |x| > \sigma. \end{cases}$$

To find such a harmonic function we use the Poisson integral formula for the upper half-plane.

$$\begin{aligned} V(x, y) &= \frac{1}{\pi} \int_{-\infty}^{\infty} V(t, 0) \frac{y}{(x - t)^2 + y^2} \, dt \\ &= \frac{V_0}{\pi} \int_{-\sigma}^{\sigma} \frac{y}{(t - x)^2 + y^2} \, dt \\ &= \frac{V_0}{\pi} \left\{ \arctan\left(\frac{\sigma - x}{y}\right) - \arctan\left(\frac{-\sigma - x}{y}\right) \right\} \\ &= \frac{V_0}{\pi} \arctan\left(\frac{2\sigma y}{x^2 + y^2 - \sigma^2}\right). \end{aligned}$$

A function $U(x, y)$ that is a harmonic conjugate of $V(x, y)$ in the upper half-plane is given by

$$U(x, y) = \frac{V_0}{\pi} \log \left| \frac{z - \sigma}{z + \sigma} \right|.$$

The level curves of U represent the curves of flux of the electric field; these curves are just the circles of Apollonius

$$|z - \sigma| = \rho |z + \sigma|, \quad 0 < \rho < \infty,$$

which are centered at the real axis, either in the interval $(-\infty, -\sigma)$ or in the interval (σ, ∞) (Fig. 4.26). $\qquad \square$

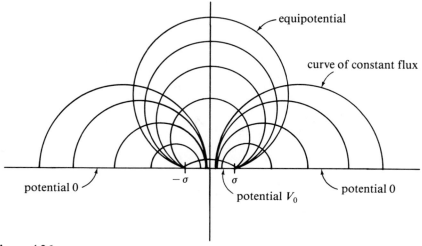

Figure 4.26

The reflection principle

The Poisson integral formula in a disc can be used to give a proof of the following theorem, which is called, for reasons that will be clear from its statement, the **reflection principle**.

> **Theorem: The reflection principle.** Suppose D is a domain in the upper half-plane $\{x + iy: y > 0\}$, which includes in its boundary a segment (a, b) of the real axis. Suppose that $f = u + iv$ is analytic in D and, moreover, is continuous at each point of (a, b). If $v(x) = 0$ for all x in the interval (a, b), then f extends to be analytic in the domain $\Omega = D \cup (a, b) \cup D^*$, where D^* is the reflection of D over the real axis:
>
> $$D^* = \{z: \bar{z} \in D\}$$
>
> Furthermore, f satisfies the relation
>
> $$f(z) = \overline{f(\bar{z})}, \ z \in \Omega.$$

Proof. The sets D, D^*, and Ω are illustrated in Figure 4.27. We begin the proof by defining a function f^* in D^* by

$$f^*(z) = \overline{f(\bar{z})}, \ z \in D^*.$$

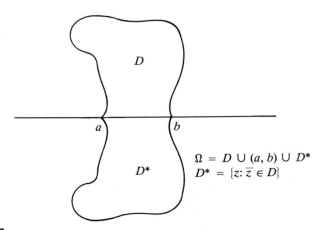

Figure 4.27

We shall show that f^* is analytic in D^* by this simple argument. If $z_0 \in D^*$, then $\bar{z}_0 \in D$. Now f is analytic in some disc $|z - \bar{z}_0| < \delta$ centered at $\bar{z}_0$ and thus, f has a power series valid in this disc:

$$f(z) = \sum_{k=0}^{\infty} A_k(z - \bar{z}_0)^k, \ |z - \bar{z}_0| < \delta.$$

Hence, in the disc $|z - z_0| < \delta$, we have

$$f^*(z) = \overline{f(\bar{z})} = \overline{\sum_{k=0}^{\infty} A_k(\bar{z} - \bar{z}_0)^k}$$

$$= \sum_{k=0}^{\infty} \bar{A}_k(z - z_0)^k.$$

Consequently, f^* is analytic for $|z - z_0| < \delta$ since it has a representation there as the sum of a convergent power series.

Next, we define a function F on Ω by

$$F(z) = \begin{cases} f(z), & z \in D \cup (a, b) \\ f^*(z), & z \in D^*. \end{cases}$$

The assumption that $v(x) = 0$ for $a < x < b$ implies that F is continuous on Ω. The functions u and v are extended by D^* by the rules $u(z) = u(\bar{z})$, $v(z) = -v(\bar{z})$, $z \in D^*$. The tricky part of the proof is to show that $F = u + iv$ is analytic across the segment (a, b). To see this, let $x_0 \in (a, b)$ and let r be a positive number so small that the disc $|z - x_0| < 2r$ lies in Ω. Let γ be the positively oriented circle $|z - x_0| = r$. On this circle the function

$$w(\theta) = \begin{cases} v(x_0 + re^{i\theta}), & \text{if } 0 \leqslant \theta \leqslant \pi \\ -v(x_0 + re^{-i\theta}), & \text{if } -\pi \leqslant \theta \leqslant 0 \end{cases}$$

is continuous and odd: $w(\theta) = -w(-\theta)$. Let v_1 be the harmonic function on the disc $|z - x_0| < r$, whose values on γ are w; v_1 is found by using the Poisson integral, appropriately scaled. Then v_1 satisfies $v_1(z) = -v_1(\bar{z})$ if $|z - x_0| < r$ (also, see Exercise 8). Thus, v_1 vanishes on the real segment $x_0 - r \leqslant x \leqslant x_0 + r$. The function $v - v_1$ is then harmonic on the half-disc $U = \{z : |z - x_0| < r \text{ and } \operatorname{Im} z > 0\}$ and vanishes on the boundary of U since $v = v_1 = w$ on the semicircle $z = x_0 + re^{i\theta}$, $0 \leqslant \theta \leqslant \pi$, and $v = v_1 = 0$ on the segment $x_0 - r \leqslant x \leqslant x_0 + r$. Hence, the maximum principle for harmonic functions [(5) in Section 1] implies that $v = v_1$ on all of U. Likewise, $v = v_1$ on all of $U^* = \{z : \bar{z} \in U\}$. Thus, $v = v_1$ on the disc $|z - x_0| < r$. But v_1 is harmonic on all this disc and hence so must be v. The function u is a harmonic conjugate of v on both U and U^*. Since v is harmonic on all the disc $|z - x_0| < r$, it has a harmonic conjugate on all this disc that must differ from u by a constant on $U \cup U^*$. Hence, u is also harmonic on the disc $|z - x_0| < r$, so F is actually analytic on this disc. $\qquad \square$

Remark: The reflection principle is frequently proved by use of Morera's Theorem.

Example 3 Show that an entire function which is real-valued on some interval (a, b) in the real axis is real-valued on all the real axis.

Solution Let f be an entire function that is real-valued on the interval (a, b).

Set

$$F(z) = \begin{cases} f(z), & \text{if } \text{Im } z \geqslant 0 \\ \overline{f(\bar{z})}, & \text{if } \text{Im } z < 0. \end{cases}$$

The reflection principle implies immediately that F is analytic on the domain Ω, which consists of the union of the upper and lower half-planes and the interval (a, b). Furthermore, f itself is of course analytic on Ω and equals F if $\text{Im } z > 0$. Thus, $f = F$ throughout Ω and, in particular,

$$f(z) = \overline{f(\bar{z})}, \quad \text{Im } z \neq 0.$$

Since f is actually continuous on the whole plane we have for any real number x,

$$f(x) = \lim_{y \to 0} f(x + iy) = \lim_{y \to 0} \overline{f(x - iy)} = \overline{f(x)},$$

which says exactly that f is real on the real axis. □

Exercises for Section 4.3

1. An arc of θ_0 radians in a circle is kept at temperature T_1, while the remainder of the circle is kept at temperature T_2. Find the temperature distribution inside the circle. Sketch some of the equitherms.

2. The segment $[-\sigma, \sigma]$ is kept at temperature T_1, while the remainder of the real axis is kept at temperature T_2. Find the temperature distribution in the upper half-plane; sketch some of the equitherms.

3. Let $w(t)$ be a bounded piecewise continuous function on the real axis with $\int_{-\infty}^{\infty} |w(t)| \, dt < \infty$ and $w(t) \to \infty$ as $|t| \to \infty$. Let $W(\zeta)$ be the harmonic function on the upper half-plane $\text{Im } \zeta > 0$, with $W(\sigma, 0) = w(\sigma)$, given by formula (11).
    (a) Show that $\lim_{|\zeta| \to \infty} W(\zeta) = 0$.
    (b) Show that if $W_1(\zeta)$ is (another) harmonic function on $\text{Im } \zeta > 0$ with $W_1(\sigma, 0) = w(\sigma)$ and if W_1 also satisfies (a), then $W_1 = W$.
    (c) Show that $\int_{-\infty}^{\infty} |W(x + iy)| \, dx \leqslant \int_{-\infty}^{\infty} |w(t)| \, dt$ for each $y > 0$.

4. Let $P_r(t)$ be the Poisson kernel, defined in formula (5). Show that
    (i) $P_r(t) > 0$ for $0 \leqslant r < 1$, $-\pi \leqslant t \leqslant \pi$,
    (ii) $\dfrac{1}{2\pi} \displaystyle\int_{-\pi}^{\pi} P_r(t) \, dt = 1, \, 0 \leqslant r < 1$,
    (iii) if $\delta > 0$, then $\lim_{r \to 1} \{\max_{\pi \geqslant |t| \geqslant \delta} P_r(t)\} = 0$.

5. We show here that (10) holds. Fix a point λ with $|\lambda| = 1$ and write $\lambda = e^{i\theta_0}$. Fill in the reasons for each of the following steps
    (i) given $\varepsilon > 0$, there is a $\delta > 0$ such that

$$|u(e^{it}) - u(e^{i\theta_0})| < \varepsilon/2 \quad \text{if } |t - \theta_0| < \delta$$

(ii) $U(re^{i\theta}) - u(e^{i\theta_0}) = \dfrac{1}{2\pi} \displaystyle\int_{-\pi}^{\pi} \{u(e^{it}) - u(e^{i\theta_0})\} P_r(\theta - t)\, dt.$

Now consider θ with $|\theta - \theta_0| < \delta/4$. Write the integral as the sum of two integrals: the first, I_1, is over those t with $|t - \theta_0| < \delta$; the second, I_2, is over all the remaining t.

(iii) $|I_1| < \dfrac{\varepsilon}{2}\dfrac{1}{2\pi} \displaystyle\int_{|t-\theta_0|<\delta} P_r(\theta - t)\, dt < \varepsilon/2$

(iv) $|I_2| < 2M\left(\max\left\{ P_r(\theta - t) : |\theta - t| \geqslant \dfrac{3\delta}{4} \right\} \right),$

where M is a upper bound on $|u(e^{it})|$, $-\pi \leqslant t \leqslant \pi$.
(v) $|I_2| < \varepsilon/2$ if r is close enough to 1
(vi) $|U(re^{i\theta}) - u(e^{i\theta_0})| < \varepsilon$ if $|\theta - \theta_0| < \delta/4$ and r is close enough to 1.

6.   Show that the function

$$u(x, y) = \frac{1}{\pi} \int_0^\infty \left\{ \frac{y}{(x-t)^2 + y^2} - \frac{y}{(x+t)^2 + y^2} \right\} f(t)\, dt$$

is harmonic in the upper half-plane and satisfies

$$u(0, y) = 0, \quad 0 < y < \infty$$
$$u(x, 0) = f(x), \; 0 < x < \infty.$$

7.   Use Exercise 6 to find an integral representation for a harmonic function $u(x, y)$ in the first quadrant $\{x + iy : x > 0, y > 0\}$ with

$$u(x, 0) = f_1(x), \; 0 < x < \infty$$
$$u(0, y) = f_2(y), \; 0 < y < \infty.$$

8.   Show that the function

$$u(re^{i\theta}) = \frac{1}{2\pi} \int_0^\pi \{P_r(t - \theta) - P_r(t + \theta)\} f(e^{it})\, dt$$

is harmonic in the half-disc $\{re^{i\theta} : 0 < r < 1, 0 < \theta < \pi\}$ and satisfies the boundary conditions

$$u(e^{i\theta}) = f(e^{i\theta}), \quad 0 < \theta < \pi$$
$$u(x) = 0, \quad\quad -1 < x < 1.$$

9.   Solve the Dirichlet problem for the exterior of the disc of radius 1 centered at the origin; that is, give an integral formula for a harmonic function $v(z)$ on the region $1 < |z| \leqslant \infty$ (including ∞) that satisfies

$$\lim_{z \to \lambda} v(z) = f(\lambda), \; |\lambda| = 1.$$

10.   Use the result of Example 2 to find the electrostatic potential in the region

$$D = \{z + iy : x^2 + y^2 > \sigma^2 \text{ and } y > 0\}$$

when the potential is 0 on both the segments $(-\infty, -\sigma)$ and (σ, ∞) and 1 on the semicircle $x^2 + y^2 = \sigma^2$, $y > 0$.

11.   The function

$$u(re^{i\theta}) = \frac{1 - r^2}{1 - 2r \cos \theta + r^2}$$

is harmonic on the disc $0 \leqslant r < 1$. Show that

$$\lim_{r \to 1} u(re^{i\theta}) = 0 \quad \text{if } \pi \geqslant |\theta| > 0.$$

Show, moreover, that $u(re^{i\theta})$ represents the temperature distribution in the disc $0 \leqslant r < 1$ due to a "hot spot" at the point $r = 1$, $\theta = 0$ (that is, the temperature there is "∞.") **(Hint:** Find the temperature distribution u_n in the disc due to the boundary temperature

$$u_n(e^{i\theta}) = \begin{cases} n, & |\theta| < \dfrac{\pi}{2n} \\[2ex] 0, & \dfrac{\pi}{2n} < |\theta| \leqslant \pi. \end{cases}$$

Then let $n \to \infty$.)

12.   Extend the result in Exercise 11 to the case when there are N "hot spots" at points $e^{i\theta_1}, \ldots, e^{i\theta_N}$ of the circle of radius 1 centered at the origin. Discuss what happens if some of the "hot spots" are instead "cold spots" (that is, the temperature there is $-\infty$).

13.   Give the temperature distribution in the upper half-plane $y > 0$ that is produced by a single "hot spot" at the point x_0 on the real axis. Do the same for several "hot/cold spots" at the points $x_1, \ldots, x_N$ on the real axis.

14.   Argue on physical grounds that if $v(re^{i\theta})$ is a nonnegative harmonic function on the disc $0 \leqslant r < 1$ with

$$v(re^{i\theta}) \leqslant \frac{1 - r^2}{1 - 2r \cos \theta + r^2} = P_r(\theta), \quad 0 \leqslant r < 1, \ 0 \leqslant \theta \leqslant 2\pi,$$

then $v(re^{i\theta}) = \lambda P_r(\theta)$ for some λ, $0 < \lambda \leqslant 1$. **(Hint:** What temperature distribution on $|z| = 1$ does v represent?)

15.   Suppose $u(z)$ is a real-valued harmonic function on the disc $\{z: |z| < 1 + \delta\}$ for some $\delta > 0$; let v be the harmonic conjugate of u on this disc, which is zero at the origin. Show that

$$v(re^{i\theta}) = \frac{1}{2\pi} \int_0^{2\pi} u(e^{it}) \frac{2r \sin(\theta - t)}{1 - 2r \cos(\theta - t) + r^2} \, dt.$$

(Hint: The function

$$f(z) = \frac{1}{2\pi} \int_0^{2\pi} u(e^{it}) \frac{e^{it} + z}{e^{it} - z} \, dt$$

is analytic on the disc $\{z : |z| < 1\}$ and its real part is $u(z)$ by (7).)

Fourier series and harmonic functions★

16.   Let $u(e^{i\theta})$ be a piecewise continuous function on the circle $\{e^{i\theta} : 0 \leqslant \theta \leqslant 2\pi\}$. Set

$$\hat{u}(n) = \frac{1}{2\pi} \int_0^{2\pi} u(e^{i\theta}) e^{-in\theta} \, d\theta, \quad n = 0, \pm 1, \pm 2, \ldots;$$

$\hat{u}(n)$ is the **nth Fourier coefficient** of u. Show that

$$|\hat{u}(n)| \leqslant \max_{0 \leqslant \theta \leqslant 2\pi} |u(e^{i\theta})|.$$

17.   Let k be an integer. Show that

$$\frac{1}{2\pi} \int_0^{2\pi} e^{ik\theta} \, d\theta = \begin{cases} 1, & k = 0 \\ 0, & k \neq 0. \end{cases}$$

18.   If $u(e^{i\theta}) = \sum_{-\infty}^{\infty} a_n e^{in\theta}$, where $\sum_{-\infty}^{\infty} |a_n| < \infty$, then $a_n = \hat{u}(n)$.

19.   Let $P_r(\theta)$ be the Poisson kernel:

$$P_r(\theta) = \frac{1 - r^2}{1 - 2r \cos \theta + r^2} = \mathrm{Re}\left(\frac{1 + re^{i\theta}}{1 - re^{i\theta}}\right).$$

Show that

$$P_r(\theta) = \sum_{-\infty}^{\infty} r^{|n|} e^{in\theta}.$$

(**Hint:** $1/(1 - z) = 1 + z + z^2 + \cdots$, if $0 \leqslant |z| < 1$.)

20.   Let $U(re^{i\theta})$ be the harmonic function on the disc $\{z : |z| < 1\}$ with $\lim_{r \to 1} U(re^{i\theta}) = u(e^{i\theta})$, except at the finitely many points of discontinuity of $u(e^{i\theta})$. Show that

$$U(re^{i\theta}) = \sum_{-\infty}^{\infty} \hat{u}(n) r^{|n|} e^{in\theta}, \quad 0 \leqslant r < 1, \ 0 \leqslant \theta \leqslant 2\pi.$$

21.   Suppose $v(e^{i\theta})$ is another piecewise continuous function on the circle $\{e^{i\theta} : 0 \leqslant \theta \leqslant 2\pi\}$ and $V(re^{i\theta})$ is its harmonic extension to the disc $\{z : |z| < 1\}$. Show that for each r, $0 < r < 1$, we have

$$\frac{1}{2\pi} \int_0^{2\pi} U(re^{i\theta}) \overline{V(re^{i\theta})} \, d\theta = \sum_{-\infty}^{\infty} r^{2n} \hat{u}(n) \overline{\hat{v}(n)}. \tag{12}$$

(**Hint:** By Exercise 18, with $U(re^{i\theta})$ in place of $u(e^{i\theta})$, we know that the nth Fourier coefficient of $U(re^{i\theta})$ is $r^{|n|}\hat{u}(n)$.)

22.   Let r increase to 1 in (12); conclude that

$$\frac{1}{2\pi} \int_0^{2\pi} u(e^{i\theta})\overline{v(e^{i\theta})}\, d\theta = \sum_{-\infty}^{\infty} \hat{u}(n)\overline{\hat{v}(n)}. \tag{13}$$

(This is a correct formula but this passage from (12) to (13) needs more mathematical justification than just letting $r \to 1$.)

23.   In Exercise 22 take $v = u$ and conclude that

$$\frac{1}{2\pi} \int_0^{2\pi} |u(e^{i\theta})|^2\, d\theta = \sum_{-\infty}^{\infty} |\hat{u}(n)|^2. \tag{14}$$

This equation is known as **Parseval's equality**.

The reflection principle

24.   Suppose that g is analytic in a domain D in the right half-plane, Re $z > 0$, which includes a segment $(i\alpha, i\beta)$ of the imaginary axis in its boundary. Show that if g is real-valued on this segment, then g extends to be analytic in the domain formed by the union of D, the reflection of D over the imaginary axis $(= \{-\bar{z}: z \in D\})$, and the interval $(i\alpha, i\beta)$.

25.   Suppose that f is an entire function that is real-valued on some segment of the real axis and also real-valued on some segment of the imaginary axis. Show that $f(-z) = f(z)$ for all z. (**Hint:** Reflect f over the imaginary axis.)

26.   Use the reflection principle to prove this result. Let D be a domain in the upper half-plane with the segment (a, b) of the real axis in its boundary. Suppose that f is analytic on D, continuous on $D \cup (a, b)$, and $f(z) = 0$ if z is in (a, b). Then $f(z) = 0$ for all $z \in D$.

27.   Find the form of a rational function M that is real on the circle $\{z: |z| = 1\}$; assume that M has no poles on this circle.

28.   Show that an entire function has only real coefficients in its power series expansion about the origin if and only if it is real-valued on the real axis.

29.   Suppose that f is analytic on the open disc $\{z: |z| < 1\}$ and continuous on the closed disc $\{z: |z| \leqslant 1\}$. Show that if $|f(z)| = 1$ whenever $|z| = 1$, then f can be extended by reflection to be analytic, except for finitely many poles, in the whole complex plane by the rule

$$f^*(z) = \frac{1}{\overline{f(1/\bar{z})}}, \quad |z| > 1.$$

Show further that f is a rational function and, in particular, has the form

$$f(z) = \lambda \frac{z - a_1}{1 - \bar{a}_1 z} \cdots \frac{z - a_M}{1 - \bar{a}_M z}, \quad |a_j| < 1; |\lambda| = 1.$$

30.   Suppose f is analytic on the disc $\{z : |z| < 1\}$, continuous on the set $\{z : |z| \leqslant 1\}$, and real-valued on the circle $\{z : |z| = 1\}$. Use the reflection principle to show that f is constant. (**Hint:** Extend f to the set $\{z : |z| > 1\}$ by $f*(z) = \overline{f(1/\bar{z})}$.)

4.4   Boundary-value problems

In a **boundary-value problem** we are given information about an unknown function on the boundary of a domain and an equation (usually a differential equation) that the function satisfies on the domain; we are challenged to find the function itself on the domain. In complex variables, the boundary-value problems typically have one of two different forms. In the first, we are to find a harmonic function u on the domain D from the knowledge of the function itself on the boundary of the domain; this is the **Dirichlet* problem** for D. In the second, we are again to find a harmonic function on D, but this time we know only the normal derivative of the function on the boundary of D; this is the **Neumann† problem** for D. Of course, there is also a mixture of these two cases, where we know the function on part of the boundary and its normal derivative elsewhere on the boundary.

There are two basic techniques in solving boundary-value problems: the first is to use an explicit integral representation formula such as the Poisson integral formula. This technique was illustrated in Section 3, where we examined the Poisson integral formula both on the disc and on the upper half-plane. The second basic technique is to transform the given region by a conformal mapping to a new region on which the boundary-value problem is easily solved. This solution can then be carried back to the original region by means of the conformal mapping, thus yielding an explicit solution to the problem. Care must be taken to be sure that the boundaries of the two regions correspond to each other in the correct way but this happens automatically in most cases.

We illustrate this second technique of conformal mapping by several examples.

Example 1   Find the electrostatic potential between two long, parallel, charged, hollow, circular cylinders if one is inside the other.

Solution   A cross-section of the cylinders perpendicular to their axes gives

---

* Peter Gustav Lejeune Dirichlet, 1805–1859.
† Carl Neumann, 1832–1925.

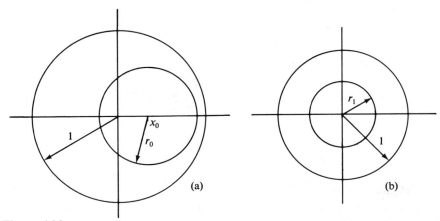

Figure 4.28

two circles, one inside the other. By a simple translation, rotation, and change of scale we may assume the outer circle is the circle $\Gamma_0 = \{z\colon |z| = 1\}$ and the inner circle is the circle $\Gamma_1 = \{z\colon |z - x_0| = r_0\}$, where $0 \leqslant x_0 < 1$ and $x_0 + r_0 < 1$ (Fig. 4.28). Let the charge on the inner and outer cylinders be C_1 and C_0, respectively. Thus, we are searching for a harmonic function on the region between the two circles Γ_0 and Γ_1, which has the value C_0 on the outer circle Γ_0 and the value C_1 on the inner circle Γ_1. If the smaller circle is centered at the origin (that is, if the two cylinders are **coaxial**) then the solution is immediate*; namely,

$$U(re^{i\theta}) = C_0 \frac{\log r_0 - \log r}{\log r_0} + C_1 \frac{\log r}{\log r_0}, \quad r_0 \leqslant r \leqslant 1$$

because we know that $w(re^{i\theta}) = \log r$ is a harmonic function (see Exercise 3 of Section 1). However, it may not be the case that the circles are concentric, so that we must first transform the region between Γ_0 and Γ_1 into the region between concentric circles. This can be accomplished by use of an appropriate linear fractional transformation. A computation (or see Exercise 16, Section 5, Chapter 3) shows that there is a real number b such that the function

$$\phi(z) = \frac{z - b}{1 - bz}$$

maps the inner circle Γ_1 onto a circle of radius r_1 centered at the origin; of course, ϕ also maps the outer circle Γ_0 onto itself. Consequently, from the solution to the problem in the coaxial case, we obtain the desired solution in this case:

$$u(z) = C_0 \frac{\log r_1 - \log |\phi(z)|}{\log r_1} + C_1 \frac{\log |\phi(z)|}{\log r_1}. \qquad \square$$

---

* The domain and the boundary data are circularly symmetric and hence the solution $U(re^{i\theta})$ should be as well. However, the only harmonic function that is independent of θ is $A + B \log r$ (see Exercise 5(b) in Section 1).

Example 2 A magnetic field is established by a north and south pole, each of large rectangular cross-section (Fig. 4.29a). Find the magnetic field in the surrounding space.

Solution We arbitrarily assign the value $+1$ to the field on the boundary of the north pole and -1 to the field on the boundary of the south pole. Our task is then to find a harmonic function in the region D outside the poles with these boundary values. (In practice, of course, the poles do not have infinite extent; this model assumes the gap between the poles is small in comparison to the width of the poles. The model is accurate for the magnetic field in the region and close to the gap, where it is of the most interest in any event.)

In Section 5, Chapter 3 we found the Schwarz–Christoffel transformation that maps the upper half plane U onto the domain D:

$$\phi(z) = \frac{2a}{\pi} \{ \sqrt{z^2 - 1} + \arcsin(1/z) \};$$

ϕ carries the ray $(-\infty, 0)$ onto the edge of the north pole and the ray $(0, \infty)$ to the edge of the south pole (Fig. 4.29b). The function $1 - (2/\pi)\operatorname{Arg} z = U(z)$ is harmonic on the upper half-plane and its boundary values are 1 on $(0, \infty)$ and -1 on $(-\infty, 0)$. Therefore, the function

$$u(w) = 1 - \frac{2}{\pi} \operatorname{Arg}(\phi^{-1}(w))$$

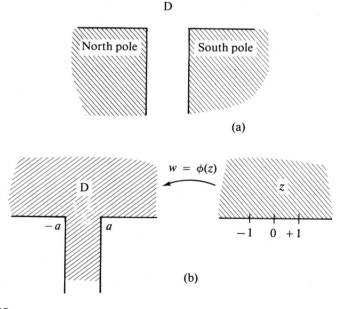

D

(a)

$w = \phi(z)$

D

$-a$ a

z

-1 0 $+1$

(b)

Figure 4.29

is the desired function on D. The curves of magnetic flux are the images under ϕ of the level lines of V, a harmonic conjugate of U: $V(z) = (2/\pi)\log|z|$. These level lines are circles centered at the origin and so the curves of magnetic flux of the field are the curves

$$\Gamma_r = \frac{2a}{\pi}\left\{\sqrt{r^2e^{2it} - 1} + \arcsin\left(\frac{e^{-it}}{r}\right): 0 \leqslant t \leqslant \pi\right\}$$

for $0 < r < \infty$. Some of these curves are shown in Figure 4.30.   □

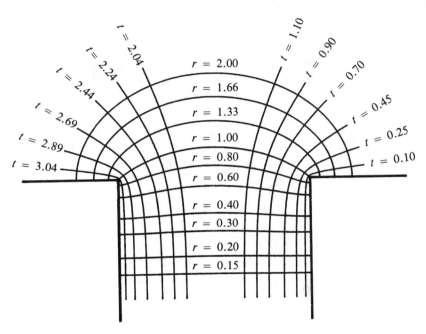

Figure 4.30   The curves of magnetic flux and equipotentials

The Neumann problem

In this boundary-value problem we are given a function g on the boundary of a domain D and we are to find a harmonic function u on D with normal derivative equal to g on the boundary of D: $\partial u/\partial n = g$ on ∂D. This is physically realizable as specifying the flux over the boundary of the region of a field or the permeability of the boundary for diffusion. This problem, like the Dirichlet problem, can be explicitly solved in the case of a disc or a half-plane. We work out the case for the disc $|z| < 1$ here and leave the other setting (which is handled similarly) for the exercises.

Let g be the value of the normal derivative at the unit circle $|z| = 1$.

We seek a function $v(re^{i\theta})$ that is harmonic on the disc $\{re^{i\theta}: 0 \leqslant r < 1\}$ and satisfies

$$\frac{\partial v}{\partial r}(e^{it}) = g(t), \quad 0 \leqslant t \leqslant 2\pi.$$

Consider the function

$$v(re^{i\theta}) = \frac{-1}{2\pi} \int_0^{2\pi} g(t) \text{Log}(1 - 2r\cos(\theta - t) + r^2) \, dt.$$

This function is harmonic since it is the real part of the analytic function

$$\frac{-1}{\pi} \int_0^{2\pi} g(t) \text{Log}(1 - ze^{-it}) \, dt, \quad z = re^{i\theta},$$

assuming $g(t)$ is real. Furthermore,

$$\frac{\partial v}{\partial r}(se^{it}) = -\frac{1}{2\pi} \int_0^{2\pi} g(t) \frac{\partial}{\partial r} \text{Log}(1 - 2r\cos(\theta - t) + r^2)\Big|_{r=s} dt$$

$$= \frac{1}{2\pi} \int_0^{2\pi} g(t) \frac{2\cos(\theta - t) - 2s}{1 - 2s\cos(\theta - t) + s^2} \, dt$$

$$= \frac{1}{2\pi s} \int_0^{2\pi} g(t) \left[\frac{1 - s^2}{1 - 2s\cos(\theta - t) + s^2} - 1 \right] dt$$

$$= \frac{1}{2\pi s} \int_0^{2\pi} g(t) P_s(\theta - t) \, dt - \frac{1}{2\pi s} \int_0^{2\pi} g(t) \, dt,$$

where $P_s(\theta - t)$ is the Poisson kernel. However, we know from Section 3 and the properties of the Poisson kernel that

$$\lim_{s \to 1} \frac{1}{2\pi} \int_0^{2\pi} g(t) P_s(\theta - t) \, dt = g(\theta).$$

Hence,

$$\lim_{s \to 1} \frac{\partial v}{\partial r}(se^{i\theta}) = g(\theta) - \frac{1}{2\pi} \int_0^{2\pi} g(t) \, dt.$$

However, the integral of g must vanish since there are no sources or sinks within the disc $\{z: |z| < 1\}$. Therefore, v is the desired solution. (Another way to see that $\int_0^{2\pi} g(t) \, dt = 0$ is to use Green's Theorem:

$$\int_0^{2\pi} g(t) \, dt = \int_\Gamma \frac{\partial v}{\partial n} \, ds = \int\int_\Omega \Delta v \, dx \, dy = 0 \quad \text{since } \Delta v \equiv 0.)$$

Example 3 A long cylinder with a circular cross-section (for example, a blood vessel) contains a fluid with a certain chemical A dissolved within it. The walls of the cylinder allow the chemical A to permeate through with a permeability that depends only on the angular location; moreover, there is a "line" source of the chemical A within the cylinder of strength exactly sufficient to keep the whole system in steady state. If the permeability function is given, find the strength of the source and the lines of flow of the chemical within a cross-section of the cylinder (Fig. 4.31).

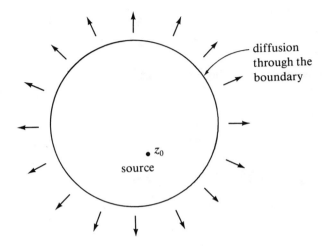

Figure 4.31

Solution   By taking a cross-section we work within a circle, which we can take to be the circle $x^2 + y^2 = 1$. The total flux of the chemical A through the circle is

$$Q = \int_0^{2\pi} g(t) \, dt,$$

where $g(t)$ is the coefficient of permeability at the point e^{it}, $0 \leqslant t \leqslant 2\pi$. Since the system is in steady state, Q is also the strength of the source. If the source is located at the point z_0, then it is given by

$$u_1(z) = \frac{Q}{2\pi} \log \left| \frac{1 - \bar{z}_0 z}{z - z_0} \right|, \quad |z| < 1.$$

We wish, then, to find a function $u(z)$ that is harmonic on the disc $|z| < 1$ except at the point z_0 and that satisfies

$$\frac{\partial u}{\partial r} (e^{it}) = g(t).$$

The function $u_1(z)$ satisfies

$$-\frac{\partial u_1}{\partial r} (e^{it}) = \frac{Q}{2\pi} \left(\frac{1 - r_0^2}{1 - 2r_0 \cos(\theta_0 - t) + r_0^2} \right), \quad z_0 = r_0 e^{i\theta_0};$$

(see Exercise 13 at the end of this section). Furthermore, the function

$$u_2(re^{i\theta}) = -\frac{1}{2\pi} \int_0^{2\pi} \left\{ g(t) - \frac{\partial u_1}{\partial r}(e^{it}) \right\} \text{Log}(1 - 2r\cos(\theta - t) + r^2)\, dt$$

is harmonic on the disc $|z| < 1$ and satisfies

$$\frac{\partial u_2}{\partial r}(e^{it}) = g(t) - \frac{\partial u_1}{\partial r}(e^{it}).$$

Consequently, $u(z) = u_1(z) + u_2(z)$ is the desired solution to the problem. In particular, in the very special case when the permeability is constant and the source is at the origin we find that

$$u(z) = -\frac{Q}{2\pi} \log |z|. \qquad \square$$

Example 4 A thin layer of a chemical solution occupies the lower half-plane $\{x + iy: y < 0\}$. The segments $(-\infty, -1]$ and $[1, \infty)$ are insulated so that there is no diffusion across these portions of the boundary but across the segment $(-1, 1)$ there is a semiporous membrane that allows diffusion at a constant rate into the upper half-plane. Find the steady-state concentration of the substance in the upper half-plane, and the lines of flow of the chemical.

Solution The concentration function $c(x, y)$ for which we are searching must satisfy

$$\frac{\partial c}{\partial y}(x, 0) = \begin{cases} 0, & \text{if } |x| > 1 \\ c_0, & \text{if } |x| < 1. \end{cases} \tag{1}$$

As a first guess we try

$$b(x, y) = \frac{c_0}{2\pi} \int_{-1}^{1} \log[(x - t)^2 + y^2]\, dt.$$

The function b has the correct normal derivative on the real axis since

$$\frac{\partial b}{\partial y} = \frac{c_0}{\pi} \int_{-1}^{1} \frac{y}{(x - t)^2 + y^2}\, dt$$

and by the Poisson integral formula for the upper-half-plane we know that

$$\lim_{y \to 0} \frac{\partial b}{\partial y}(x, y) = \begin{cases} c_0, & |x| < 1 \\ 0, & |x| > 1 \end{cases}$$

(also see Exercise 1 in this section). However, this function is not the solution we wish because it is clearly unbounded for large values of $x^2 + y^2$. Instead, we consider

$$c(x, y) = \frac{c_0}{2\pi} \left[\int_{-1}^{1} \log[(x - t)^2 + y^2]\, dt - 2\log(x^2 + y^2) \right].$$

This function is still harmonic for $y > 0$ and certainly satisfies (1). Furthermore, if $x^2 + y^2$ is large, $c(x, y)$ is virtually zero. Thus, $c(x, y)$ given above is the desired solution.   □

Example 5   Groundwater flows out of an underground aquifer bounded on the bottom by a layer of impervious rock and on the top by a semi-infinite layer of impervious rock; the remainder of the top is fully permeable (Fig. 4.32). Find the flow and its streamlines if the discharge per unit width of the aquifer is q.

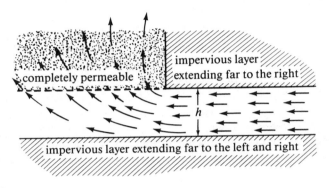

Figure 4.32

Solution   By the introduction of a coordinate system with the x-axis along the bottom face of the aquifer and the y-axis passing through the left end of the top face we have the situation shown in Figure 4.33a. The flow has a source "at ∞," that is, far to the right, of total strength $Q = qh$. We now consider the domain D, which is the horizontal strip $\{x + iy: -\infty < x < \infty, 0 < y < h\}$. See Figure 4.33b. On D we seek a harmonic function $v(x, y)$ (the potential) with these boundary conditions:

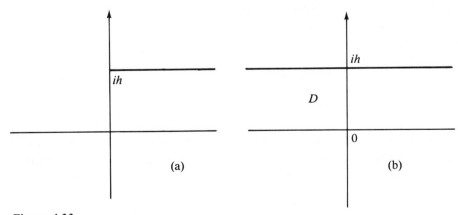

Figure 4.33a

$$\frac{\partial v}{\partial n}(x,\, y) = 0 \ \text{ if } \begin{cases} y = 0 \text{ and } -\infty < x < \infty \\ y = h \text{ and } \quad\ 0 < x < \infty \end{cases}$$

$$v(x,\, y) = 0 \text{ if } \ y = h \text{ and } -\infty < x < 0.$$

The first of these conditions says the boundary is impermeable and the second says that the groundwater is free to go out of the aquifer from that portion of the top that is not blocked by the impervious rock. The flow must, of course, have a source of strength Q far to the right. Therefore, we are faced with a mixed Dirichlet–Neumann problem; we solve it by finding a conformal mapping of the strip D onto a domain in which the corresponding problem is solved by inspection.

The function $\zeta = (-1 + \cos w)/2$ maps the vertical half-strip $S = \{w:$ $\text{Im } w > 0$ and $-\pi < \text{Re } w < 0\}$ onto the upper half-plane, $\text{Im } \zeta > 0$. In turn, the function $z = (h/\pi)\text{Log } \zeta$ maps the upper half-plane onto D. Thus,

$$z = f(w) = \frac{h}{\pi} \text{Log}\!\left(\frac{-1 + \cos w}{2}\right)$$

is a one-to-one analytic mapping of the strip S onto D. Furthermore, the mapping has been carefully arranged so that the left vertical edge of the strip S is mapped to the segment $\{x + ih: 0 < x < \infty\}$, the right vertical edge of S is sent to the real axis, and the horizontal bottom edge of S is sent to the segment $\{x + ih: -\infty < x < 0\}$.

On S the function $(Q/\pi)\text{Im } w$ has all the required properties: it is zero on the bottom edge, its normal derivative (that is, its partial derivative with respect to Re w) is zero on both the vertical edges and the flow given by the gradient of $(Q/\pi)\text{Im } w$ has a source at ∞ at strength Q. Since

$$w = \arccos[1 + 2\,\exp(z\pi/h)],$$

we see that

$$v(x,\, y) = \frac{Q}{\pi}\,\text{Im}\{\arccos[1 + 2\,\exp(z\pi/h)]\}$$

is the required potential. The complex potential is, then,

$$G(z) = -i\,\frac{Q}{\pi}\,\arccos[1 + 2\,\exp(z\pi/h)]. \qquad \square$$

Exercises for Section 4.4

1. Derive this formula for a solution to the Neumann problem in the upper half-plane $y > 0$:

$$u(x,\, y) = \frac{1}{2\pi} \int_{-\infty}^{\infty} g(t)\log[(x - t)^2 + y^2]\, dt + c$$

(where we assume that $|g(t)| \leqslant Ct^{-\nu}$ for all large t and for some $\nu > 1$ to make the integral converge).

2.   Suppose that $g(-t) = -g(t)$ for all t; show that

$$u(x, y) = \frac{1}{2\pi} \int_0^\infty g(t) \log \frac{(t-x)^2 + y^2}{(t+x)^2 + y^2} \, dt + c$$

is a harmonic function in the upper half-plane $y > 0$ with $u(0, y) = 0$ and $(\partial u / \partial y) = g$ on the real line.

3.   Suppose that $g(t) = 0$ if $|t| > M$; let

$$u(x, y) = \frac{1}{2\pi} \int_{-M}^M g(t) \log[(x-t)^2 + y^2] \, dt - A \log(x^2 + y^2),$$

where $A = (1/2\pi) \int_{-M}^M g(t) \, dt$. Show that $u(x, y)$ is harmonic in the upper half-plane, $u(x, y) \to 0$ as $x^2 + y^2 \to \infty$ and

$$\lim_{\delta \downarrow 0} \frac{\partial u}{\partial y}(x, \delta) = g(x), \quad -\infty < x < \infty.$$

4.   Show that the function

$$u(x, y) = \frac{1}{\pi} \int_0^\infty g(t) \left\{ \frac{y}{(x-t)^2 + y^2} + \frac{y}{(t+x)^2 + y^2} \right\} dt$$

is harmonic in the first quadrant and satisfies

$$u(x, 0) = g(x), \quad 0 < x < \infty$$

$$\frac{\partial u}{\partial x}(0, y) = 0.$$

5.   Modify the integral formula in Exercise 4 to find an integral formula for a function $u(x, y)$ harmonic in the first quadrant $x > 0$, $y > 0$, which satisfies

$$u(0, y) = f(y)$$

$$\frac{\partial u}{\partial y}(x, 0) = 0.$$

6.   Find the temperature distribution $T(x, y)$ in the upper half-plane if $T(x, 0) = T_1$ for $x < -\sigma$, $T(x, 0) = T_2$ for $x > \sigma$ and the segment $-\sigma < x < \sigma$ is insulated (that is, $(\partial T / \partial y)(x, 0) = 0$ for $-\sigma < x < \sigma$).

7.   Find the temperature distribution in the first quadrant if $T(0, y) = T_1$, $0 < y < \infty$, $T(x, 0) = T_2$ for $\sigma < x < \infty$, and the segment $(0, \sigma)$ on the real axis is insulated.

8.   Find the temperature distribution in an infinite straight rod if the boundary temperatures are as shown in Figure 4.34. (**Hint:** Assume that the width is π and use $w = e^z$ to map the strip onto the upper half-plane.)

9.   Find the electrostatic potential between two long parallel cylinders of

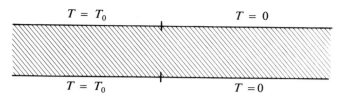

$T = T_0$ $T = 0$

$T = T_0$ $T = 0$

Figure 4.34

circular cross-section if one is inside the other and they touch along one line (the line of contact is assumed to be insulated).

10. Find the electrostatic potential in the region exterior to two cylinders (neither inside the other). (See Figure 4.35 for a cross-section.) (**Hint:** Invert over a point inside one of the cylinders; you may also need to use Exercise 16, Section 5, Chapter 3.)

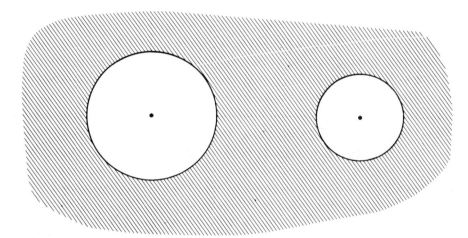

Figure 4.35

11. Find the temperature distribution inside the half-disc $\{x^2 + y^2 < 1,$ $y > 0\}$, if the temperature is T_1 along the bottom and T_2 along the semicircle. (**Hint:** Use a conformal mapping onto the first quadrant.)

12. Find the temperature distribution in the lens-shaped region within both the circles $(x - 1)^2 + y^2 = 1$ and $x^2 + (y + 1)^2 = 1$ if one edge of the lens is held at temperature T_1 and the other at temperature T_2.

13. Let $|a| < 1$; show that the function $g(z) = \log |(1 - \bar{a}z)/(z - a)|$ satisfies

$$-\frac{\partial g}{\partial r}(e^{it}) = \frac{1 - |a|^2}{1 - 2|a|\cos t + |a|^2}.$$

14. Justify the use of conformal mapping to solve the Neumann problem by showing that a Neumann condition on a segment of the boundary

of one domain is transformed to a Neumann condition on the corresponding segment in the boundary of the second domain provided that the mapping is actually conformal not only inside the domain but on these segments in the boundaries as well.

15.   Let D be the region obtained from the strip $\{x + iy: 0 < y < \pi\}$ by deleting the half-line $\{x + (i\pi/2): 0 \leqslant x < \infty\}$ (Fig. 4.36).

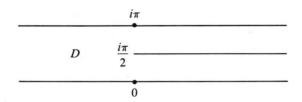

Figure 4.36

Exercise 14 in Section 5, Chapter 3 gives the Schwarz–Christoffel transformation of the upper half-plane U onto D. Use this to determine the potential in D if the top and bottom surfaces are held at potential zero and the middle (semi-infinite) surface is held at potential $V_0 \neq 0$.

4.5   Impulse functions and the Green's function of a domain

It is often very useful in the analysis of physical systems to imagine an "impulse function" with these two properties: (1) the impulse function is zero everywhere except at one point where it has the value ∞ and (2) the total area (or volume) under the graph of the function is 1. Such "functions" represent an idealization of an impact of short duration or a short, strong driving pulse, for instance.* We can understand an impulse function, say on a domain D in the plane, by the following limit argument. Fix a point p in D, the point where the impulse will be felt, and let r be a small positive number. Set

$$h_r(z) = \begin{cases} 0, & \text{if } |z - p| > r, \; z \in D \\ (\pi r^2)^{-1}, & \text{if } |z - p| \leqslant r. \end{cases}$$

Then the total volume under the graph of h_r is 1 and, for each continuous function f on D we have

$$\lim_{r \to 0} \int \int_D f(z) h_r(z) \, dx \, dy = f(p). \tag{1}$$

---

* In mechanics, this type of function is also referred to as a "concentrated force." In electrostatics, a unit point charge is like an impulse function.

Equation (1) is valid by the following argument. Since the volume under the graph of h_r is 1, we have

$$\int\int_D f(z)h_r(z)\ dx\ dy - f(p) = \frac{1}{\pi r^2}\int\int_{|z-p|\leq r} \{f(z) - f(p)\}\ dx\ dy.$$

The function f is continuous at p so that if $\varepsilon > 0$ is given, then r can be chosen with

$$|f(z) - f(p)| < \varepsilon \text{ whenever } |z - p| < r.$$

Hence

$$\left|\int\int_D f(z)h_r(z)\ dx\ dy - f(p)\right| \leq \varepsilon \text{ if } r \text{ is small}$$

and this is what we wished to show. Thus, in some sense, the impulse function δ_p equals $\lim_{r\to 0} h_r$.

In complex variables, and in particular, in the applications of complex variables, we are interested in harmonic functions u, that is, in solutions of the equation $\Delta u = 0$. Usually, as well, the values of u are prescribed on the boundary of the domain D. Let us consider here the function G, which is the solution of the boundary-value problem

$$\begin{cases} \Delta G = \delta_p \text{ on } D \\ \quad G = 0 \quad \text{on the boundary of } D, \end{cases}$$

where δ_p is the impulse function for $p \in D$. Thus, G is harmonic on the open set $D\backslash\{p\}$ and vanishes on the boundary of D.* Green's formula (formula (9) of Section 6, Chapter 1) gives us the relation

$$\int_\Gamma \left(\frac{\partial G}{\partial n} f - G \frac{\partial f}{\partial n}\right) ds = \int\int_D \{(\Delta G)f - G(\Delta f)\}\ dx\ dy,$$

where Γ is the boundary of D. If we assume that f is continuous on $D \cup \Gamma$ and harmonic on D, then we find that

$$\int_\Gamma f \frac{\partial G}{\partial n}\ ds = \int\int_D (\Delta G)f\ dx\ dy = \int\int_D \delta_p f\ dx\ dy = f(p).$$

Consequently, the normal derivative of G is precisely the function that takes the boundary values f and produces the value at p of the harmonic function with these boundary values; that is, $\partial G/\partial n$ solves the Dirichlet problem for the domain D. Of course, the derivation carried out above is only heuristic, but it can be made mathematically sound (see the exercises).

We now give the "official" definition of the Green's function and then some examples that illustrate the Green's function and its use.

---

* That is, G is the **response** to the impulse at p for a system governed by the Laplacian, with zero boundary data.

The definition of the Green's function of a domain

Suppose that D is a bounded domain whose boundary is a finite number of disjoint smooth simple closed curves (Fig. 4.37).

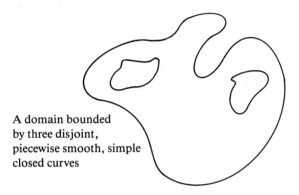

A domain bounded
by three disjoint,
piecewise smooth, simple
closed curves

Figure 4.37

Let p be a point in D. The **Green's function for D with singularity at p** is the function $G(z; p)$ with these three properties

(a) $G(z; p)$ is harmonic for z in D, $z \neq p$,
(b) $G(z; p) + \log |z - p|$ is harmonic near p,   **(2)**
(c) $G(z; p) = 0$ if z is in the boundary of D.

Such a function $G(z; p)$ can be found in the following way. Let v be the harmonic function on D with $v(z) = \log |z - p|$ if z is in the boundary of D; then set

$$G(z; p) = v(z) - \log|z - p|.$$

Exercise 1 at the end of this section shows that $G(z; p)$ is unique.

The Green's function for an unbounded domain is defined exactly as in (a), (b), and (c) above if p is not the point at ∞. If ∞ is a point in the domain D, then the Green's function for D with singularity at ∞ is the function $G(z; \infty)$ satisfying

(a)′ $G(z; \infty)$ is harmonic on $D\backslash\{\infty\}$,
(b)′ $G(z; \infty) - \log|z|$ is harmonic for $|z|$ large,   **(3)**
(c)′ $G(z; \infty) = 0$ if z is in the boundary of D.

Domains that include the point at ∞, and their Green's functions, are important in the subject of potential theory.

Example 1   Let D be the disc $\{z: |z| < 1\}$ and fix $p \in D$. Set

$$G(z; p) = \log\left|\frac{1 - \bar{p}z}{z - p}\right|, \quad |z| < 1.$$

Then G is harmonic except at p, $G = 0$ on the boundary of D, since if $|z| = 1$, then $|(1 - \bar{p}z)/(z - p)| = 1$; on the disc D, we have

$$G(z; p) + \log|z - p| = \log|1 - \bar{p}z|.$$

Note that $\log|1 - \bar{p}z|$ is harmonic on D. Furthermore, the (inward) normal derivative of G at the boundary is just minus the partial derivative of G with respect to r and this is

$$-\frac{\partial}{\partial r}\left\{\log\left|\frac{1 - \bar{p}re^{i\theta}}{re^{i\theta} - p}\right|\right\}\Bigg|_{r=1} = \frac{1 - |p|^2}{|e^{i\theta} - p|^2}.$$

You should recognize this function immediately as the Poisson kernel for the point p, $|p| < 1$.

Example 2   Let D be a simply connected domain, bounded or unbounded, with at least two points in its boundary and fix a point $p \in D$. Let ϕ be the Riemann mapping of D onto the disc $\Delta = \{w: |w| < 1\}$ with $\phi(p) = 0$, $\phi'(p) > 0$; such a mapping exists and is unique by Theorem 1, Section 5 Chapter 3. Show that

$$G(z; p) = -\log|\phi(z)|, \quad z \in D$$

is the Green's function for D with singularity at p.

Solution   Let us suppose first that p is not the point at ∞ (if this point is even in D.) On $D\backslash\{p\}$, the function ϕ is analytic and nonzero so that $\log|\phi|$ is harmonic by Exercise 2 of Section 1. Furthermore, ϕ has a zero of order 1 at p so that

$$\phi(z) = (z - p)g(z),$$

where $g(z)$ is never zero on D. Thus, we have

$$\log|\phi(z)| = \log|z - p| + \log|g(z)|, \quad z \in D$$

and $\log|g|$ is harmonic on all of D. This shows that $\log|\phi|$ satisfies (a) and (b) of (2); to see that $\log|\phi|$ satisfies (c) we need to use this fact:

> if $\{z_n\}$ is a sequence of points in D with $\lim_{n \to \infty} z_n = \lambda$
>
> for some λ in the boundary of D, then $\lim_{n \to \infty} |\phi(z_n)| = 1$.

(See Exercise 19, Section 5, Chapter 3). From this, it is evident that (c) holds and so $-\log|\phi(z)|$ is, in fact, the Green's function for D with singularity at p.

The case when p is the point at ∞ is left for the exercises but it is done in much the same way.   □

Example 3   Let D be the complement of the segment $[a, b]$, including the point at ∞. Find the Green's function for D with singularity at ∞ and find

$$\kappa = \lim_{|z| \to \infty} \{G(z; \infty) - \log|z|\}.$$

Solution   Let

$$\phi_1(z) = \frac{2z}{b - a} + \frac{a + b}{a - b}.$$

$\phi_1(z)$ is a one-to-one analytic function that maps the segment $[a, b]$ onto $[-1, 1]$ and D onto the domain D_1, which is the complement of $[-1, 1]$. Let $\phi_2(w) = w - \sqrt{w^2 - 1}$; $\phi_2(w)$ is an analytic function that maps the domain D_1 one-to-one onto the disc $\Delta = \{\zeta \colon |\zeta| < 1\}$; ϕ_2 is the inverse of the function $\psi(\zeta) = \frac{1}{2}(\zeta + \zeta^{-1})$, which maps the unit disc Δ onto D_1 (see Example 2, Section 1, Chapter 3). The composition, $\phi(z) = \phi_2(\phi_1(z))$ is the Riemann mapping of D onto Δ with $\phi(\infty) = 0$. According to Example 2, the Green's function for D with singularity at ∞ is

$$G(z; \infty) = -\log|\phi(z)|$$
$$= -\log|(Az + B) - \sqrt{(Az + B)^2 - 1}|,$$

where $A = 2/(b - a)$ and $B = (a + b)/(a - b)$. Next, for $|w|$ large, we have

$$\sqrt{w^2 - 1} = w\sqrt{1 - w^{-2}} = w\left\{1 - \frac{1}{2w^2} - \frac{c_2}{w^4} - \frac{c_3}{w^6} \cdots\right\},$$

where $c_2, c_3, \ldots$ are constants (see Exercises 34 to 36 of Section 2, Chapter 2). Hence,

$$w - \sqrt{w^2 - 1} = \frac{1}{2w} + \frac{c_2}{w^3} + \frac{c_3}{w^5} + \cdots,$$

and consequently

$$-\log|\phi_2(w)| = -\log\left|\frac{1}{2w} + \frac{c_2}{w^3} + \cdots\right|$$

$$= \log|2w| - \log\left|1 + \frac{2c_2}{w^2} + \cdots\right|.$$

Replacing w by $Az + B$ gives

$$G(z; \infty) - \log|z| = \log\left|2\left(A + \frac{B}{z}\right)\right| - \log\left|1 + \frac{2c_2}{(Az + B)^2} + \cdots\right|$$

and so finally

$$\kappa = \lim_{|z| \to \infty} \{G(z; \infty) - \log|z|\} = \log(2A)$$

$$= \log\left(\frac{4}{b-a}\right).$$

Exercises for Section 4.5

1. Let $p \in D$. Show that the Green's function of the domain D with singularity at p is unique. (**Hint:** If both $G(z; p)$ and $G_1(z; p)$ satisfy the definition of the Green's function, show that the difference, $G - G_1$, is harmonic and bounded in D, and vanishes on the boundary of D.)

2. Let D_1 and D_2 be two domains with Green's functions G_1 and G_2, respectively. Suppose that ϕ is a one-to-one analytic function mapping D_1 onto D_2. Show that $G_1(z; p) = G_2(\phi(z); \phi(p))$.

Use Exercise 2 or the definition of Green's function to explicitly find the Green's function for the domain D:

3. $D = \{z = x + iy: y > 0\}$.

4. $D = \{z: |z - z_0| < R\}$; z_0, R fixed, $R > 0$.

5. $D = \{z = x + iy: x > 0 \text{ and } y > 0\}$.

6. $D = \{z = x + iy: y > 0 \text{ and } |z| < 1\}$.

Use Example 2 to find the Green's function for the following domains

7. $D = \{z = x + iy: 0 < y < \pi\}$.

8. $D = \{z = x + iy: -\pi/2 < x < \pi/2 \text{ and } y > 0\}$.

9. $D = \{z = x + iy: x \notin (-\infty, 0]\}$.

10. Show that the Green's function for a domain D with singularity at p is always positive on $D\backslash\{p\}$. (**Hint:** Show that $G(z; p)$ is positive on the circle $|z - p| = \varepsilon$ for all small ε; then use the maximum/minimum principle on the domain $D\backslash\{z: |z - p| \leqslant \varepsilon\}$.)

11. Argue on physical grounds that the Green's function is symmetric: $G(q; p) = G(p; q)$ for all $p \neq q$ in D.

12. Let D be a bounded domain whose boundary, Γ, consists of a finite number of disjoint piecewise smooth simple closed curves. It is a fact that the normal derivative of the Green's function exists and is continuous on Γ. Use this to show that

$$\frac{1}{2\pi} \int_\Gamma u(z) \frac{\partial}{\partial n} \{G(z; p)\} \, ds = u(p)$$

for any function u that is continuous on $D \cup \Gamma$ and harmonic on D. (**Hint:** Apply Green's formula on the domain $D_\varepsilon = \{z \in D:$

$|z - p| \geqslant \varepsilon \}$; you'll have to show that the term

$$\frac{1}{2\pi} \int_{|z-p|=\varepsilon} \left\{ \frac{\partial G}{\partial n} u - \frac{\partial u}{\partial n} G \right\} ds$$

approaches $u(p)$ as ε decreases to 0. This is accomplished by noting that

$$\frac{\partial G}{\partial n} = \frac{1}{|z-p|} + \text{bounded term}$$

on this circle. Next, estimate $|G(z; p)|$ by $\log(1/|z - p|)$ and recall that $\varepsilon \log \varepsilon$ goes to zero as $\varepsilon \to 0$.)

13.   Let D be an unbounded domain including the point at ∞ and let E be the complement of D; let $G(z; \infty)$ be the Green's function for D with singularity at ∞. The number $c(E)$ defined by

$$-\log c(E) = \lim_{|z| \to \infty} \{ G(z; \infty) - \log|z| \}$$

is called the **logarithmic capacity** of E. Find the logarithmic capacity of E if (a) $E = \{z: |z| \leqslant 1\}$; (b) $E = \{z: |z - z_0| \leqslant r_0\}$; (c) $E = [a, b]$, $a < b$ (see Example 3); (d) $E = \{x + iy: (x^2/a^2) + (y^2/b^2) \leqslant 1\}$.

14.   Complete Example 2 by showing that if D is a simply connected domain containing the point at ∞, then

$$G(z; \infty) = -\log|\phi(z)|,$$

where $\phi(z)$ is the Riemann mapping of D onto the disc $\{w: |w| < 1\}$ with $\phi(\infty) = 0$.

Impulse functions

15.   Show that the Poisson kernel

$$P_r(\theta) = \frac{1 - r^2}{1 - 2r \cos \theta + r^2}$$

converges to the impulse function at $\theta = 0$ as $r \to 1$ in the sense that $\lim_{r \to 1} 1/2\pi \int_{-\pi}^{\pi} h(e^{i\theta}) P_r(e^{i\theta}) \, d\theta = h(1)$ for each function that is continuous on the circle $\{e^{i\theta}: -\pi \leqslant \theta \leqslant \pi\}$.

16.   Let $[a, b]$ be an interval in the line and let $a < p < b$. Show that the "impulse function" δ_p for $[a, b]$ has the property that

$$\int_a^b f(x) \, \delta_p(x) \, dx = f(p)$$

for every continuous function $f(x)$ on $[a, b]$. (**Hint:** Let $u_n(x)$ be zero for $|x - p| > 1/(2n)$ and $u_n(x)$ be n for $|x - p| < 1/(2n)$. Show that

$$\lim_{n \to \infty} \int_a^b f(x)u_n(x) \, dx = f(p).)$$

17. Let $H(x) = \begin{cases} 0, & a \leqslant x < p \\ 1, & p < x \leqslant b. \end{cases}$ Use integration by parts to "show" that $H' = \delta_p$. (**Hint:** if $f(b) = 0$, then

$$\int_a^b f'(x)H(x) \, dx = - \int_a^b f(x)H'(x) \, dx.$$

But $\int_a^b f'(x)H(x) \, dx = \int_p^b f'(x) \, dx = f(b) - f(p) = -f(p).)$

Further reading

The following two books present a substantial number of applications of complex variables to engineering and science problems.

Henrici, P. *Applied and computational complex analysis.* Vol. 1. New York: John Wiley, 1974.

Kyrala, A. *Applied functions of a complex variable.* New York: John Wiley, 1972.

More specialized books that I have found useful in dealing with the topics in this chapter include

Chorin, L., Marsden J. *A mathematical introduction to fluid mechanics.* New York: Springer-Verlag, 1979.

Crank, J. *The mathematics of diffusion,* New York: Oxford University Press, 1956.

Kellogg, O. D. *Foundations of potential theory.* New York: Dover, 1953.

Milne-Thompson, L. M. *Theoretical aerodynamics.* 4th ed. New York: Mac-Millan, 1966.

Milne-Thompson, L. M. *Theoretical hydrodynamics.* 5th ed. New York: Mac-Millan, 1968.

Muskhelishvili, N. I. *Some basic problems of the mathematical theory of elasticity.* Groningen, Netherlands: P. Noordhoff, 1963.

Verruijt, A. *Theory of groundwater flow.* New York: MacMillan, 1970.

5

Transform methods

5.1 The Fourier* transform: Basic properties

The Fourier transform is a powerful tool in both applied and theoretical mathematics. Its utility is due in no small part to the simple relationships between a function and its Fourier transform. However, a careful mathematical justification of these relationships is not elementary. For that reason we postpone our discussion of these relationships until Section 2. In this section we define the Fourier transform, demonstrate some of its basic properties, and show some of its applications. The relationships between the function and its transform are stated in Section 2 and several examples illustrate their uses. At the end of Section 2 we sketch a portion of the mathematics establishing these relationships.

We begin with the notion of a piecewise smooth function on the real line. Suppose t_0 is a real number and u is a function defined for t near t_0; then we say that u has a **limit at t_0 from the right** if

$$\lim_{h \downarrow 0} u(t_0 + h) = u^+(t_0), \quad h > 0$$

exists; we say that u has a **limit at t_0 from the left** if

$$\lim_{h \downarrow 0} u(t_0 - h) = u^-(t_0), \quad h > 0$$

exists. We do not make any requirement about the possible equality of $u^+(t_0)$ and $u^-(t_0)$. If it happens that these two limits exist and are both equal to $u(t_0)$, then (and only then) u is continuous at t_0.

Example 1 The function

$$u(t) = \begin{cases} 1, & |t| < \sigma \\ 0, & |t| > \sigma \end{cases}$$

has both left and right limits at all points t_0. The left and right limits coincide except at $t_0 = \pm\sigma$; at $t_0 = -\sigma$, the limit from the left is zero and the limit from the right is 1; at $t_0 = \sigma$, exactly the reverse situation occurs.

---

* Jean Baptiste Joseph Fourier, 1768–1830.

A function u is **piecewise smooth** if the following conditions are met: there are a finite number of points $t_1 < t_2 < \cdots < t_N$ in the real line such that

(a) both u and u' are continuous on all the intervals $(-\infty, t_1), (t_1, t_2), \ldots, (t_N, \infty)$.

(b) both u and u' have limits from the right *and* left at each of the points $t_1, \ldots, t_N$.

We do not require that the right and left limits of either u or u' coincide at any of the points $t_1, \ldots, t_N$, although this may happen.

Example 2   The function

$$u(t) = \begin{cases} 1, & |t| < \sigma \\ 0, & |t| > \sigma \end{cases}$$

from Example 1 is piecewise smooth.

Example 3   The function

$$u(t) = \begin{cases} 1 - t^2, & |t| < 1 \\ t^2 - 1, & |t| > 1 \end{cases}$$

is piecewise smooth; u is continuous for all t but u' is discontinuous at $t_0 = \pm 1$.

Example 4   If u is piecewise smooth and $U'(t) = u(t)$ for all $t \neq t_j, j = 1, \ldots, N$, then U is piecewise smooth. The elementary demonstration of this is left to the exercises.

Example 5   The function

$$u(t) = \begin{cases} 0, & t \leqslant 0 \\ \sin(\pi/t), & t > 0 \end{cases}$$

is *not* piecewise smooth, since u has no limit from the right at $t_0 = 0$. This follows, for example, because $u(1/n) = 0$, $n = 1, 2, \ldots$, while $u(2/(1 + 4n)) = 1$, $n = 1, 2, \ldots$.

The Fourier transform

Let u be a piecewise smooth function on the real line and assume that

$$\int_{-\infty}^{\infty} |u(t)| \, dt < \infty. \tag{1}$$

Condition (1) will certainly hold if $u(t) = 0$ for $|t| > M$ or if $|u(t)| \leqslant C/t^2$ for all large t. We define the **Fourier transform** of u by the rule

$$\hat{u}(x) = \int_{-\infty}^{\infty} u(t)e^{-itx} \, dt, \quad -\infty < x < \infty. \tag{2}$$

The condition (1), together with the fact that $|e^{-itx}| = 1$ if both t and x are real, assures us that the integral defining $\hat{u}(x)$ makes good sense (see Exercises 16 to 19 in Section 2).

Example 6 Find the Fourier transform of the function

$$u(t) = \begin{cases} 1, & |t| < \sigma \\ 0, & |t| > \sigma. \end{cases}$$

Solution By the definition of the Fourier transform, we have

$$\hat{u}(x) = \int_{-\sigma}^{\sigma} e^{-itx} \, dt = \frac{-1}{ix}\left[e^{-itx} \Big|_{t=-\sigma}^{t=\sigma} \right]$$

$$= \frac{2}{x}\left[\frac{e^{i\sigma x} - e^{-i\sigma x}}{2i} \right]$$

$$= 2\,\frac{\sin \sigma x}{x}. \qquad \square$$

Example 7 Find the Fourier transform of $u(t) = e^{-t^2}$.

Solution The Fourier transform of u is given by

$$\hat{u}(x) = \int_{-\infty}^{\infty} e^{-t^2} e^{-itx} \, dt = \int_{-\infty}^{\infty} e^{-(t+ix/2)^2} e^{-x^2/4} \, dt$$

$$= \{\exp(-x^2/4)\} \int_{-\infty}^{\infty} \exp\left[-\left(t + \frac{ix}{2} \right)^2 \right] dt.$$

To evaluate this last integral we make use of some complex analysis. Let R be a large positive number; the integral

$$\int_{-R}^{R} \exp\left\{ -\left(t + \frac{ix}{2} \right)^2 \right\} dt$$

converges to

$$\int_{-\infty}^{\infty} \exp\left\{ -\left(t + \frac{ix}{2} \right)^2 \right\} dt$$

as $R \to \infty$. However, the integral from $-R$ to R is nothing but the integral

$$\int_{\gamma_R} \exp(-z^2) \, dz, \quad z = t + \frac{ix}{2}, \quad -R \leqslant t \leqslant R,$$

where γ_R is the line segment joining $-R + (ix/2)$ to $R + (ix/2)$. Consider now the rectangle Γ_R with vertices at $-R, R, R + (ix/2)$, and $-R + (ix/2)$ (Fig. 5.1).

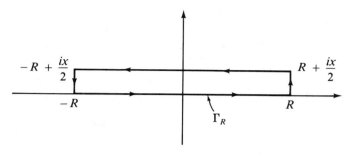

Figure 5.1

By Cauchy's theorem the line integral of $f(z) = \exp(-z^2)$ about Γ_R is zero. Furthermore, on the vertical edges of Γ_R, we have the estimate

$$|f(z)| = |\exp(-z^2)|$$

$$= \exp\left[-R^2 + \frac{s^2}{4}\right], \quad 0 \leqslant s \leqslant x/2,$$

and so the integral of f over the two vertical edges of Γ_R is very small if R is very large. The integral of $f(z)$ as z runs from $-R$ to R cannot be computed exactly, but as $R \to \infty$ it converges to

$$\int_{-\infty}^{\infty} e^{-t^2}\, dt = \sqrt{\pi}.$$

(The value of this integral should be familiar from calculus; also see Exercise 20, Section 3, Chapter 2.) Hence, we have shown that

$$\int_{-\infty}^{\infty} \exp(-t^2 - ixt)\, dt = \sqrt{\pi}\, \exp[-x^2/4]$$

and so

$$\hat{u}(x) = \sqrt{\pi}\, \exp[-x^2/4].$$

That is, the Fourier transform of e^{-t^2} is $\sqrt{\pi}e^{-x^2/4}$.   □

Example 8   Find the Fourier transform of

$$u(t) = \frac{1}{1 + t^2}.$$

Solution   We have

$$\hat{u}(x) = \int_{-\infty}^{\infty} \frac{e^{-ixt}}{1 + t^2}\, dt.$$

To evaluate this integral we use the residue theorem. Let

$$f(z) = \frac{e^{-ixz}}{1 + z^2}$$

and set $z = \sigma + i\tau$. Thus,

$$|f(z)| \leqslant \frac{e^{x\tau}}{|z|^2 - 1}, \quad |z| \text{ large}$$

$$\leqslant \frac{1}{|z|^2 - 1} \quad \text{if } x\tau \leqslant 0.$$

The necessity that $x\tau$ be less than or equal to zero requires us to consider two cases. If x is negative, we integrate f over the contour shown in Figure 5.2a; if x is positive, we integrate f over the contour in Figure 5.2b. We can apply the residue theorem (or just use Cauchy's Theorem). For the case where $x \leqslant 0$, the pole is at i and we find that

$$\int_{\Gamma_R} f(z)\, dz = 2\pi i \, \frac{e^{-ix(i)}}{2i} = \pi e^x, \quad x \leqslant 0.$$

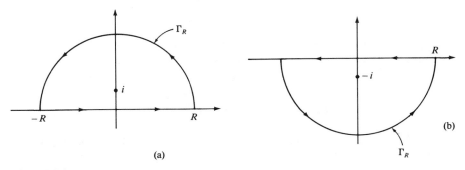

Figure 5.2

For the case where $x \geqslant 0$, the pole is at $-i$ and we have

$$\int_{\Gamma_R} f(z)\, dz = 2\pi i \, \frac{e^{-ix(-i)}}{-2i} = -\pi e^{-x}, \quad x \geqslant 0.$$

Since the integral of f over the semicircular part of Γ_R goes to zero as $R \to \infty$, we find that

$$\hat{u}(x) = \int_{-\infty}^{\infty} \frac{e^{ixt}}{1 + t^2}\, dt = \begin{cases} \pi e^x, & x \leqslant 0 \\ \pi e^{-x}, & x \geqslant 0. \end{cases}$$

Consequently,

$$\hat{u}(x) = \pi e^{-|x|}, \quad x \text{ real.} \qquad \square$$

Basic properties of the Fourier transform

The operation of taking the Fourier transform of a function has certain convenient and useful properties, which we list here (in all cases we assume that (1) holds).

Property 1

If u_1 and u_2 are functions and λ_1 and λ_2 are complex numbers, then

$$(\lambda_1 u_1 + \lambda_2 u_2)\hat{}(x) = \lambda_1 \hat{u}_1(x) + \lambda_2 \hat{u}_2(x). \tag{3}$$

Property 2

If a is a real number and v is the function defined by

$$v(t) = u(a + t), \quad -\infty < t < \infty,$$

then

$$\hat{v}(x) = e^{iax}\hat{u}(x), \quad -\infty < x < \infty. \tag{4}$$

Property 3

If b is a real number, $b \neq 0$, and w is the function defined by

$$w(t) = u(bt), \quad -\infty < t < \infty,$$

then

$$\hat{w}(x) = \frac{1}{b}\hat{u}(x/b). \tag{5}$$

Property 4

If u is differentiable everywhere on the real axis, and if u and u' both satisfy (1), then

$$\widehat{u'}(x) = ix\hat{u}(x), \quad -\infty < x < \infty. \tag{6}$$

Property 5

$$\frac{d}{dx}\hat{u}(x) = -i(tu(t))\hat{}(x) \tag{7}$$

assuming that

$$\int_{-\infty}^{\infty} |tu(t)| \, dt < \infty.$$

The demonstrations of (3) through (7) are simple and left for the exercises; (4) and (5) are done by a change of variables and (6) by an integration by parts. (See Exercise 20, Section 2 of this chapter, to conclude that $\lim_{|t| \to \infty} u(t) = 0$.)

Example 9 Find the Fourier transform of

(i) $u_1(t) = (t^2 + 4t + 5)^{-1}$.

(ii) $u_2(t) = \exp[-t^2/2]$.

(iii) $u_3(t) = \begin{cases} 1+t, & -1 \leqslant t \leqslant 0 \\ 1-t, & 0 \leqslant t \leqslant 1 \\ 0, & |t| \geqslant 1. \end{cases}$

Solution   (i) u_1 is given by

$$u_1(t) = \frac{1}{(t+2)^2 + 1} = u(t+2),$$

where u is the function in Example 8. Hence,

$$\hat{u}_1(x) = e^{2ix}\hat{u}(x)$$

$$= \pi e^{2ix}e^{-|x|}, \quad -\infty < x < \infty.$$

(ii)   Next, $u_2(t)$ is exactly $u(t/\sqrt{2})$, where u is the function from Example 7. Consequently, with $b = 1/\sqrt{2}$ we have

$$\hat{u}_2(x) = \sqrt{2}\hat{u}(\sqrt{2}x)$$
$$= \sqrt{2\pi}\,\exp[-2x^2/4]$$
$$= \sqrt{2\pi}u_2(x).$$

Thus, u_2 is a constant multiple of its Fourier transform.

(iii)   Finally, the function $u_3(t)$ has a very simple derivative

$$u_3'(t) = \begin{cases} 1, & -1 < t < 0 \\ -1, & 0 < t < 1 \\ 0, & |t| > 1. \end{cases}$$

The Fourier transform of u_3' is easily computed to be

$$\widehat{u_3'(x)} = 2(\cos x - 1)/ix.$$

Thus,

$$\hat{u}_3(x) = \frac{1}{ix}\widehat{u_3'(x)} = 2\frac{1 - \cos x}{x^2}. \qquad \square$$

There is one more very important property of the Fourier transform but we need a definition in order to introduce it. Suppose u and v are two functions satisfying (1); the **convolution of u and v** is the function $u * v$ defined by

$$(u * v)(t) = \int_{-\infty}^{\infty} u(s)v(t-s)\,ds, \quad -\infty < t < \infty. \tag{8}$$

Example 10   Let

$$u(t) = v(t) = \begin{cases} 1, & |t| < \sigma \\ 0, & |t| > \sigma. \end{cases}$$

Find $u * v$.

Solution

$$(u * v)(t) = \int_{-\infty}^{\infty} u(s)v(t - s) \, ds.$$

Now $v(t - s) = 0$ unless $|t - s| < \sigma$; since $u(s) = 1$ only when $|s| < \sigma$, we know that

$$u(s)v(t - s) = \begin{cases} 1, & \text{if } |t - s| < \sigma \text{ and } |s| < \sigma \\ 0, & \text{otherwise.} \end{cases}$$

Thus, $u(s)v(t - s) = 0$ for all s if $|t| > 2\sigma$ and so $(u * v)(t) = 0$ if $|t| > 2\sigma$. Moreover, if $0 \leqslant t < 2\sigma$, then

$$u(s)v(t - s) = \begin{cases} 1, & \text{if } t - \sigma < s < \sigma \\ 0, & \text{otherwise,} \end{cases}$$

and, for $-2\sigma < t \leqslant 0$,

$$u(s)v(t - s) = \begin{cases} 1, & \text{if } -\sigma < s < t + \sigma \\ 0, & \text{otherwise.} \end{cases}$$

Thus,

$$(u * v)(t) = \begin{cases} 0, & t \leqslant -2\sigma \\ t + 2\sigma, & -2\sigma \leqslant t \leqslant 0 \\ 2\sigma - t, & 0 \leqslant t \leqslant 2\sigma \\ 0, & t \geqslant 2\sigma. \end{cases}$$

A sketch of the graphs of u, v, and $u * v$ is shown in Figure 5.3a and 5.3b.   □

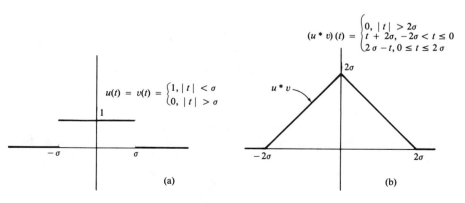

(a)   (b)

Figure 5.3

Example 11 Let u be a bounded continuous function on $(-\infty, \infty)$, satisfying (1), and, for $y > 0$, let

$$K_y(t) = \frac{1}{\pi} \frac{y}{t^2 + y^2}.$$

Then the function

$$u(x, y) = (u * K_y)(x) = \frac{1}{\pi} \int_{-\infty}^{\infty} u(t) \frac{y}{(x - t)^2 + y^2} \, dt$$

is exactly the harmonic function of $x + iy$ defined in the upper half-plane $\{x + iy: y > 0\}$, which satisfies

$$\lim_{x+iy \to s} u(x, y) = u(s), \quad -\infty < s < \infty;$$

see Section 3, Chapter 4.

In the exercises the reader is asked to show that $u * v$ is the same as $v * u$; that is, the order of the convolution is not important. It is also true that convolution "distributes" across addition just as multiplication does: $u * (v + w) = u * v + u * w$.

The critical connection between the Fourier transform and convolution is that the Fourier transform of the convolution of u and v is just the *product* of the Fourier transforms of u and v, respectively. In symbols,

$$(u * v)\,\hat{}\,(x) = \hat{u}(x)\hat{v}(x), \quad -\infty < x < \infty. \tag{9}$$

This is not too hard to show (if one takes on faith the reversibility of the order of integration in an iterated integral):

$$(u * v)\,\hat{}\,(x) = \int_{-\infty}^{\infty} (u * v)(t)e^{-itx} \, dt$$

$$= \int_{-\infty}^{\infty} \left\{ \int_{-\infty}^{\infty} u(s)v(t - s) \, ds \right\} e^{-itx} \, dt$$

$$= \int_{-\infty}^{\infty} u(s) \left\{ \int_{-\infty}^{\infty} v(t - s)e^{-i(t-s)x} \, dt \right\} e^{-isx} \, ds$$

$$= \left(\int_{-\infty}^{\infty} u(s)e^{-isx} \, ds \right) \hat{v}(x)$$

$$= \hat{u}(x)\hat{v}(x).$$

The Fourier transform has powerful applications in the area of ordinary and partial differential equations; we illustrate with two examples. Both examples make use of the fact that if u_1 and u_2 are two functions with $\hat{u}_1 = \hat{u}_2$, then $u_1 = u_2$. This is proved in Section 2.

Example 12 At a particular time a very long thin rod has a given temperature distribution $f(s)$, $-\infty < s < \infty$. Find the temperature at all future times.

Solution   The rod is one-dimensional (it just has length) and so the temperature is a function of the position s on the rod at the time t; we let $u(t, s)$ be the temperature at time t and position s. The function $u(t, s)$ must satisfy the **heat equation**

$$\frac{\partial u}{\partial t} = \frac{\partial^2 u}{\partial s^2}$$

with **initial condition**

$$u(0, s) = f(s), \quad -\infty < s < \infty.$$

We now use the Fourier transform (on the s variable) to solve this equation. We begin by setting

$$\phi(t, x) = \int_{-\infty}^{\infty} u(t, s)e^{-isx}\, ds, \quad -\infty < x < \infty.$$

Then

$$\frac{\partial \phi}{\partial t} = \int_{-\infty}^{\infty} \frac{\partial u}{\partial t}(t, s)e^{-isx}\, ds$$

$$= \int_{-\infty}^{\infty} \frac{\partial^2 u}{\partial s^2}(t, s)e^{-isx}\, ds.$$

In this second integral we use (6) (or integrate twice by parts); we also make the natural assumption that $u(t, s)$ and $(\partial u/\partial s)(t, s)$ both tend to zero as $s \to \pm\infty$ for each t. The net result, then, is that

$$\frac{\partial \phi}{\partial t}(t, x) = -x^2 \int_{-\infty}^{\infty} u(t, s)e^{-isx}\, ds$$

$$= -x^2 \phi(t, x).$$

Furthermore, we also have

$$\phi(0, x) = \int_{-\infty}^{\infty} u(0, s)e^{-isx}\, ds$$

$$= \int_{-\infty}^{\infty} f(s)e^{-isx}\, ds$$

$$= \hat{f}(x).$$

Consequently, for each fixed x, $\phi(x, t)$ satisfies the ordinary differential equation

$$\frac{d\phi}{dt} = -x^2\phi, \quad \phi(0, x) = \hat{f}(x).$$

Hence, we must have

$$\phi(t, x) = \hat{f}(x)\exp[-tx^2].$$

However, we recognize (and this is critical!) that $\exp[-x^2t]$ is the Fourier transform of

$$K(s) = \frac{\sqrt{\pi}}{\sqrt{t}} e^{-s^2/4t}, \ t > 0.$$

Since $\phi(t, x)$ is the Fourier transform of $u(t, s)$, we use the convolution theorem to discover that $u(t, s)$ must be the convolution of f and K:

$$u(t, s) = \int_{-\infty}^{\infty} f(y) \frac{\sqrt{\pi}}{\sqrt{t}} e^{-(s-y)^2/4t} \, dy$$

$$= \frac{\sqrt{\pi}}{\sqrt{t}} \int_{-\infty}^{\infty} f(y) e^{-(s-y)^2/4t} \, dy.$$

This explicitly displays the dependence of the solution $u(t, s)$ on the initial temperature distribution f. The function K given above is the **heat kernel**.   □

Example 13   (Biharmonic functions in the upper half-plane)   In this example we show how the Fourier transform can be used to give an integral representation formula for a function $u(x, y)$, which is biharmonic on the upper half-plane, $U = \{x + iy: y > 0\}$, and which has given values on the real axis. The formula will be in many ways analogous to the Poisson integral formula for the upper half-plane. The reader is referred to Section 2, Chapter 4, for a physical problem whose solution is biharmonic.
We wish to find a function $u(x, y)$ satisfying

(i) $\Delta(\Delta u) = \dfrac{\partial^4 u}{\partial x^4} + 2 \dfrac{\partial^4 u}{\partial x^2 \, \partial y^2} + \dfrac{\partial^4 u}{\partial y^4} = 0$ on U

(ii) $u(x, 0) = f(x), \ -\infty < x < \infty.$

Let $U(s, y)$ be the Fourier transform of $u(x, y)$ on the x-variable:

$$U(s, y) = \int_{-\infty}^{\infty} u(x, y) e^{-isx} \, dx.$$

We hold s fixed and treat U as a function of y. Differentiation yields

$$\frac{d^4 U}{dy^4} = \int_{-\infty}^{\infty} \frac{\partial^4 u}{\partial y^4} e^{-isx} \, dx$$

$$= \int_{-\infty}^{\infty} \left\{ -\frac{\partial^4 u}{\partial y^2 \, \partial x^2} - \frac{\partial^4 u}{\partial x^4} \right\} e^{-isx} \, dx.$$

Using (6), or integrating by parts, we find that

$$\frac{d^4 U}{dy^4} = s^2 \int_{-\infty}^{\infty} \frac{\partial^2 u}{\partial y^2} e^{-isx}\, dx - s^4 \int_{-\infty}^{\infty} u(x, y)e^{-isx}\, dx$$

$$= s^2 \frac{d^2 U}{dy^2} - s^4 U.$$

Hence, U satisfies the fourth-order linear ordinary differential equation

$$\frac{d^4 U}{dy^4} - s^2 \frac{d^2 U}{dy^2} + s^4 U = 0$$

with

$$U(s, 0) = \int_{-\infty}^{\infty} u(x, 0)e^{-isx}\, dx$$

$$= \int_{-\infty}^{\infty} f(x)e^{-isx}\, dx$$

$$= \hat{f}(s).$$

The linearly independent solutions of $v^{(4)} - s^2 v'' + s^4 v = 0$ are e^{sy}, ye^{sy}, e^{-sy}, and ye^{-sy}. In our problem, y is positive and the solution U must be a Fourier transform. This leads to the conclusion that

$$U(s, y) = A(s)e^{-y|s|} + B(s)ye^{-y|s|}, \quad -\infty < s < \infty,$$

where $A(s)$ and $B(s)$ are functions of s (see Exercise 21 at the end of this section). The condition $U(s, 0) = \hat{f}(s)$ yields $A(s) = \hat{f}(s)$. Furthermore, $e^{-s|y|}$ is the Fourier transform of

$$u_1(t) = \frac{1}{\pi} \frac{y}{s^2 + y^2}$$

(see Example 8 and formula (5)). Likewise, $ye^{-y|s|}$ is the Fourier transform of

$$u_2(t) = \frac{1}{\pi} \frac{y^2}{s^2 + y^2}.$$

The convolution result, equation (9), then leads to the conclusion that

$$u(x, y) = \frac{1}{\pi} \int_{-\infty}^{\infty} f(t) \frac{y}{(x - t)^2 + y^2}\, dt + \frac{1}{\pi} \int_{-\infty}^{\infty} g(t) \frac{y^2}{(x - t)^2 + y^2}\, dt.$$

The Poisson integral formula shows that the first term on the right is the harmonic function on the upper half-plane U whose boundary values on the real axis are exactly f. The second term is just $yG(x, y)$, where G is the harmonic function on U whose boundary values on the real axis are g. Thus, as y decreases to zero, $yG(x, y)$ converges to $0 \cdot g(x) = 0$. Since G is harmonic, $yG(x, y)$ is biharmonic. The function g is arbitrary, within the limitations that the integral defining $G(x, y)$ makes good sense. For instance g could be any piecewise continuous, bounded function.

Exercises for Section 5.1

In Exercises 1 to 12 find $\hat{u}$ from the given u

1. $u(t) = \begin{cases} t, & |t| < b \\ 0, & |t| > b \end{cases};$ b a positive constant.

2. $u(t) = \begin{cases} -1, & -\sigma < t < 0 \\ 1, & 0 < t < \sigma; \\ 0, & |t| > \sigma \end{cases}$ σ a positive constant.

3. $u(t) = \begin{cases} (1 + t)^2, & -1 \leqslant t \leqslant 0 \\ (1 - t)^2, & 0 \leqslant t \leqslant 1 \\ 0, & |t| \geqslant 1. \end{cases}$

4. $u(t) = 16e^{-4t^2}$.

5. $u(t) = (20 + 8t + t^2)^{-1}$.

6. $u(t) = (a^2 + b^2 t^2)^{-1}$; a and b positive constants.

7. $u(t) = \displaystyle\int_{-\sigma}^{\sigma} \frac{1}{1 + (t - s)^2}\, ds$; σ a positive constant.

8. $u(t) = \displaystyle\int_{-\infty}^{\infty} \frac{1}{1 + s^2} \frac{1}{1 + (t - s)^2}\, ds$.

9. $u(t) = \begin{cases} e^{-t^2}, & |t| > \sigma \\ e^{-t^2} - e^{-\sigma^2}, & |t| < \sigma \end{cases};$ σ a positive constant.

    (**Hint:** Let $u_1(t) = e^{-t^2}$ and $u_2 = u - u_1$. Find $\hat{u}_1$ and $\hat{u}_2$.)

10. u has the graph shown in Figure 5.4.

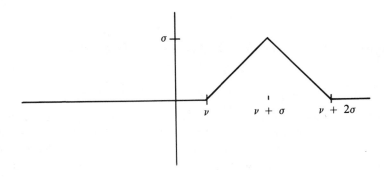

Figure 5.4

11. u has the graph shown in Figure 5.5.

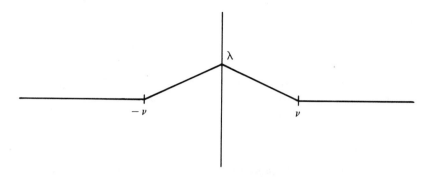

Figure 5.5

12. $u(t) = \begin{cases} A, & |t| < a \\ B, & a < |t| < b; \\ 0, & |t| > b \end{cases}$ a, b, A, and B constants, and $0 < a < b$.

(**Hint:** Write $u = u_1 + u_2$, where each of u_1 and u_2 has just one step.)

13. Give the details to show that (i) (3) is valid, (ii) (4) is valid, (iii) (5) is valid, and (iv) (6) is valid.

14. Let u be piecewise smooth and $U'(t) = u(t)$. Show that U is also piecewise smooth.

15. Suppose that $u(t) = 0$ if $|t| > A$. Use Morera's Theorem to show that $\hat{u}(z)$ is an entire function of the complex variable z; moreover, $\hat{u}$ is bounded on the real axis and, in general, satisfies $|\hat{u}(z)| \leqslant 2AMe^{A|y|}$ if $z = x + iy$, $M = \text{maximum}\{|u(t)|: |t| \leqslant A\}$.

16. Find u if $\hat{u}(x) = e^{-2|x|}$.

17. Let $*$ be the operation of convolution defined in (8). Show that $u * v = v * u$ and that $u * (v_1 + v_2) = u * v_1 + u * v_2$.

18. Let $\delta(t)$ be the "impulse function" at 0 discussed in Section 5, Chapter 4, Exercise 16. Show that $\delta * \delta = \delta$ and that $\hat{\delta}(x) = 1$ for all x.

19. Suppose that f is piecewise continuous on $(-\infty, \infty)$ and satisfies

$$\int_{-\infty}^{\infty} |f(t)|\, dt = M.$$

Let

$$u(x, y) = \frac{1}{\pi} \int_{-\infty}^{\infty} f(t)\, \frac{y}{(x-t)^2 + y^2}\, dt, \quad y > 0.$$

Show that

$$\int_{-\infty}^{\infty} |u(x, y)|\, dx \leqslant M, \quad \text{for each } y > 0.$$

20.   Suppose that $u(x, y)$ satisfies

$$\int_{-\infty}^{\infty} |u(x, y)| \, dx \leqslant M$$

for each y, $y \geqslant 0$. Let

$$U(s, y) = \int_{-\infty}^{\infty} u(x, y)e^{-isx} \, dx.$$

Show that $|U(s, y)| \leqslant M$ for all $y \geqslant 0$ and all s, $-\infty < s < \infty$.

21.   Let $U(s, y)$ be a bounded function on the region $-\infty < s < \infty$, $y > 0$. Suppose that U can be written as

$$U(s, y) = [A(s) + yB(s)]e^{sy} + [C(s) + yD(s)]e^{-sy}.$$

Show that

$$A(s) = B(s) = 0 \quad \text{if} \quad s > 0$$

and

$$C(s) = D(s) = 0 \quad \text{if} \quad s < 0.$$

(**Hint:** e^{sy} is unboundedly large as $y\uparrow\infty$ if $s > 0$, and e^{-sy} and ye^{-sy} both go to zero as $y\uparrow\infty$ if $s > 0$.)

22.   Let u and v satisfy (1) and let w be their convolution:

$$w(x) = \int_{-\infty}^{\infty} u(x - t)v(t) \, dt.$$

Show that

$$\int_{-\infty}^{\infty} |w(x)| \, dx \leqslant \left(\int_{-\infty}^{\infty} |u(t)| \, dt\right)\left(\int_{-\infty}^{\infty} |v(t)| \, dt\right).$$

In Exercises 23 to 26 use the Fourier transform in the manner of Examples 12 and 13 to solve the given partial differential equations.

23.   Solve the **wave equation**

$$u_{tt} = u_{xx}, \quad t \geqslant 0, \quad -\infty < x < \infty$$

subject to

$$u(x, 0) = 0$$
$$u_t(x, 0) = g(x), \quad -\infty < x < \infty.$$

24.   Solve the Dirichlet problem in the upper half-plane:

$$u_{xx} + u_{yy} = 0, \qquad u(x, 0) = f(x), \quad -\infty < x < \infty.$$

(At one point, you will need to make use of the conclusion of Exercise 21.) Compare your answer to the solution given in Section 3, Chapter 4.

25.   Solve the Neumann problem in the upper half-plane:

$$u_{xx} + u_{yy} = 0, \qquad \lim_{t \downarrow 0} \frac{\partial u}{\partial y}(x, t) = g(x), \quad -\infty < x < \infty.$$

Compare your answer to the solution given in Section 4, Chapter 4. (As in Exercise 24, at one point you will need to use the conclusion of Exercise 21.)

26.   Solve the **modified heat equation**

$$\frac{\partial u}{\partial t} = \frac{\partial^2 u}{\partial x^2} + g''(x); \quad -\infty < x < \infty, \, t \geqslant 0$$

with

$$u(0; x) = f(x), \quad -\infty < x < \infty.$$

Here g is a function of x alone; we assume that both g and g' satisfy (1).

27.   **(The telegraph equation.*)** Current flows along a long cable according to the equation

$$\frac{\partial^2 u}{\partial x^2} = CL \frac{\partial^2 u}{\partial t^2} + (RC + SL) \frac{\partial u}{\partial t} + RSu,$$

where $u(x; t)$ is the current at point x, $-\infty < x < \infty$, and time t, $t \geqslant 0$. The constants C, L, R, and S are, respectively, the capacitance, inductance, resistance, and leakage, all per unit length. Suppose that the initial conditions are $u(x; 0) = f(x)$ and $(\partial u/\partial t)(x; 0) = g(x)$, $-\infty < x < \infty$. We shall take steps toward solving this equation. Set $a = RS/CL$, $b = \frac{1}{2}[(R/L) + (S/C)]$, and $c = (CL)^{-1/2}$. The equation becomes

$$c^2 \frac{\partial^2 u}{\partial x^2} = \frac{\partial^2 u}{\partial t^2} + 2b \frac{\partial u}{\partial t} + au.$$

Define

$$U(t) = \int_{-\infty}^{\infty} u(x; t)e^{-ixs} \, dx, \quad s \text{ fixed.}$$

(i) Show that U satisfies the ordinary differential equation

$$U'' + 2bU' + (a + s^2c^2)U = 0$$

and solve for U.

(ii) In the special case where $RC = SL$, complete the work by solving for $u(x; t)$.

---

* The telegraph equation was discovered by Oliver Heaviside; see Section 3.

5.2   Formulas relating u and $\hat{u}$

The relationships between a function u and its Fourier transform $\hat{u}$ are set forth in the following two formulas. The first, the **Fourier inversion formula**, is

$$u(t) = \frac{1}{2\pi} \int_{-\infty}^{\infty} \hat{u}(x)e^{ixt} \, dx \tag{1}$$

if the function u is *continuous* at t. Since u could possibly be discontinuous at a finite number of points, a more general version of (1) is

$$\tfrac{1}{2}(u^+(t) + u^-(t)) = \frac{1}{2\pi} \int_{-\infty}^{\infty} \hat{u}(x)e^{ixt} \, dx. \tag{1$'$}$$

Either of these formulas shows that the function u can be recovered from its Fourier transform $\hat{u}$ by means of a formula almost identical to the formula that defines $\hat{u}$ in terms of u.

    The second formula states that 2π times the integral over the real axis of $|u|^2$ is exactly the same as the integral of $|\hat{u}|^2$:

$$2\pi \int_{-\infty}^{\infty} |u(t)|^2 \, dt = \int_{-\infty}^{\infty} |\hat{u}(x)|^2 \, dx. \tag{2}$$

These two formulas are at the heart of the theory of the Fourier transform. As mentioned in the opening paragraph of Section 1, a thorough mathematical justification of formulas (1) and (2) is not elementary.

    The equality in (2) is called **Parseval's equality**. In physical terms it states that the total energy of a system is preserved by the Fourier transform— the transformed system has exactly as much energy as the original system. Furthermore, equation (2) is equivalent to the seemingly more general equality

$$2\pi \int_{-\infty}^{\infty} u(t)\overline{v(t)} \, dt = \int_{-\infty}^{\infty} \hat{u}(x)\overline{\hat{v}(x)} \, dx. \tag{3}$$

Indeed, (3) follows from (2) by replacing u by $u + v$ and using the known equality (2) for both $|u|^2$ and $|v|^2$ (see the exercises at the end of this section).

Example 1   Find the Fourier transform of $f(t) = (\sin t)/t$.

Solution   From Example 6, Section 1, we know that $2(\sin x)/x$ is the Fourier transform of the function

$$u(t) = \begin{cases} 1, & |t| < 1 \\ 0, & |t| > 1. \end{cases}$$

Thus, by the Fourier inversion formula, we have

$$\hat{f}(x) = \begin{cases} \pi, & |x| < 1 \\ 0, & |x| > 1. \end{cases} \qquad \square$$

You might try to compute $\hat{f}(x)$ directly to better appreciate the strength of the Fourier inversion formula that we are using here.

Example 2   Compute

$$\int_{-\infty}^{\infty} \frac{(1 - \cos x)^2}{x^4}\, dx.$$

Solution   The function

$$f(x) = 2\, \frac{1 - \cos x}{x^2}$$

is the Fourier transform of the function

$$u(t) = \begin{cases} 1 + t, & -1 \leqslant t \leqslant 0 \\ 1 - t, & 0 \leqslant t \leqslant 1 \\ 0, & |t| > 1 \end{cases}$$

(see the function u_3 in Example 9, Section 1). The integral we want is one-quarter the integral of the square of $f = \hat{u}$ and, by (2), this is equal to

$$(\tfrac{1}{4})2\pi \int_{-\infty}^{\infty} (u(t))^2\, dt = \frac{\pi}{3}.$$

Hence,

$$\int_{-\infty}^{\infty} \frac{(1 - \cos x)^2}{x^4}\, dx = \frac{\pi}{3}. \qquad \square$$

The complex Fourier transform

Let us look back at formula (2) in Section 1, which defines the Fourier transform of a function u, but now replace the real variable x by the complex variable $z = x + iy$:

$$\hat{u}(z) = \int_{-\infty}^{\infty} u(t) e^{-izt}\, dt. \tag{4}$$

We need to concern ourselves first with the question of whether the integral makes sense and, if so, for which z. We have

$$|u(t)e^{-izt}| = |u(t)|e^{yt}.$$

First take $-\infty < t \leqslant 0$; in order that the integral on $(-\infty, 0)$ converges, we assume that

$$|u(t)| \leqslant M_1 e^{-\tau_1 t}, \ t \leqslant 0 \tag{5}$$

for some constant M_1 and some number τ_1. It follows from this that

$$|u(t)e^{-izt}| \leqslant M_1 e^{t(y-\tau_1)}, \ t \leqslant 0$$

and so

$$\int_{-\infty}^{0} u(t)e^{-izt} \, dt$$

converges for all z with $y = \mathrm{Im}\, z > \tau_1$ (see Exercises 16 to 19 at the end of this section). Likewise, if u satisfies the growth condition

$$|u(t)| \leqslant M_2 \, e^{-\tau_2 t} \ \text{for} \ t \geqslant 0, \tag{6}$$

then the integral

$$\int_{0}^{\infty} u(t)e^{-izt} \, dt$$

is absolutely convergent for all z with $-\tau_2 + y < 0$. Thus, if both (5) and (6) hold, then the integral defining $\hat{u}(z)$ is convergent for all z with

$$\tau_1 < \mathrm{Im}\, z < \tau_2. \tag{7}$$

In this way we see that the complex Fourier transform $\hat{u}(z)$ is defined for all z in a strip of the form (7). Moreover, in this strip $\hat{u}(z)$ is an analytic function of z with derivative

$$[\hat{u}]'(z) = -i \int_{-\infty}^{\infty} u(t)t \, e^{-izt} \, dt. \tag{8}$$

The analyticity of $\hat{u}(z)$ can validated more fully by use of Morera's Theorem (see the exercises at the end of this section). Thus, for example, if $u(t) = 0$ for all t with $|t| \geqslant A > 0$, then surely both (5) and (6) hold for every τ_1 and τ_2 and so $\hat{u}(z)$ is analytic for all z; that is, $\hat{u}$ is an entire function.

Example 3 Let $u(t) = e^{-|t|}$. Then $\tau_1 = -1$ and $\tau_2 = 1$ so $\hat{u}(z)$ is analytic within the strip $|\mathrm{Im}\, z| < 1$. In fact,

$$\hat{u}(z) = 2 \left(\frac{1}{1+z^2} \right)$$

as Example 8, Section 1, and the Fourier inversion formula show. Thus, the strip of analyticity of $\hat{u}$ is as wide as possible: $\hat{u}$ is not analytic in any wider (open) strip.

Example 4 Let us look at the Example 3 the other way around. Take $u(t) = 2(1 + t^2)^{-1}$; then $\hat{u}(x) = 2\pi e^{-|x|}$ and this function has no analytic extension to a strip of the form $\tau_1 < \mathrm{Im}\, z < \tau_2$ no matter how τ_1 and τ_2 are chosen $\tau_1 \leqslant 0 \leqslant \tau_2$. Thus, the estimates of growth on $|u(t)|$ given by (5) and (6) are not at all superfluous.

To invert the complex Fourier transform we use the usual formula but in a new location. Let y_0 be any number with $\tau_1 < y_0 < \tau_2$; then

$$|u(t)e^{-it(x+iy_0)}| = |u(t)|e^{ty_0}$$

$$\leqslant \begin{cases} M_1 e^{-\varepsilon_1 t} & \text{if } t \geqslant 0 \\ M_2 e^{+\varepsilon_2 t} & \text{if } t \leqslant 0, \end{cases}$$

where M_1, M_2, ε_1, and ε_2 are positive constants (see (5) and (6)). Let

$$u_1(t) = u(t)e^{ty_0}.$$

Then for $-\infty < x < \infty$, we have

$$\hat{u}_1(x) = \hat{u}(x + iy_0).$$

Consequently, by the Fourier inversion formula,

$$u(t) = e^{-ty_0}u_1(t)$$

$$= e^{-ty_0}\frac{1}{2\pi}\int_{-\infty}^{\infty}\hat{u}_1(x)e^{itx}\,dx$$

$$= \frac{1}{2\pi}\int_{-\infty}^{\infty}\hat{u}(x+iy_0)e^{it(x+iy_0)}\,dx$$

$$= \frac{1}{2\pi}\int_{-\infty+iy_0}^{\infty+iy_0}\hat{u}(z)e^{itz}\,dz.$$

The lower and upper limits on the integral mean that z traverses the entire horizontal line $y = y_0$ from left to right. Since y_0 is any number in the interval (τ_1, τ_2) we can write

$$u(t) = \frac{1}{2\pi}\int_{-\infty+iy}^{\infty+iy}\hat{u}(z)e^{itz}\,dz, \quad \tau_1 < y < \tau_2.$$

$\hat{u}(z)$ is analytic and obviously so is e^{itz}, and thus it is frequently possible to use the residue theorem, or some other similar device, to evaluate this integral.

Example 5   Find u if $\hat{u}(z) = 1/z^2$.

Solution $\hat{u}(z)$ is analytic except at the origin so we may use the residue theorem to evaluate the integral

$$\int_{-\infty+iy}^{\infty+iy}\frac{e^{izt}}{z^2}\,dz.$$

We choose y_0 to be a positive number. If t is positive we integrate over the semicircle shown in Figure 5.6a, while if t is negative, we integrate over the semicircle in Figure 5.6b. In both cases the integral over the curved portion goes to zero as $R \to \infty$. Cauchy's Theorem then gives

$$u(t) = \begin{cases} 0, & t > 0 \\ t, & t < 0. \end{cases} \qquad \square$$

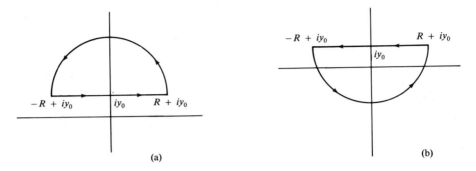

Figure 5.6

A mathematical justification of formulas (1) and (2)

The correctness of what is done below depends on the validity of two steps that, although clearly believable, are themselves in need of further mathematical justification. We will point these out and comment further on them after the derivation of the formulas.

Let u and v be two piecewise smooth functions that satisfy the conditions

$$\int_{-\infty}^{\infty} |u(t)| \, dt < \infty; \qquad \int_{-\infty}^{\infty} |u(t)|^2 \, dt < \infty; \qquad \lim_{|t| \to \infty} u(t) = \lim_{|t| \to \infty} u'(t) = 0$$

and the same with v in place of u. Then

$$\int_{-\infty}^{\infty} \hat{u}(x)v(x)e^{isx} \, dx = \int_{-\infty}^{\infty} v(x) \left\{ \int_{-\infty}^{\infty} u(t)e^{-ixt} \, dt \right\} e^{isx} \, dx$$

$$= \int_{-\infty}^{\infty} u(t) \left\{ \int_{-\infty}^{\infty} v(x)e^{-ix(t-s)} \, dx \right\} dt$$

$$= \int_{-\infty}^{\infty} u(t)\hat{v}(t-s) \, dt = \int_{-\infty}^{\infty} u(t+s)\hat{v}(t) \, dt.$$

Now take $v(x) = e^{-\varepsilon x^2}$, where ε is a small positive number. We know, from Example 7 in Section 1 and (5) in Section 1 that

$$\hat{v}(t) = \sqrt{\pi} \, \frac{e^{-t^2/4\varepsilon^2}}{\varepsilon}.$$

Hence,

$$\int_{-\infty}^{\infty} e^{-\varepsilon x^2} \hat{u}(x)e^{isx} \, dx = \sqrt{\pi} \int_{-\infty}^{\infty} u(t+s) \frac{e^{-t^2/4\varepsilon^2}}{\varepsilon} \, dt$$

$$= \sqrt{\pi} \int_{-\infty}^{\infty} u(\varepsilon y + s)e^{-y^2/4} \, dy.$$

Now let ε decrease to zero; the left-hand side converges to

$$\int_{-\infty}^{\infty} \hat{u}(x)e^{isx}\,dx.$$

The right-hand side converges to

$$\sqrt{\pi} \int_{-\infty}^{\infty} u(s)e^{-y^2/4}\,dy = u(s)2\pi$$

if u is continuous at s or to $(2\pi)\tfrac{1}{2}(u^+(s) + u^-(s))$ if s happens to be one of the finite number of points of discontinuity of u. This gives (1) if u is continuous at s or (1)′ if u is discontinuous at s.*

To prove (2) let $\varepsilon > 0$ and set

$$v_\varepsilon(t) = \overline{\hat{u}(t)}\,e^{-\varepsilon^2 t^2},$$

$$w_\varepsilon(t) = \frac{1}{2\sqrt{\pi}\,\varepsilon}\,e^{-t^2/4\varepsilon^2},$$

and

$$p(t) = \overline{u(-t)}.$$

A simple computation gives $\overline{\hat{u}(t)} = \hat{p}(t)$ and so

$$v_\varepsilon(t) = \hat{p}(t)\hat{w}_\varepsilon(t) = (p * w_\varepsilon)\hat{\;}(t).$$

Using (1) we find that

$$\frac{1}{2\pi}\hat{v}_\varepsilon(x) = (p * w_\varepsilon)(-x)$$

$$= \int_{-\infty}^{\infty} p(-x - s)w_\varepsilon(s)\,ds$$

$$= \frac{1}{\sqrt{\pi}} \int_{-\infty}^{\infty} p(-x - 2\varepsilon t)e^{-t^2}\,dt$$

$$= \frac{1}{\sqrt{\pi}} \int_{-\infty}^{\infty} \overline{u(x + 2\varepsilon t)}\,e^{-t^2}\,dt.$$

In the formula

$$\int_{-\infty}^{\infty} \hat{u}(x)v(x)e^{isx}\,dx = \int_{-\infty}^{\infty} u(x + s)\hat{v}(x)\,dx$$

derived above, we take $s = 0$ and $v = v_\varepsilon$. This produces the relation

---

* An alternative justification of (1) is given in Exercises 22 to 25.

$$\int_{-\infty}^{\infty} |\hat{u}(x)|^2 e^{-\varepsilon^2 x^2}\, dx = \int_{-\infty}^{\infty} u(x) \left\{ \int_{-\infty}^{\infty} (2\sqrt{\pi})\overline{u(x + 2\varepsilon t)}\, e^{-t^2}\, dt \right\} dx$$

$$= 2\sqrt{\pi} \int_{-\infty}^{\infty} e^{-t^2} \left\{ \int_{-\infty}^{\infty} u(x)\overline{u(x + 2\varepsilon t)}\, dx \right\} dt.$$

Upon letting ε decrease to zero, we obtain (2).

A few comments now on the mathematics above. To prove (1) we wrote down the integral of $\hat{u}(x)v(x)$, then replaced $\hat{u}(x)$ by its integral formulation and then interchanged the order of integration in the resulting iterated integral. This is mathematically justifiable here but only with some effort. We then made the choice $v(x) = \exp(-\varepsilon x^2)$ and, after a change of variables we then asserted that

$$\lim_{\varepsilon \to 0} \int_{-\infty}^{\infty} (\quad) = \int_{-\infty}^{\infty} \lim_{\varepsilon \to 0} (\quad),$$

where the expression (   ) depends on ε as well as the variable of integration. Again, this is reasonable but it needs mathematical justification.

Exercises for Section 5.2

1.   Use (3) to show that

$$\int_{-\infty}^{\infty} \frac{\sin \alpha x \sin \beta x}{x^2}\, dx = \pi \min\{\alpha, \beta\}, \quad \alpha \text{ and } \beta \text{ positive.}$$

2.   Show that

$$\int_0^{\infty} e^{-t}\, \frac{\sin \alpha t}{t}\, dt = \arctan \alpha, \quad \alpha \text{ positive.}$$

(**Hint:** Rewrite the integral as

$$\tfrac{1}{2} \int_{-\infty}^{\infty} e^{-|t|}\, \frac{\sin \alpha t}{t}\, dt$$

and use (3).)

3.   Compute the value of

$$\int_{-\infty}^{\infty} \left(\frac{1 - \cos \sigma x}{x} \right)^2 dx$$

using (2) and Exercise 2, Section 1.

4.   Compute the value of

$$\int_{-\infty}^{\infty} \frac{(x - \sin x)^2}{x^6}\, dx$$

using (2) and Exercise 3, Section 1.

5.   Using (2), compute the value of

$$\int_{-\infty}^{\infty} \frac{dx}{(1 + x^2)^2}.$$

6.   Find $\hat{u}$ if

$$u(t) = \int_{-\infty}^{\infty} e^{-|t-s|} \frac{\sin \sigma s}{s} \, ds.$$

7.   Find u if

$$\hat{u}(x) = e^{-|x|}\left(\frac{\sin 4x}{x}\right).$$

8.   Use Fourier transforms to solve the following equation for ϕ:

$$e^{-x^2} = \int_{-\infty}^{\infty} \phi(t)e^{-|x-t|} \, dt.$$

9.   Use (3) to find the value of

$$\int_{-\infty}^{\infty} \frac{(x - \sin x)(\sin \sigma x)}{x^4} \, dx.$$

10.   Use (3) to show that

$$\int_{-\infty}^{\infty} \frac{(\sin^2 \alpha x)(\sin^2 \beta x)}{x^2} \, dx = \frac{\pi}{2} \min\{\alpha, \beta\}.$$

(**Hint:** $2 \sin^2 \theta = 1 - \cos 2\theta$.)

11.   Suppose that $u(t) = 0$ for $|t| \geqslant A$ and that $\hat{u}(x) = 0$ for $|x| \geqslant B$, A and $B > 0$. Show that u is identically zero.

12.   Suppose that u_1 and u_2 are two functions satisfying (1) of Section 1 and u_1 and u_2 have the same Fourier transform. Show that $u_1 = u_2$.

13.   Give the details of how (2) implies (3).

14.   Suppose that $\int_{-\infty}^{\infty} |u(t)| \, dt < \infty$; show that $\hat{u}(x)$ is a continuous function of x. (**Hint:** We can write

$$(\hat{u}(x_2) - \hat{u}(x_1)) = \int_{-\infty}^{\infty} u(t)\{e^{-ix_2t} - e^{-ix_1t}\} \, dt = \int_{-M}^{M} + \int_{M}^{\infty} + \int_{-\infty}^{-M}.$$

The second and third integrals are small for all x_1 and x_2 if only M is big, while for M fixed the first integral is small if x_1 is near x_2.)

15.   Suppose (1) of Section 1 holds. Fill in the details of the following steps to show that the **Riemann–Lebesgue lemma** holds:

$$\lim_{x \to \infty} \hat{u}(x) = 0.$$

(a) $\hat{u}(x) = -\int_{-\infty}^{\infty} e^{-ixt} u(t + (\pi/x)) \, dt$

(b) $\hat{u}(x) = \int_{-\infty}^{\infty} e^{-ixt} \frac{1}{2}\{u(t) - u(t + \pi/x)\} \, dt$

(c) $|\hat{u}(x)| \leq \{\int_{-M}^{M} + \int_{M}^{\infty} + \int_{-\infty}^{-M}\} w(t) \, dt$,
where $w(t) = |u(t) - u(t + \pi/x)|$. Show that the second and third integrals are small for any x if M is big; show that the first integral is small for M fixed if x is big and u is continuous.

(d)   Modify step (c) above to include the case when u is merely piecewise continuous.

Improper integrals on $(-\infty, \infty)$

A piecewise continuous function u on the real axis has an **improper integral** on $(-\infty, \infty)$ if

$$\lim_{\substack{M_1 \to \infty \\ M_2 \to \infty}} \int_{-M_2}^{M_1} u(t) \, dt$$

exists no matter *how* M_1 and M_2 go to ∞. $u(t)$ is **absolutely integrable** if $|u(t)|$ has an improper integral. We denote the improper integral of u on $(-\infty, \infty)$ by $\int_{-\infty}^{\infty} u(t) \, dt$.

16.   Show that u is absolutely integrable if and only if there is a constant C such that

$$\int_{-M}^{M} |u(t)| \, dt \leq C$$

for all choices of M. Show that the numbers $\int_{-M}^{M} |u(t)| \, dt$ increase to $\int_{-\infty}^{\infty} |u(t)| \, dt$ as $M \uparrow \infty$.

17.   Let u be absolutely integrable. Show that

$$\lim_{B \to \infty} \int_{-\infty}^{-B} |u(t)| \, dt = \lim_{A \to \infty} \int_{A}^{\infty} |u(t)| \, dt = 0.$$

18.   Let u be absolutely integrable. Show that u has an improper integral on $(-\infty, \infty)$ and that

$$\left| \int_{-\infty}^{\infty} u(t) \, dt \right| \leq \int_{-\infty}^{\infty} |u(t)| \, dt.$$

(**Hint:** For any $A < B$, we have

$$\left| \int_{A}^{B} u(t) \, dt \right| \leq \int_{A}^{B} |u(t)| \, dt.$$

Next show that for M_1 and M_2 sufficiently large and $M_1' > M_1$, $M_2' > M_2$, we have

$$\left| \int_{M_1}^{M_1'} u(t) \, dt \right| \le \int_{M_1}^{M_1'} |u(t)| \, dt < \varepsilon$$

$$\left| \int_{-M_2'}^{-M_2} u(t) \, dt \right| \le \int_{-M_2'}^{-M_2} |u(t)| \, dt < \varepsilon.)$$

19. Let $|u(t)| \le |v(t)|$ for $-\infty < t < \infty$ and suppose that v is absolutely integrable on $(-\infty, \infty)$. Show that u is also absolutely integrable on $(-\infty, \infty)$.

20. Suppose that $\int_{-\infty}^{\infty} |u'(t)| \, dt < \infty$ and $\int_{-\infty}^{\infty} |u(t)| \, dt < \infty$. Show that $\lim_{x \to \infty} u(x) = 0$. (**Hint:** Given a small $\varepsilon > 0$ choose M so large that $\int_M^{\infty} |u'(t)| \, dt < \varepsilon$ (see Exercise 17). For any $s > 0$, we then have

$$|u(s + M) - u(M)| = \left| \int_M^{M+s} u'(t) \, dt \right| \le \int_M^{M+s} |u'(t)| \, dt < \varepsilon.$$

Now let $s \to \infty$. This shows that $\lim_{x \to \infty} u(x)$ exists. Now show that this limit must be zero.)

21. Use Morera's Theorem (Theorem 2, Section 4, Chapter 2) to prove that $\hat{u}(z)$ is analytic for $\tau_1 < \text{Im } z < \tau_2$ if u satisfies (5) and (6).

An alternate proof of the Fourier inversion formula (1)★

We shall show that

$$\lim_{M \to \infty} \frac{1}{2\pi} \int_{-M}^{M} \hat{u}(s) e^{ixs} \, ds = \tfrac{1}{2}(u^+(x) + u^-(x)).$$

22. Show that

$$\int_{-M}^{M} \hat{u}(s) \, e^{ixs} \, ds = 2 \int_{-\infty}^{\infty} u(t) \, \frac{\sin M(x - t)}{x - t} \, dt$$

$$= 2 \int_{-\infty}^{\infty} u(x + t) \, \frac{\sin Mt}{t} \, dt.$$

23. Use Exercise 15 to show that for each x and each $\delta > 0$, we have

(i) $\displaystyle \lim_{M \to \infty} \int_{\delta}^{\infty} u(x + t) \, \frac{\sin Mt}{t} \, dt = 0$

and

(ii) $\displaystyle \lim_{M \to \infty} \int_{-\infty}^{-\delta} u(x + t) \, \frac{\sin Mt}{t} \, dt = 0.$

24. Choose δ so small that $u(x + t) - u^+(x)$ is continuous on $[0, \delta]$ and $u(x + t) - u^-(x)$ is continuous on $[-\delta, 0]$. Show that

(iii) $\displaystyle \lim_{M \to \infty} \int_0^\delta (u(x + t) - u^+(x)) \frac{\sin Mt}{t} \, dt = 0$

(iv) $\displaystyle \lim_{M \to \infty} \int_{-\delta}^0 (u(x + t) - u^-(x)) \frac{\sin Mt}{t} \, dt = 0.$

25.   Conclude from Exercises 23 and 24 that

$$\lim_{M \to \infty} \int_{-\infty}^\infty u(x + t) \frac{\sin Mt}{t} \, dt = (u^+(x) + u^-(x)) \int_0^\infty \frac{\sin x}{x} \, dx$$

$$= \frac{\pi}{2} (u^+(x) + u^-(x))$$

and hence that

$$\lim_{M \to \infty} \frac{1}{2\pi} \int_{-M}^M \hat{u}(s) \, e^{ixs} \, ds = \tfrac{1}{2}(u^+(x) + u^-(x)).$$

5.3   The Laplace transform

The Laplace transform is just the Fourier transform rotated by 90° and applied to a function that vanishes on the ray $(-\infty, 0)$. Specifically, suppose that u is a piecewise continuous function on $[0, \infty)$ such that both u and u' have limits from the right at $t = 0$. The **Laplace transform** of u is

$$(\mathcal{L}u)(s) = \int_0^\infty u(t)e^{-st} \, dt. \tag{1}$$

Initially we can think of s as a real variable, but with the knowledge we already have about the complex Fourier transform, it is natural to investigate immediately for which complex numbers s the integral in (1) makes sense. We shall henceforth suppose that the function u has **exponential growth** a; that is, there is a constant M such that

$$|u(t)| \leqslant M \, e^{at}, \; t > 0. \tag{2}$$

Let the complex number s be $s = \sigma + i\tau$; then

$$|u(t)e^{-st}| \leqslant M \, e^{at} \, e^{-\sigma t}$$
$$= M \, e^{t(a - \sigma)}.$$

Hence, if $a - \sigma < 0$, then the integral in (1) is absolutely convergent. We see, therefore, that the Laplace transform of u is defined at least for all s with Re $s > a$, where a is the exponential order of growth of u from (2). Furthermore, within the half-plane Re $s > a$, the Laplace transform of u is an analytic function of the complex variable s. One way to see that $\mathcal{L}u(s)$ is analytic is to write down the elementary relationship between the Laplace and Fourier transforms of the function u:

$$(\mathscr{L}u)(s) = \hat{u}(-is), \quad \text{Re } s > a. \tag{3}$$

Since $\hat{u}(s)$ is analytic in the half-plane Im $s < -a$ it follows that $\mathscr{L}u(s)$ is analytic for Re $s > a$.*

Why should we bother with the Laplace transform when it is so much like the Fourier transform? There are several reasons. One is that in a physical system there is typically no output until there is input. That is, the function that represents the output must vanish on the ray $(-\infty, 0)$ (assuming that the input begins at $t = 0$); such functions are termed **causal**. Another reason is that by beginning the integration at $t = 0$, when we integrate by parts the boundary terms will incorporate the numbers $u(0)$, $u'(0)$, etc., and hence the Laplace transform is an effective tool for handling differential equations on the ray $(0, \infty)$, which include **initial conditions**.

Finally the Laplace transform is a better tool than the Fourier transform for dealing with the impulse function $\delta(t)$ and the **Heaviside† function** H, defined by

$$H(t) = \begin{cases} 0, & t < 0 \\ 1, & t > 0. \end{cases}$$

As was pointed out in Exercise 17, Section 5, Chapter 4, H is in some sense an indefinite integral of $\delta(t)$. Furthermore, the Fourier transform of $H(t_0 - t)$ make no sense for any t_0, whereas the Laplace transform is defined for all t_0:

$$\int_0^\infty H(t_0 - t)e^{-st}\, dt = \int_0^{t_0} e^{-st}\, dt$$

$$= \frac{1}{s}[1 - e^{-st_0}] \quad \text{if } t_0 > 0$$

and

$$\int_0^\infty H(t_0 - t)e^{-st}\, dt = 0, \text{ if } t_0 \leqslant 0.$$

Example 1   Find the Laplace transform of each of the following functions (all the functions vanish for $t < 0$):

(a) $u_1(t) = 1$;

(b) $u_2(t) = t^k$, k a positive integer;

(c) $u_3(t) = \sin At$, $A \neq 0$;

(d) $u_4(t) = \cosh Bt$, $B \neq 0$;

(e) $u_5(t)$ is the impulse function at the point t_0, $t_0 > 0$.

---

* It is possible to prove that there is a number s_0 such that the Laplace transform of u converges if Re $z > s_0$ and diverges if Re $z < s_0$; s_0 is much like the radius of convergence of a power series.

† Oliver Heaviside, 1850–1925.

Solution

(a) $\mathscr{L}u_1(s) = \displaystyle\int_0^\infty e^{-st}\, dt = \dfrac{1}{s}.$

(b) $\mathscr{L}u_2(s) = \displaystyle\int_0^\infty t^k\, e^{-st}\, dt = k!/s^{k+1};$ integrate by parts k times.

(c) $\mathscr{L}u_3(s) = \displaystyle\int_0^\infty (\sin At)\, e^{-st}\, dt = \dfrac{1}{2i}\int_0^\infty (e^{iAt} - e^{-iAt})\, e^{-st}\, dt$

$$= \dfrac{1}{2i}\left\{ \dfrac{-1}{iA - s} + \dfrac{1}{-iA - s} \right\} = \dfrac{A}{A^2 + s^2}.$$

(d) $\mathscr{L}u_4(s) = \displaystyle\int_0^\infty (\cosh Bt)\, e^{-st}\, dt = \tfrac{1}{2}\int_0^\infty (e^{Bt} + e^{-Bt})e^{-st}\, dt$

$$= \dfrac{1}{2}\left\{ -\dfrac{1}{B - s} - \dfrac{1}{-B - s} \right\} = \dfrac{s}{s^2 - B^2}.$$

(e) $\mathscr{L}u_5(s) = \displaystyle\int_0^\infty \delta(t - t_0)\, e^{-st}\, dt = e^{-st_0}.$ □

Example 2 Find the Laplace transform of u', u'', ..., in terms of the Laplace transform of u and the initial data $u(0)$, $u'(0)$, ...

Solution We assume that u and all its derivatives have exponential growth a. For Re $s > a$, we have

$$(\mathscr{L}u')(s) = \int_0^\infty u'(t)\, e^{-st}\, dt = u(t)\, e^{-st}\Big|_0^\infty + s\int_0^\infty u(t)\, e^{-st}\, dt$$

$$= -u(0) + s\,\mathscr{L}u(s).$$

Likewise,

$$(\mathscr{L}u'')(s) = \int_0^\infty u''(t)\, e^{-st}\, dt = u'(t)\, e^{-st}\Big|_0^\infty + s\int_0^\infty u'(t)\, e^{-st}\, dt$$

$$= -u'(0) + s(\mathscr{L}u')(s) = -u'(0) - su(0) + s^2\,\mathscr{L}u(s).$$

Continuing in this way we obtain the formula

$$(\mathscr{L}u^{(n)})(s) = s^n\,\mathscr{L}u(s) - s^{n-1}u(0) - s^{n-2}u'(0) - \cdots - u^{(n-1)}(0),$$

which is valid for all complex numbers s with Re $s > a$.   □

Further properties of the Laplace transform

It is clear that the Laplace transform is **linear**:

$$\mathscr{L}(\lambda_1 u_1 + \lambda_2 u_2)(s) = \lambda_1\mathscr{L}u_1(s) + \lambda_2\mathscr{L}u_2(s), \quad \text{Re } s > a,$$

for any two functions u_1, u_2, both of exponential growth a, and complex numbers λ_1, λ_2. In addition, like the Fourier transform, the Laplace transform of the convolution of two functions is just the product of their Laplace transforms. Since the functions vanish on $(-\infty, 0)$, we have

$$(u * v)(t) = \int_0^t u(x)v(t - x)\, dx, \ t > 0, \tag{4}$$

and

$$[\mathscr{L}(u * v)](s) = (\mathscr{L}u(s))(\mathscr{L}v(s)). \tag{5}$$

Another useful formula relates the derivative of the Laplace transform of u to the transform of $tu(t)$. The formula is

$$\frac{d}{ds}(\mathscr{L}u)(s) = -\mathscr{L}(tu(t))(s). \tag{6}$$

The derivation of (6) is immediate (see Exercise 18).

Finally, if the Laplace transform of u is identically zero, then u is also identically zero. Consequently, if u_1 and u_2 are two functions with the same Laplace transform, then u_1 and u_2 are the same function.

Furthermore, because of the close connection between the Fourier and Laplace transforms, we can quickly and easily obtain the inversion formula for the Laplace transform. Suppose that $u(t)$ vanishes for $t < 0$ and is of exponential growth a; let σ be any number larger than a. The Laplace and Fourier transforms are related by

$$\hat{u}(-is) = (\mathscr{L}u)(s), \ \text{Re } s > a.$$

By the Fourier inversion formula we know that

$$u(t) = \frac{1}{2\pi} \int_{-\infty + i\sigma}^{\infty + i\sigma} \hat{u}(z)\, e^{izt}\, dz.$$

In this integral set $z = -is$; this produces the formula

$$u(t) = \frac{1}{2\pi i} \int_{\sigma - i\infty}^{\sigma + i\infty} (\mathscr{L}u)(s)\, e^{st}\, ds, \tag{7}$$

where σ is any number larger than a, and the integration is over the entire vertical line $x = \sigma$. Formula (7) is the **inversion formula for the Laplace transform**. Typically the integral is evaluated by means of the residue theorem, as the following examples show.

Example 3   Find u if

$$\mathscr{L}u(s) = \frac{1}{s^2 - 2s + 3}.$$

Solution   The function u is found from the Laplace inversion formula (7); with

$\mathscr{L}u$ as given, we obtain

$$u(t) = \frac{1}{2\pi i} \int_{\sigma - i\infty}^{\sigma + i\infty} \frac{e^{st}}{s^2 - 2s + 3} \, ds.$$

The function $(s^2 - 2s + 3)^{-1}$ has poles at $1 \pm i\sqrt{2}$ and so is analytic for Re $s > 1$; hence, we choose $\sigma > 1$ for the vertical line of integration. We distinguish two cases: $t \geqslant 0$ and $t < 0$. For $t \geqslant 0$, we integrate

$$f(z) = \frac{e^{zt}}{z^2 - 2z + 3}$$

over the curve Γ shown in Figure 5.7. On the semicircular part of Γ, we have $z = \sigma + Re^{i\theta}$, $\pi/2 \leqslant \theta \leqslant 3\pi/2$; this gives the estimate (for $t \geqslant 0$)

$$|f(z)| \leqslant e^{\sigma t} \frac{e^{tR \cos \theta}}{R^2/4 - 2R - 3}, \quad \text{if } R \text{ is large},$$

$$\leqslant \frac{4e^{\sigma t}}{R^2 - 8R - 12}.$$

Thus, the integral of f over the semicircular part of Γ is no larger than

$$\frac{4e^{\sigma t}}{R^2 - 8R - 12} \pi R \to 0 \quad \text{as } R \to \infty$$

(see formula (3), Section 6, Chapter 1). The function f has simple poles at $1 \pm i\sqrt{2}$ with residues

$$\text{Res}(f; 1 + i\sqrt{2}) = \frac{e^t e^{it\sqrt{2}}}{2i\sqrt{2}}$$

and

$$\text{Res}(f; 1 - i\sqrt{2}) = \frac{e^t e^{-it\sqrt{2}}}{-2i\sqrt{2}}.$$

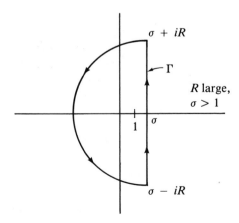

Figure 5.7

An application of the residue theorem and passage to the limit as $R \to \infty$ then yields

$$u(t) = \text{Res}(f; 1 + i\sqrt{2}) + \text{Res}(f; 1 - i\sqrt{2})$$
$$= e^t \sin(\sqrt{2}\,t), \; t \geq 0.$$

On the other hand, if $t < 0$, the integration is taken over the curve Γ shown in Figure 5.8.

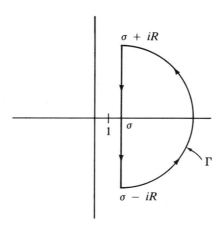

Figure 5.8

Estimates such as those just completed show that the integral of f over the semicircular part of Γ goes to zero as $R \to \infty$. However, because f is analytic on and within Γ, Cauchy's Theorem (and passage to the limit as $R \to \infty$) yields $u(t) = 0$ for $t < 0$. Hence,

$$u(t) = \begin{cases} e^t \sin(\sqrt{2}t), & t \geq 0 \\ 0, & t < 0. \end{cases}$$

□

Example 4   Find u if

$$\mathscr{L}u(s) = \frac{s}{(s^2 + A^2)^2}.$$

Solution   We recall that

$$\mathscr{L}(\sin At)(s) = \frac{A}{s^2 + A^2}.$$

Hence,

$$\frac{d}{ds}\,\mathscr{L}(\sin At)(s) = \frac{-2sA}{(s^2 + A^2)^2}.$$

However, by formula (6), we also have

$$- \mathcal{L}(t \sin At) = \frac{d}{ds} \mathcal{L}(\sin At)(s) = \frac{-2sA}{(s^2 + A^2)^2}.$$

Consequently,

$$u(t) = \frac{1}{2A} t \sin At.$$

Another technique to find u is to use the residue theorem (see the exercises). $\square$

Example 5   Let u be a piecewise smooth function that satisfies the equation

$$u(t) = u(t + T), \quad t \geqslant 0$$

for some fixed positive number T. Find the Laplace transform of u.

Solution   For Re $s > 0$,

$$(\mathcal{L}u)(s) = \int_0^\infty u(t) \, e^{-st} \, dt$$

$$= \int_0^T + \int_T^{2T} + \int_{2T}^{3T} + \cdots$$

$$= \int_0^T u(t) \, e^{-st} \, dt + \int_0^T u(t) \, e^{-s(t+T)} \, dt + \int_0^T u(t) \, e^{s(t+2T)} \, dt + \cdots$$

$$= \int_0^T u(t) \, e^{-st} \{1 + e^{-sT} + e^{-2sT} + \cdots\} \, dt$$

$$= \frac{\int_0^T u(t) \, e^{-st} \, dt}{1 - e^{-sT}}.$$

The number T is called a **period** of u and the smallest such T is called the period of u. $\square$

Exercises for Section 5.3

In Exercises 1 to 10, find the Laplace transform of the given function u; in all cases $u(t) = 0$ if $t < 0$.

1. $u(t) = \cos At$.

2. $u(t) = \sinh At$.

3. $u(t) = e^{-Bt} \sin At$.

4. $u(t) = e^{-Bt} \cos At$.

5. $u(t) = \begin{cases} 1, & a < t < b, \ a > 0 \\ 0, & \text{otherwise.} \end{cases}$

6. $u(t) = \begin{cases} 1, & 0 < t < \sigma \\ -1, & \sigma < t < 2\sigma \\ 0, & t > 2\sigma. \end{cases}$

7. $u(t) = \begin{cases} t, & 0 < t < \sigma \\ 0, & t > \sigma. \end{cases}$

8. $u(t) = \begin{cases} 1, & 0 < t < \sigma \\ 0, & \sigma < t < 2\sigma \end{cases}$ and $u(t) = u(t + 2\sigma), \quad t > 0.$

9. $u(t) = \displaystyle\int_0^t H(s)H(t - s)\, ds, \quad H = \text{Heaviside function.}$

10. $u(t) = \displaystyle\int_0^t \cos(t - u)\sin u\, du.$

11. Use term-by-term integration to show that the Laplace transform of
$$u(t) = \begin{cases} 0, & t < 0 \\ \dfrac{\sin t}{t}, & t > 0 \end{cases}$$
is $\arctan(1/s)$, $s > 1$. This can also be done by using formula (6).

12. Use term-by-term integration to show that the Laplace transform of the 0th Bessel function J_0 is $1/\sqrt{s^2 + 1}$, $s > 1$. (**Hint:** Compute $\mathscr{L}J_0$ and then square it.)

In Exercises 13 to 17 find u from $\mathscr{L}u$ by means of the residue theorem or another technique.

13. $\mathscr{L}u(s) = s/(s^2 + A^2)^2.$

14. $\mathscr{L}u(s) = 1/(s - 1)^2.$

15. $\mathscr{L}u(s) = 1/(s - 1)^4.$

16. $\mathscr{L}u(s) = (s - 2)e^{-s}/(s^2 - 4s + 3).$

17. $\mathscr{L}u(s) = 1/(1 - is).$ (**Hint:** $1/(1 - is) = 1/(1 + s^2) + (is)/(1 + s^2).$)

18. Establish the formula (6):
$$\frac{d}{ds}(\mathscr{L}u)(s) = -\mathscr{L}(tu(t))(s).$$

(**Hint:** The differentiation on the s variable can be brought inside the integral; this can be proved by techniques from real analysis.)

19. Show that if $u(t) = 0$ for $t \geqslant A$, $A > 0$, then $\mathscr{L}u(s)$ is an entire function of the complex variable s and
$$(\mathscr{L}u)(s) = \sum_{n=0}^{\infty} s^n \left\{ \frac{(-1)^n}{n!} \int_0^{\infty} u(t)t^n\, dt \right\}$$

is its power series expansion about $s = 0$.

20.   Find the Laplace transform of the function

$$u(t) = \begin{cases} 0, & t \leqslant 0 \\ \dfrac{1}{\sqrt{t}}\, e^{-c/t}, & t > 0, \end{cases}$$

where c is a positive constant.

$$\left(\text{Answer: } \mathscr{L}u(s) = \sqrt{\frac{\pi}{s}}\, e^{-2\sqrt{cs}}, \; s > 0. \right)$$

Further Integral Transforms★: The Gamma and Beta functions

The **Gamma function** is defined by

$$\Gamma(z) = \int_0^\infty e^{-t} t^{z-1} \, dt, \; \text{Re } z > 0. \tag{8}$$

For Re $z = x > 0$, we have

$$\left| e^{-t} t^{z-1} \right| = e^{-t} t^{x-1},$$

and so the integral is absolutely convergent.

21.   Show that $t^{z-1} = e^{(z-1)\log t}$ is analytic on the half-plane $\{z: \text{Re } z > 0\}$.

22.   Use Morera's Theorem and an interchange of the order of integration to show that $\Gamma(z)$ is analytic on the half-plane $\{z: \text{Re } z > 0\}$.

23.   Let $z = x$ be real and bigger than 1; use integration by parts to show that

$$\Gamma(x) = (x - 1)\Gamma(x - 1).$$

24.   Conclude from Exercise 23 that $\Gamma(z) = (z - 1)\Gamma(z - 1)$ if Re $z > 1$.

25.   Show that $\Gamma(n) = (n - 1)!$ if n is a positive integer.

26.   Derive the formula

$$\frac{\Gamma(x)\Gamma(y)}{\Gamma(x + y)} = \int_0^\infty \frac{t^{y-1}}{(1 + t)^{x+y}} \, dt. \tag{9}$$

$\Bigg($ **Hint:** Write

$$\Gamma(x)\Gamma(y) = \int_0^\infty t^{x-1}\, e^{-t}\, dt \int_0^\infty u^{y-1}\, e^{-u}\, du.$$

Now let $u = tv$; we obtain

$$\Gamma(x)\Gamma(y) = \int_0^\infty t^{x+y-1} \int_0^\infty e^{-t(v+1)} v^{y-1} \, dv \, dt$$

Then set $s = (1 + v)t$ to obtain

$$\Gamma(x)\Gamma(y) = \Gamma(x + y) \int_0^\infty \frac{v^{y-1}}{(1 + v)^{x+y}} \, dv.\Big)$$

The **Beta function** is defined by

$$\beta(x, y) = \frac{\Gamma(x)\Gamma(y)}{\Gamma(x + y)}$$

for x and y positive.

27. In the integral in (9) substitute $u = 1/(1 + t)$ and conclude that

$$\beta(x, y) = \frac{\Gamma(x)\Gamma(y)}{\Gamma(x + y)} = \int_0^1 u^{x-1}(1 - u)^{y-1} \, du$$

$$= 2 \int_0^{\pi/2} (\sin \theta)^{2x-1}(\cos \theta)^{2y-1} \, d\theta.$$

28. Set $x = y = 1/2$ and use $\Gamma(1) = 1$ to get $\Gamma(1/2) = \sqrt{\pi}$.

29. Put $y = 1 - x, 0 < x < 1$, and obtain

$$\Gamma(x)\Gamma(1 - x) = \int_0^\infty \frac{u^{x-1}}{1 + u} \, du.$$

Now employ formula (8), Section 6, Chapter 2, to see that

$$\Gamma(x)\Gamma(1 - x) = \frac{\pi}{\sin(\pi x)}, \quad 0 < x < 1.$$

30. Use the conclusion of Exercise 29 to deduce that

$$\Gamma(z)\Gamma(1 - z) = \frac{\pi}{\sin(\pi z)} \quad \text{for } 0 < \text{Re } z < 1.$$

Now use the fact that $\Gamma(z)$ is analytic for Re $z > 0$ to define $\Gamma(w)$ for those w with Re $w \leqslant 0$:

$$\Gamma(w) = \frac{\pi}{\sin(\pi(1 - w))} \cdot \frac{1}{\Gamma(1 - w)}, \quad \text{Re } w \leqslant 0.$$

More on Bessel functions⋆

31. Let v be a real number, $v \geqslant 0$; show that one solution of the differential equation

$$z^2 f''(z) + z f'(z) + (z^2 - v^2) f(z) = 0$$

is

$$J_v(z) = \sum_{n=1}^\infty (-1)^n \frac{z^{2n+v}}{2^{2n+v} n! \Gamma(v + n + 1)}, \quad \text{Re } z > 0,$$

where Γ is the Gamma function.

5.4   Applications of the Laplace transform to differential equations

The Laplace transform finds a major application in the solution of linear ordinary differential equations; it is frequently used to solve certain partial differential equations as well. Its advantage as a tool for solving such equations are twofold: first, the initial conditions and the nonhomogeneous term are incorporated from the start and thus the solution is generated without any "constants of integration," which need further evaluation; second, an ordinary differential equation is converted into an algebraic equation. The technique is illustrated by means of several examples.

Example 1   Solve the equation

$$u'(t) + u(t) = \begin{cases} 1, & \text{if } 0 < t < 1 \\ 0, & \text{if } 1 < t < \infty \end{cases}$$

with initial condition $u(0) = 1$.

Solution   We take the Laplace transform of both sides of the equation; this yields

$$\mathscr{L}u'(s) + \mathscr{L}u(s) = \frac{1}{s}(1 - e^{-s}).$$

However, $\mathscr{L}u'(s) = -u(0) + s\mathscr{L}u(s)$ by Example 2, Section 3, so that we have the equation

$$-1 + s\mathscr{L}u(s) + \mathscr{L}u(s) = \frac{1}{s}(1 - e^{-s}),$$

and hence

$$-1 + (s + 1)(\mathscr{L}u)(s) = \frac{1 - e^{-s}}{s}$$

or

$$\mathscr{L}u(s) = \frac{1}{s} - e^{-s}\left\{\frac{1}{s} - \frac{1}{s+1}\right\}.$$

The function e^{-s} is the Laplace transform of the impulse function at $t_0 = 1$; the functions $1/s$ and $1/(s + 1)$ are the Laplace transforms of 1 and e^{-t}, respectively. Hence, by the convolution result,

$$\frac{1}{s}e^{-s} \quad \text{and} \quad \frac{1}{s+1}e^{-s}$$

are the Laplace transforms of the convolution of the impulse function at $t_0 = 1$

with 1 and e^{-t}, respectively. These are, respectively,

$$\int_0^t 1\, \delta(1-x)\, dx = \begin{cases} 0, & \text{if } 0 < t < 1 \\ 1, & \text{if } 1 \leqslant t < \infty \end{cases}$$

and

$$\int_0^t e^{-(t-x)}\, \delta(1-x)\, dx = \begin{cases} 0, & \text{if } 0 < t < 1 \\ e^{-(t-1)}, & \text{if } 1 \leqslant t < \infty. \end{cases}$$

Consequently,

$$u(t) = \begin{cases} 1, & \text{if } 0 < t < 1 \\ e^{-(t-1)}, & \text{if } 1 \leqslant t < \infty. \end{cases} \qquad \square$$

You should take note that the solution $u(t)$, although continuous on $[0, \infty)$, fails to be differentiable at $t_0 = 1$, exactly because the function on the right-hand side of the original equation is not continuous at $t_0 = 1$.

Example 2   Solve each of the following differential equations by means of the Laplace transform
(a) $u'' + u = f$; $u(0) = 1, u'(0) = 0$ and

$$f(t) = \begin{cases} 1, & 0 < t < \pi \\ 0, & \pi < t < 2\pi \end{cases}, \ f(t) = f(t + 2\pi).$$

(b) $u'' + 2u' + 5u = \delta(t - \pi)$; $u(0) = u'(0) = 0$.

Solutions
(a) The Laplace transform of the 2π-periodic function $f(t)$ is

$$\mathscr{L} f(s) = \frac{\int_0^{2\pi} f(t)\, e^{-st}\, dt}{1 - e^{-2\pi s}} = \frac{1}{s} \frac{1 - e^{-\pi s}}{1 - e^{-2\pi s}} = \frac{1}{s(1 + e^{\pi s})}.$$

Hence, the Laplace transform of the equation gives us

$$s^2 \mathscr{L}u(s) - s + \mathscr{L}u(s) = \frac{1}{s(1 + e^{\pi s})}$$

or

$$\mathscr{L}u(s) = \frac{s}{1 + s^2} + \frac{1}{1 + s^2} \cdot \frac{1}{s(1 + e^{\pi s})}.$$

The function $s/(1 + s^2)$ is the Laplace transform of $\cos t$ and $1/(1 + s^2)$ is the transform of $\sin t$. The convolution theorem then gives

$$u(t) = \cos t + (f * \sin)(t)$$

$$= \cos t + \int_0^t f(x)\sin(t - x)\, dx$$

$$= \begin{cases} 1, & 0 < t \leqslant \pi \\ -\cos t, & \pi \leqslant t \leqslant 2\pi \\ 1 - 2\cos t, & 2\pi \leqslant t \leqslant 3\pi \\ -3\cos t, & 3\pi \leqslant t \leqslant 4\pi. \\ \text{etc.} \end{cases}$$

(b) Here, after taking Laplace transforms, we obtain

$$s^2 \mathcal{L}u(s) + 2s\mathcal{L}u(s) + 5\mathcal{L}u(s) = e^{-\pi s}.$$

Thus,

$$\mathcal{L}u(s) = \frac{e^{-\pi s}}{s^2 + 2s + 5}.$$

The function $1/(s^2 + 2s + 5)$ is the Laplace transform of $(1/2)\, e^{-t} \sin 2x$ so that the convolution theorem gives us the result that

$$u(t) = \int_0^t \delta(x - \pi)\tfrac{1}{2}\, e^{-(t-x)}\sin 2(t - x)\, dx$$

$$= \begin{cases} 0, & 0 < t \leqslant \pi \\ \tfrac{1}{2}\, e^{-(t-\pi)}\sin 2t, & \pi \leqslant t < \infty. \end{cases} \qquad \square$$

Example 3 Use the Laplace transform to solve the partial differential equation

$$u_{xx} = u_{tt} \text{ subject to } \begin{cases} u(0, t) = u(\pi, t) = 0, & t > 0 \\ u(x, 0) = \sin x, & 0 < x < \pi \\ u_t(x, 0) = 0. \end{cases}$$

Solution Let $U(x, s)$ be the Laplace transform of $u(x, t)$ on the t variable:

$$U(x, s) = \int_0^\infty u(x, t)\, e^{-st}\, dt.$$

Then $U(0, s) = U(\pi, s) = \int_0^\infty 0 \cdot e^{-st}\, dt = 0$ for each s. Treating U as a function of x alone (holding s fixed) we find that

$$\frac{d^2 U}{dx^2} = \int_0^\infty u_{xx}(x, t)\, e^{-st}\, dt$$

$$= \int_0^\infty u_{tt}(x, t)\, e^{-st}\, dt.$$

We now integrate twice by parts and use the conditions $u_t(x, 0) = 0$, $u(x, 0) = \sin x$. The end result is the equation

$$\frac{d^2 U}{dx^2} = s^2 U - s \sin x, \quad 0 < x < \pi$$

$$U(0) = U(\pi) = 0.$$

This is an elementary linear second-order equation whose solution is easily found to be

$$U(x, s) = \frac{s}{s^2 + 1} \sin x.$$

The term $\sin x$ is constant with respect to s and $s/(s^2 + 1)$ is the Laplace transform of $\cos t$. Hence, the solution is

$$u(x, t) = \cos t \sin x. \qquad \square$$

The knowledgeable reader will be aware that in the previous example the original equation is easily solved by separation of variables. However, the example is designed to illustrate, with a minimum of extraneous details, the use of the Laplace transform in solving partial differential equations. The following two examples are further illustrations of this technique, this time of a more complicated nature.

Example 4 A string of length 1 is clamped at its left end and initially is straight and at rest. At time $t = 0$ the right end begins to move up and down, with height $\sin t$ at time t. Find the resulting displacement at position x, $0 \leqslant x \leqslant 1$, and time t, $t > 0$.

Solution Let $u(x, t)$ be the displacement at position x and time t, $0 \leqslant x \leqslant 1$, $t \geqslant 0$. The equation that determines u is

$$u_{xx} = u_{tt} \text{ subject to } \begin{cases} u(x, 0) = u_t(x, 0) = 0, & 0 < x < 1 \\ u(0, t) = 0, & t > 0 \\ u(1, t) = \sin t, & t > 0. \end{cases}$$

Let $U(x, s)$ be the Laplace transform of $u(x, t)$ on the t variable:

$$U(x, s) = \int_0^\infty u(x, t) e^{-st} \, dt.$$

We treat U as a function of x alone and differentiate twice. This gives

$$\frac{d^2 U}{dx^2} = \int_0^\infty u_{xx}(x, t) e^{-st} \, dt$$

$$= \int_0^\infty u_{tt}(x, t) e^{-st} \, dt$$

$$= s^2 \int_0^\infty u(x, t) e^{-st} \, dt$$

after two integrations by parts (or the use of the formula in Example 2, Section 3) and the use of the initial conditions $u(x, 0) = u_t(x, 0) = 0$. The boundary conditions on u yield

$$U(0) = \int_0^\infty u(0, t) e^{-st} \, dt = \int_0^\infty 0 \cdot e^{-st} \, dt = 0$$

and

$$U(1) = \int_0^\infty u(1, t) e^{-st} \, dt = \int_0^\infty (\sin t) e^{-st} \, dt = \frac{1}{1 + s^2}.$$

Hence, U satisfies the elementary second-order equation

$$\frac{d^2 U}{dx^2} = s^2 U \text{ with } U(0) = 0, \ U(1) = \frac{1}{1 + s^2}.$$

This equation has solution

$$U(x, s) = \frac{1}{1 + s^2} \frac{\sinh xs}{\sinh s}.$$

Our final task, then, is to invert U and recover u:

$$u(x, t) = \frac{1}{2\pi i} \int_{\sigma_0 - i\infty}^{\sigma_0 + i\infty} \frac{1}{1 + s^2} \frac{\sinh xs}{\sinh s} e^{st} \, ds.$$

The function U has simple poles at $\pm i$ and $\pm in\pi$, $n = 1, 2, \ldots$. The residues of the integrand $U(s, x) e^{st}$ are

$$\text{Res}(Ue^{st}; \pm i) = \pm \frac{1}{2i} \frac{\sin x}{\sin 1} e^{\pm it}$$

and

$$\text{Res}(Ue^{st}; i\pi n) = \frac{i}{1 - \pi^2 n^2} \frac{\sin(\pi n x)}{(-1)^n} e^{i\pi nt}, \ n = \pm 1, \pm 2, \ldots.$$

Hence, the solution is

$$u(x, t) = \frac{\sin x}{\sin 1} \sin t + \sum_{n=1}^\infty \frac{(-1)^{n+1}}{\pi^2 n^2 - 1} \{\sin(\pi n x)\}\{\sin(\pi n t)\}. \qquad \square$$

Example 5   A steel wire of length 1 is clamped at both ends. The wire is straight and at rest when a sinusoidal magnetic field is activated. Find the resulting displacement of the wire at each future time.

Solution   Let $u(x, t)$, $0 \leqslant x \leqslant 1$, $0 \leqslant t$, be the displacement at the position x and the time t from the rest position. The partial differential equation that determines the function $u(x, t)$ is

$$u_{xx} = u_{tt} - \sin \omega t, \ 0 < x < 1, \ t > 0.$$

with boundary and initial conditions:

$$u(0, t) = u(1, t) = 0, \quad t > 0 \qquad \text{(the ends are clamped)}$$
$$u(x, 0) = u_t(x, 0) = 0, \ 0 < x < 1 \ \text{(the wire is initially straight and at rest).}$$

Let $U(x, s)$ be the Laplace transform of $u(x, t)$ on the time variable t:

$$U(x, s) = \int_0^\infty u(x, t) \, e^{-st} \, dt.$$

We hold s fixed and treat U as a function of x. Then

$$\frac{d^2U}{dx^2} = \int_0^\infty u_{xx}(x, t) \, e^{-st} \, dt$$

$$= \int_0^\infty u_{tt}(x, t) \, e^{-st} \, dt - \int_0^\infty (\sin \omega t) \, e^{-st} \, dt$$

$$= s^2 U - \frac{\omega}{s^2 + \omega^2}.$$

Furthermore, we have the boundary conditions $U(0) = U(1) = 0$. This is an ordinary differential equation for U whose solution is easily found to be

$$U(x, s) = \frac{\omega}{s^2(s^2 + \omega^2)} \frac{(e^s - e^{sx})(1 - e^{-sx})}{1 + e^s}$$

$$= \frac{2\omega}{s^2(s^2 + \omega^2)} \frac{\sinh(sx/2)\sinh((s/2)(1 - x))}{\cosh(s/2)}.$$

The frequency ω of the magnetic field plays a critical role in the nature of the solution u. We shall first analyze the case when ω is not an odd integer multiple of π. We must find $u(x, t)$ from $U(x, s)$; that is, we must invert the Laplace transform. The inversion formula (7), Section 3, gives

$$u(x, t) = \frac{1}{2\pi i} \int_{\sigma - i\infty}^{\sigma + i\infty} \frac{\omega}{s^2(s^2 + \omega^2)} \frac{(e^s - e^{sx})(1 - e^{-sx})}{1 + e^s} e^{st} \, ds.$$

The integrand has simple poles at $s = \pm i\omega$ and at $s = \pm i\pi, \pm 3i\pi, \ldots$; there is no pole at $s = 0$ since $(e^s - e^{sx})(1 - e^{-sx})$ has a double zero at $s = 0$. The residue at $s = ik\pi, k = \pm 1, \pm 3, \ldots$, is

$$A_k = \frac{2i\omega}{\pi^4} \frac{\sin k\pi x}{k^2(k^2 - (\omega/\pi)^2)} e^{ik\pi t},$$

so that

$$A_k + A_{-k} = -\frac{4\omega}{\pi^4} \frac{\sin k\pi x \sin k\pi t}{k^2(k^2 - (\omega/\pi)^2)}, \quad k = 1, 3, 5, \ldots.$$

The residues at $s = i\omega$ and $s = -i\omega$ sum to

$$B(x, t) = \frac{\sin \omega t}{\omega^2} \left\{ \frac{(e^{i\omega} - e^{i\omega x})(1 - e^{-i\omega x})}{1 + e^{i\omega}} \right\}.$$

This gives the solution

$$u(x, t) = B(x, t) - \frac{4\omega}{\pi^4} \sum_{k \text{ odd}} \frac{(\sin k\pi x)(\sin k\pi t)}{k^2(k^2 - (\omega/\pi)^2)}.$$

When ω is an odd integer multiple of π there is a new difficulty; we shall examine the case when $\omega = \pi$. We are still faced with the job of inverting U:

$$u(x, t) = \frac{1}{2\pi i} \int_{\sigma - i\infty}^{\sigma - i\infty} \frac{\pi}{s^2(s^2 + \pi^2)} \frac{(e^s - e^{sx})(1 - e^{-sx})}{1 + e^s} e^{st} \, ds.$$

The integrand again has simple poles at $s = \pm 3i\pi, \pm 5i\pi, \ldots$, and again there is no pole at $s = 0$. However, there are poles of order 2 at $s = \pm i\pi$. The residue at $s = ik\pi, k = \pm 3, \pm 5, \ldots$, is

$$A_k = \frac{2i}{\pi^3} \frac{\sin k\pi x}{k^2(k^2 - 1)} e^{ik\pi t}, \quad k = \pm 3, \pm 5, \ldots,$$

and, as before,

$$A_k + A_{-k} = -\frac{4}{\pi^3} \frac{\sin k\pi x \sin k\pi t}{k^2(k^2 - 1)}, \quad k = 3, 5, \ldots.$$

The residue at both $i\pi$ and $-i\pi$ is more complicated because these points are poles of order 2. Diligent computation shows these residues to be

$$A_1 = t \, e^{i\pi t} \frac{\sin \pi x}{\pi^2} + i \left[\frac{2\pi + (\cos \pi x)(4\pi x - 2\pi) - 10 \sin \pi x}{4\pi^4} \right]$$

and

$$A_{-1} = t \, e^{-i\pi t} \frac{\sin \pi x}{\pi^2} - i \left[\frac{2\pi + (\cos \pi x)(4\pi x - 2\pi) - 10 \sin \pi x}{4\pi^4} \right].$$

Their sum is

$$A_1 + A_{-1} = \frac{2t \cos \pi t \sin \pi x}{\pi^2},$$

and so the solution is given by

$$u(x, t) = \frac{2t \cos \pi t \sin \pi x}{\pi^2} - \frac{4}{\pi^3} \sum_{n=1}^{\infty} \frac{(\sin(2n + 1)\pi x)(\sin(2n + 1)\pi t)}{(2n + 1)^2[(2n + 1)^2 - 1]}.$$

Note that the second term of the solution remains bounded for all t but the first term can become unboundedly large for large t. This is the phenomenon of **resonance**.   □

Exercises for Section 5.4

In Exercises 1 to 5, use the Laplace transform to solve the given differential equation.

1. $u'' - 2u' + 2u = 6e^{-t}$; $u(0) = 0$, $u'(0) = 1$.

2. $u'' + 9u = \begin{cases} 1, & 0 < t < \pi \\ 0, & \pi < t \end{cases}$; $u(0) = 1$, $u'(0) = 0$.

3. $u''(t) + tu'(t) + u(t) = 0$; $u(0) = 1$, $u'(0) = 0$.

4. $u'' + 2u' + 2u = \delta(t - 2\pi)$; $u(0) = 1$, $u'(0) = 0$.

5. $u'' + u = \begin{cases} 0, & 0 \leqslant t \leqslant \pi \\ \cos t, & \pi < t < \infty \end{cases}$; $u(0) = 0$, $u'(0) = 1$.

6. Solve the **Volterra* integral equation**

$$\phi(t) + \int_0^t k(t - s)\phi(s) \, ds = f(t)$$

for the unknown function ϕ by taking Laplace transforms. Express the answer in terms of the Laplace transform of ϕ.

In Problems 7 to 11, use the Laplace transform to solve the given partial differential equation.

7. **(The traveling wave)** Solve the wave equation

$$u_{tt} = u_{xx}, \quad x > 0, \, t > 0,$$

subject to

$$\begin{cases} u(x, 0) = u_t(x, 0) = 0, & x > 0 \\ u(0, t) = f(t), & t \geqslant 0, \, f(0) = 0 \\ \lim_{x \to \infty} u(x, t) = 0, & t \geqslant 0. \end{cases}$$

8. **(The falling string)** Solve the modified wave equation

$$u_{tt} = u_{xx} - g, \quad x > 0, \, t > 0,$$

g is a positive constant, subject to

$$\begin{cases} u(x, 0) = u_t(x, 0) = 0, & x > 0 \\ u(0, t) = 0, & t \geqslant 0 \\ \lim_{x \to \infty} u_x(x, t) = 0, & t \geqslant 0. \end{cases}$$

9. Solve the wave equation

$$u_{tt} = u_{xx}, \text{ subject to } \begin{cases} u(x, 0) = u_t(x, 0) = 0, & 0 < x < 1 \\ u(0, t) = 0, & t \geqslant 0 \\ u(1, t) = \sin \omega t, & t \geqslant 0. \end{cases}$$

---

* Vito Volterra, 1860–1940.

The value of ω will affect the nature of the solution.

10.   Solve the wave equation

$$u_{tt} = u_{xx}, \text{ subject to } \begin{cases} u(x, 0) = u_t(x, 0) = 0, & 0 < x < 1 \\ u(0, t) = 0, & t \geq 0 \\ u_x(1, t) = c_0, & t \geq 0. \end{cases}$$

Here c_0 is a constant.

11.   Solve the heat equation

$$\frac{\partial u}{\partial t} = k \frac{\partial^2 u}{\partial x^2}, \ x > 0, \ t > 0,$$

subject to the conditions

$$u(x, 0) = 0, \qquad\qquad x > 0$$

$$-k \frac{\partial u}{\partial x} (0, t) = \phi_0, \text{ a constant, } \ t > 0$$

$$\lim_{x \to \infty} u(x, t) = 0, \qquad\qquad t > 0.$$

To invert $U(x, s)$ you will need to use the conclusion of Exercise 20, Section 3. k is a constant.

5.5   The Z-transform

The Z-transform provides a technique to deal with sequences, and solve difference equations involving sequences, in much the same way that the Laplace transform provides a technique to solve differential equations involving functions. The Z-transform is a valuable tool in areas such as numerical analysis where a typical problem is to solve a difference equation involving discrete data.

Let $\{a_j\}$ be a sequence of complex numbers satisfying the growth condition

$$|a_j| \leq Mr_0^j, \ j = 0, 1, 2, \ldots \tag{1}$$

for some positive numbers M and r_0. The **Z-transform of** $\{a_j\}$ is the function

$$Z(\{a_j\}) = \sum_{j=0}^{\infty} a_j/z^j.$$

The series converges absolutely for all z with $|z| = r > r_0$ because

$$|a_j/z^j| = |a_j|/r^j \leq M\left(\frac{r_0}{r}\right)^j$$

and the series $\sum (r_0/r)^j$ converges since the ratio r_0/r is less than one. Consequently, the Z-transform of $\{a_j\}$ is an analytic function in the exterior of the circle $|z| = r_0$, including at ∞.

Example 1   Find the Z-transform of the sequence $\{a_j\}$ if $a_j = 3$ for all j.

Solution   If $a_j = 3$ for all j, then

$$Z(\{3\}) = \sum_{j=0}^{\infty} 3/z^j = \frac{3}{1 - 1/z} = \frac{3z}{z - 1}.$$

This function is analytic outside the circle $|z| = 1$. □

Example 2   Find the Z-transform of the sequence $a_j = 1/j!$

Solution   We have

$$Z\left(\left\{\frac{1}{j!}\right\}\right) = \sum_{j=0}^{\infty} \frac{1}{j!z^j} = e^{1/z}, \ |z| > 0.$$

This function is analytic for all z with $|z| > 0$. □

Basic properties of the Z-transform

The Z-transform is a discrete version of the Laplace transform. To see this let δ be a positive number and set $x_n = n\delta$, $n = 1, 2, 3, \ldots$. The Laplace transform of a function u is given by the integral $\int_0^{\infty} u(t) \, e^{-st} \, dt$. This integral is approximated by the sum $\sum_{n=0}^{\infty} u(x_n) \, e^{-sx_n} \, \Delta x_n = \delta \sum_{n=0}^{\infty} u(n\delta) \, e^{-s\delta n}$. The function $z = F(s) = e^{s\delta}$ maps the half-plane $\text{Re}(s) > a$ onto the exterior of the circle $|z| = e^{a\delta} = 1/r_0$. If we set $a_n = \delta u(n\delta)$, then we see that

$$(\mathscr{L}u)(s) = \int_0^{\infty} u(t) \, e^{-st} \, dt$$

is approximated by

$$\sum_0^{\infty} \frac{a_n}{z^n} = Z(\{a_n\}), \ z = e^{s\delta}.$$

Note that the condition $|u(t)| \leqslant M \, e^{-at}$ is translated into the condition $|a_n| \leqslant \delta M r_0^n$, $n = 0, 1, 2, \ldots$. This is a reason for saying that the Z-transform is a "discrete Laplace transform."

The Z-transform shares many of the general properties of the Laplace transform. First, it is linear:

if $\{a_j\}$ and $\{b_j\}$ are two sequences both satisfying (1) and if λ is a complex number, then $Z(\{\lambda a_j + b_j\}) = \lambda Z(\{a_j\}) + Z(\{b_j\})$. (2)

Second, like the Laplace transform, the Z-transform turns a convolution into a product. Convolution is defined in this way. If $\{a_j\}$ and $\{b_j\}$ are two sequences of complex numbers, then the **convolution of** $\{a_j\}$ **and** $\{b_j\}$ is the sequence $\{c_n\}$ given by

$$c_n = \sum_{k=0}^{n} a_k b_{n-k}, \quad n = 0, 1, 2, \dots. \tag{3}$$

We make note of the fact that if $\{a_k\}$ and $\{b_k\}$ both satisfy (1), then $|c_n| \leqslant MM'(n+1)r_0^n < M''r^n$ for each $r > r_0$. Hence, the Z-transform of $\{c_n\}$ is well defined on the domain $|z| > r_0$.

The Z-transform of $\{c_n\}$ is given in the following:

> If $\{c_n\}$ is the convolution of $\{a_j\}$ and $\{b_j\}$,
> then $Z(\{c_n\}) = Z(\{a_j\})Z(\{b_j\})$. $\tag{4}$

The demonstration of (4) is easy. We have

$$Z(\{a_j\})Z(\{b_j\}) = \left(\sum_{j=0}^{\infty} a_j/z^j \right)\left(\sum_{k=0}^{\infty} b_k/z^k \right)$$

$$= \sum_{n=0}^{\infty} \frac{1}{z^n}\left(\sum_{k=0}^{n} a_k b_{n-k} \right)$$

$$= \sum_{n=0}^{\infty} c_n/z^n = Z(\{c_n\}).$$

The passage from the first line to the second occurs simply by multiplying out the two series and collecting equal powers of z (see Section 2, Chapter 2).

The convolution result in (4) is enormously useful; we give one elementary example here to illustrate this fact.

Example 3   Consider the (infinite) system of linear equations

$$\begin{aligned}
b_0 &= a_0 \\
b_1 &= a_1 - 2a_0 \\
b_2 &= a_2 - 2a_1 + a_0 \\
b_3 &= a_3 - 2a_2 + a_1 \\
&\cdots
\end{aligned}$$

In general, $b_n = a_n - 2a_{n-1} + a_{n-2}$, $n \geqslant 2$. Solve this system for the sequence $\{a_j\}$ in terms of the sequence $\{b_j\}$.

Solution   We have

$$\{b_j\} = \{a_j\} * \{f_j\},$$

where

$$f_0 = 1, f_1 = -2, f_2 = 1, \text{ and } f_j = 0 \text{ for } j \geqslant 3.$$

Note that $\{b_n\}$ satisfies (1) with constant $\left(1 + \dfrac{1}{r_0}\right)^2$ M. After taking Z-transforms, we have

$$Z(\{b_j\}) = Z(\{a_j\})Z(\{f_j\}) = Z(\{a_j\})\left(1 + \frac{-2}{z} + \frac{1}{z^2}\right)$$

$$= \left(\frac{z-1}{z}\right)^2 Z(\{a_j\}).$$

Division then yields this equation for the Z-transform of $\{a_j\}$:

$$Z(\{a_j\}) = \left(\frac{z}{z-1}\right)^2 Z(\{b_j\}).$$

The convolution result then implies that

$$\{a_j\} = \{g_j\} * \{b_j\},$$

where $\{g_j\}$ is the sequence whose Z-transform is $(z/(z-1))^2$. To find $\{g_j\}$ we must expand $(z/(z-1))^2$ in powers of $1/z$.

$$\frac{z^2}{(z-1)^2} = \left(1 + \frac{1}{z-1}\right)^2 = 1 + \frac{2}{z-1} + \frac{1}{(z-1)^2}$$

$$= 1 + \frac{2}{z-1} - \left(\frac{1}{z-1}\right)'$$

$$= 1 + 2\left\{\frac{1}{z} + \frac{1}{z^2} + \cdots\right\} - \left\{\frac{1}{z} + \frac{1}{z^2} + \cdots\right\}'$$

$$= 1 + \frac{2}{z} + \frac{2}{z^2} + \cdots + \frac{1}{z^2} + \frac{2}{z^3} + \frac{3}{z^4} + \cdots$$

$$= 1 + \frac{2}{z} + \frac{3}{z^2} + \frac{4}{z^3} + \frac{5}{z^4} + \cdots.$$

Hence, $g_j = j + 1$, $j = 0, 1, 2, \ldots$, and so

$$a_j = \sum_{n=0}^{j} (n+1)b_{j-n}, \ j = 0, 1, 2, \ldots.$$

In particular, $a_0 = b_0$, $a_1 = b_1 + 2b_0$, $a_2 = b_2 + 2b_1 + 3b_0$, etc. □

Shifting

If the sequence $\{a_j\}$ is shifted by one unit to form the new sequence $\{b_j\}$, $b_j = a_{j+1}, j = 0, 1, 2, \ldots$, then

$$Z(\{b_j\}) = z[Z(\{a_j\}) - a_0]. \tag{5}$$

This follows because

$$Z(\{b_j\}) = b_0 + \frac{b_1}{z} + \frac{b_2}{z^2} + \cdots = a_1 + \frac{a_2}{z} + \frac{a_3}{z^2} + \cdots$$

$$= z\left[a_0 + \frac{a_1}{z} + \frac{a_2}{z^2} + \cdots - a_0\right]$$

$$= z[Z(\{a_j\}) - a_0].$$

More generally, if N is a fixed positive integer and

$$b_j = a_{j+N}, \ j = 0, 1, 2, \ldots,$$

then

$$Z(\{b_j\}) = z^N \left[Z(\{a_j\}) - \sum_{i=0}^{N-1} \frac{a_i}{z^i} \right].\tag{6}$$

The result in (5) or (6) is referred to as the **theorem on shifting**.

The Z-transform is a particularly effective tool for studying the solution of a **linear difference equation with constant coefficients**. Such an equation has the form

$$\sum_{j=0}^{P} A_j y_{j+n} = \sum_{j=0}^{Q} B_j x_{j+n}, \quad n = 0, 1, 2, \ldots.\tag{7}$$

Here P and Q are fixed nonnegative integers, $A_0, \ldots, A_P$ and $B_0, \ldots, B_Q$ are known real or complex numbers, the coefficients of the system. The numbers $\{x_j\}$ (which satisfy (1) for some r_0), called the **forcing function** or the **input**, are known. We wish to solve for the numbers $\{y_j\}$, the **response** or the **output of the system** to the forcing function. The numbers $y_0, \ldots, y_{P-1}$ are **initial conditions** and are assumed known as well. (If $P = 0$, then there are no initial conditions.)

We begin by taking the Z-transform of both sides of this equation. This yields

$$\sum_{j=0}^{P} A_j z^j \left[Y(z) - \sum_{k=0}^{j-1} y_k z^{-k} \right] = \sum_{j=0}^{Q} B_j z^j \left[X(z) - \sum_{k=0}^{j-1} x_k z^{-k} \right],$$

where

$$Y(z) = \sum_{j=0}^{\infty} y_j/z^j, \quad X(z) = \sum_{j=0}^{\infty} x_j/z^j$$

are the Z-transforms of $\{y_j\}$ and $\{x_j\}$, respectively. We now solve for $Y(z)$:

$$Y(z) = \frac{\displaystyle\sum_{j=0}^{Q} B_j z^j [X(z) - \sum_{k=0}^{j-1} x_k z^{-k}] + \sum_{j=0}^{P} A_j z^j \sum_{k=0}^{j-1} y_k z^{-k}}{\displaystyle\sum_{j=0}^{P} A_j z^j}.\tag{8}$$

The solution is completed by applying the inverse Z-transform or by expanding the right-hand side in powers of $1/z$ (which is the same thing).* One of the advantages of this form of the answer is that it prominently displays the denominator, which is critical in the determination of the stability of the system. This is discussed in detail in Section 5.1.

Example 4 Solve the linear difference equation

$$2y_n + y_{n+1} = x_n - x_{n+2}, \quad n = 0, 1, \ldots; \ y_0 = 1.$$

---

* Here the material at the end of Section 5, of Chapter 2, can be useful.

Solution   Here $P = 1$ with $A_0 = 2$ and $A_1 = 1$, $Q = 2$ with $B_0 = 1$, $B_1 = 0$, and $B_2 = -1$. The formula for $Y(z)$ yields

$$Y(z) = \frac{X(z) - z^2[X(z) - (x_0 + x_1/z)] + 2 + z}{2 + z}$$

$$= \frac{(x_0 + x_1/z + x_2/z^2 + \cdots) - (x_2 + x_3/z + x_4/z^2 + \cdots) + 2 + z}{2 + z}.$$

The expansion of $1/(2 + z)$ in powers of $1/z$, valid for $|z| > 2$, is

$$\frac{1}{2 + z} = \left(\frac{1}{z}\right)\left(\frac{1}{1 + 2/z}\right)$$

$$= \left(\frac{1}{z}\right)(1 - (2/z) + (2/z)^2 - (2/z)^3 + \cdots)$$

$$= \frac{1}{z} - \frac{2}{z^2} + \frac{4}{z^3} - \frac{8}{z^4} + \cdots.$$

This yields

$$Y(z) = 1 + \left(\frac{1}{2 + z}\right)\left(x_0 - x_2 + \frac{x_1 - x_3}{z} + \frac{x_2 - x_4}{z^2} + \cdots\right)$$

$$= 1 + \left(\frac{1}{z} - \frac{2}{z^2} + \frac{4}{z^3} - \frac{8}{z^4} + \cdots\right)\left(x_0 - x_2 + \frac{x_1 - x_3}{z} + \frac{x_2 - x_4}{z^2} + \cdots\right)$$

$$= 1 + \frac{x_0 - x_2}{z} + \frac{-2(x_0 - x_2) + x_1 - x_3}{z^2}$$

$$+ \frac{x_2 - x_4 - 2(x_1 - x_3) + 4(x_0 - x_2)}{z^3} + \cdots.$$

Hence, $y_1 = x_0 - x_2$, $y_2 = -2x_0 + x_1 + 2x_2 - x_3$, and in general,

$$y_n = \sum_{j=1}^{n} (-2)^{j-1}(x_{n-j} - x_{n+2-j}), \quad n = 1, 2, 3, \ldots. \qquad \square$$

Example 5   Find the current i_k in each step of the ladder network shown in Figure 5.9.

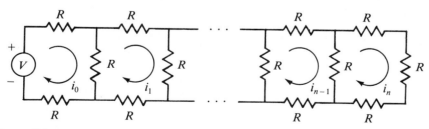

Figure 5.9

Solution   In the first loop

$$3Ri_0 - Ri_1 = V$$

and in the other loops

$$-Ri_k + 4i_{k+1} - Ri_{k+2} = 0.$$

Let $I(z) = \sum_{k=0}^{\infty} i_k/z^k$ be the Z-transform of $\{i_k\}$. Using the difference equations given above and (6) we obtain

$$i_1 = 3i_0 - V/R$$

and

$$I(z) - 4z(I(z) - i_0) + z^2(I(z) - i_0 - i_1/z) = 0.$$

Upon substituting for i_1 and solving we find that

$$I(z) = i_0 \frac{z^2 - z\left(1 + \dfrac{V}{i_0 R}\right)}{z^2 - 4z + 1}.$$

The numbers i_1, i_2, ... are then obtained by expanding this rational function in powers of $1/z$. One way to accomplish this is by using equations (9) to (11) of Section 5, Chapter 2. We first write

$$I(z) = i_0\left[1 + \frac{-1 + z\left(3 - \dfrac{V}{i_0 R}\right)}{z^2 - 4z + 1}\right].$$

The rational function

$$f(z) = \frac{-1 + z\left(3 - \dfrac{V}{i_0 R}\right)}{z^2 - 4z + 1}$$

has poles of order one at $z_1 = 2 - \sqrt{3}$ and $z_2 = 2 + \sqrt{3}$ with residues

$$r_1 = \text{Res}(f; z_1) = \frac{-1 + (2 - \sqrt{3})\left(3 - \dfrac{V}{i_0 R}\right)}{-2\sqrt{3}}$$

and

$$r_2 = \text{Res}(f; z_2) = \frac{-1 + (2 + \sqrt{3})\left(3 - \dfrac{V}{i_0 R}\right)}{2\sqrt{3}}.$$

Hence, for $|z| > z_2$ we obtain the expansion

$$I(z) = \sum_0^{\infty} \frac{i_k}{z^k}$$

where

$$i_k = r_1 z_1^{k-1} + r_2 z_2^{k-1}, \quad k = 1, 2, 3, \ldots.$$

Exercises for Section 5.5

In Exercises 1 to 7, find the Z-transform of the given sequence $\{a_j\}$ and the region in which $Z(\{a_j\})$ is analytic.

1. $a_j = \begin{cases} 1, & j \text{ even} \\ 0, & j \text{ odd.} \end{cases}$

2. $a_j = \begin{cases} 0, & j \text{ even} \\ (-1)^k \dfrac{1}{(2k+1)!}, & j = 2k+1, \ k = 0, 1, 2, \ldots. \end{cases}$

3. $a_j = 2^j / j!$

4. $a_j = \begin{cases} 0, & j \text{ odd} \\ \dfrac{1}{j!}, & j \text{ even.} \end{cases}$

5. $a_j = 1/(j+1).$

6. $\{a_j\} = \{b_j\} * \{b_j\}$, where $b_j = 1$ for all j.

7. $a_j = \begin{cases} 4^j, & j \text{ even} \\ 3^{-j}, & j \text{ odd.} \end{cases}$

In Exercises 8 to 14, find the sequence $\{a_j\}$ given its Z-transform $Z(\{a_j\})$.

8. $Z(\{a_j\}) = \cos(1/z).$

9. $Z(\{a_j\}) = \sinh(2/z).$

10. $Z(\{a_j\}) = \text{Log}\left(1 + \dfrac{1}{z}\right), \ |z| > 1.$

11. $Z(\{a_j\}) = z(e^{1/z} - 1).$

12. $Z(\{a_j\}) = (z^2 + z + 1)\left(\cos\left(\dfrac{1}{z}\right) - 1\right).$

13. $Z(\{a_j\}) = \left(\dfrac{1}{z-1}\right)^2.$

14. $Z(\{a_j\}) = \dfrac{1}{z^2 + z - 2}.$

15. Show that if $\{a_j\} * \{b_j\} = \{0\}$, then either $a_j = 0$ for all j or $b_j = 0$ for all j. (**Hint:** We have

$$\left(\sum_0^\infty \frac{a_j}{z^j}\right)\left(\sum_0^\infty \frac{b_j}{z^j}\right) = 0 \text{ for all } z \text{ with } |z| > r_0^{-1}.$$

Hence, one or the other of the Z-transforms must be identically zero.)

16.   Let $P(z) = A_0 + A_1 z + \cdots + A_N z^N$ be a polynomial of degree 1 or more all of whose zeros lie within the disc $|z| < 1$. Let $\{y_j\}$ be any solution of the system.

$$\sum_{j=0}^N A_j y_{j+n} = x_n, \quad n = 0, 1, 2, \ldots.$$

Show that if the forcing term $\{x_n\}$ is **bounded** (that is, $|x_n| \leq M$ for all n), then $Y(z)$, the Z-transform of $\{y_j\}$, is analytic in the region $\{z: |z| > 1\}$.

Solve the following linear systems:

17. $y_n - y_{n+1} = x_n + x_{n+1} + x_{n+2}; \, y_0 = 1.$

18. $y_n + 2y_{n+1} + y_{n+2} = x_n - x_{n+1}; \, y_0 = 1, \, y_1 = 0.$

19. $y_n + y_{n+2} = x_n; \, y_0 = 0, \, y_1 = 1.$

20. $y_n - y_{n+1} - 4y_{n+2} + 4y_{n+3} = x_n + x_{n+1}; \, y_0 = y_1 = y_2 = 0.$

21. $y_n - 2y_{n+1} = x_n - 2x_{n+1} + x_{n+2}; \, y_0 = 1.$

5.5.1   The stability of a discrete linear system

Let us look back at the linear difference equation in (7) and make three simplifying assumptions: first, that the initial conditions are all zero, $y_0 = \cdots = y_{P-1} = 0$; second, that in the forcing term (or input) we have $x_0 = \cdots = x_{Q-1} = 0$, and third, that $P \geq Q$. Then we obtain a simple explicit representation for the Z-transform, $Y(z)$, of the solution given by (8):

$$Y(z) = F(z)X(z),$$

where

$$F(z) = \frac{\sum_0^Q B_j z^j}{\sum_0^P A_j z^j} = \sum_{j=0}^\infty f_j/z^j. \tag{9}$$

Equivalently,

$$y_n = \sum_{k=0}^n f_k x_{n-k}, \quad n = 0, 1, 2, \ldots. \tag{10}$$

That is, the output $\{y_n\}$ of the system is given in terms of the input $\{x_j\}$ by the convolution equation (10). The rational function F given in (9) is called the **system transfer function**. The linear system is **stable** if bounded input always produces bounded output. The stability of the system obviously depends on the system transfer function F and we shall discover a bit later in this section exactly what properties of F correspond to stability. But first we shall work out a few examples to get an understanding of some of the mechanics of stability.

Example 1 $x_0 = y_0, x_1 - x_0 = y_1, x_2 - x_1 = y_2, \ldots$, and, in general,

$$x_{n+1} - x_n = y_{n+1}, \; n = 0, 1, 2, \ldots.$$

This is the system (7) with $B_0 = -1$, $B_1 = 1$, and $A_0 = 0$, $A_1 = 1$. Hence, the system transfer function is

$$F(z) = \frac{-1 + z}{z} = \frac{z - 1}{z} = 1 - 1/z.$$

The system is clearly stable since $|x_n| \leqslant M$ for all n gives $|y_n| \leqslant 2M$ for all n.

Example 2 $x_0 = y_0, y_1 - y_0 = x_1, y_2 - y_1 = x_2$, and, in general,

$$y_{n+1} - y_n = x_{n+1}, \; n = 0, 1, 2, \ldots.$$

This is the system (7) with $B_0 = 0$, $B_1 = 1$, and $A_0 = -1$, $A_1 = 1$. The system transfer function is

$$F(z) = \frac{z}{-1 + z} = 1 + \frac{1}{z} + \frac{1}{z^2} + \cdots$$

and the solution is

$$y_n = x_0 + x_1 + \cdots + x_n, \; n = 0, 1, 2, \ldots.$$

This system is clearly *not* stable since the bounded input $x_n = 1$ for all n produce the unbounded output $y_n = n + 1$ for all n.

Example 3 $x_0 = y_0, y_1 - \frac{1}{2}y_0 = x_1, y_2 - \frac{1}{2}y_1 = x_2$ and in general

$$y_{n+1} - \tfrac{1}{2}y_n = x_{n+1}, \; n = 0, 1, 2, \ldots.$$

Once again comparing with (7) we find that $A_1 = 1$, $A_0 = -1/2$, and $B_0 = 0$, $B_1 = 1$. The system transfer function here is

$$F(z) = \frac{z}{-1/2 + z} = \frac{2z}{2z - 1} = 1 + \frac{1}{2z} + \frac{1}{4z^2} + \cdots.$$

The solution is $y_0 = x_0$ and

$$y_n = x_n + \tfrac{1}{2}x_{n-1} + \tfrac{1}{4}x_{n-2} + \cdots + \frac{1}{2^n}x_0, \; n = 0, 1, 2, \ldots.$$

The system is stable since $|x_n| \leqslant M$ for all n yields

$$|y_n| \leqslant M + \tfrac{1}{2}M + \tfrac{1}{4}M + \cdots + \frac{1}{2^n}M < 2M \text{ for all } n.$$

Example 4 $x_0 = y_0$, $y_1 + y_0 = x_1$, $y_2 + y_1 = x_2$; generally,

$$y_{n+1} + y_n = x_{n+1}, \; n = 0, 1, 2, \ldots.$$

The system corresponds to (7) with $A_0 = A_1 = 1$ and $B_0 = 0$, $B_1 = 1$. The system transfer function is

$$F(z) = \frac{z}{1+z} = 1 - \frac{1}{z} + \frac{1}{z^2} - \frac{1}{z^3} + \cdots$$

and the solution is

$$y_n = x_n - x_{n-1} + x_{n-2} - \cdots + (-1)^n x_0, \; n = 0, 1, 2, \ldots.$$

This particular system is not stable since the bounded input $x_j = (-1)^{j+1}$ produces the unbounded output $y_n = (-1)^{n+1}(n+1)$.

Conditions for stability

The fundamental question on stability is completely answered in the following theorem.

> **Theorem:**   The linear difference equation (10) with system transfer function F given in (9) is stable if and only if all the poles of the rational function F lie in the open disc $\{z: |z| < 1\}$.

Proof. Suppose first that all the poles of F lie in the open disc $\{z: |z| < 1\}$. Then there is a number $r_0 < 1$ such that all the poles of F lie in the disc $\{z: |z| \leqslant r_0\}$; this is because F has only a finite number of poles. Hence, F is analytic on the set $\{z: |z| > r_0\}$, including at ∞, and so has a power series valid there, centered at ∞:

$$F(z) = A_0 + \frac{A_1}{z} + \frac{A_2}{z^2} + \cdots, \; |z| > r_0.$$

However, F has the power series representation (9) so that $A_j = f_j$ for $j = 0, 1, 2, \ldots$. Thus, the series $\sum f_j/z^j$ converges exterior to the circle $|z| = r_0$ and, by the basic properties of power series, is absolutely convergent on any circle of radius larger than r_0. In particular, this holds on the circle $|z| = 1$. Consequently,

$$S = \sum_{n=0}^{\infty} |f_n|$$

is finite. Suppose now that $|x_k| \leqslant M$ for all k. Then

$$|y_n| = \left| \sum_{k=0}^{n} x_k f_{n-k} \right| \leqslant \sum_{k=0}^{n} |x_k| |f_{n-k}|$$

$$\leqslant M \sum_{k=0}^{n} |f_{n-k}| \leqslant MS$$

for each $n = 0, 1, 2, \ldots$. Therefore, the linear system is stable.

Conversely, suppose that the linear system (10) is stable. We note first that if $\{y_n\}$ is a bounded sequence, say $|y_n| \leq M$ for all n, then the Z-transform $Y(z)$ of $\{y_n\}$ satisfies

$$|Y(z)| = \left| \sum_{n=0}^{\infty} y_n/z^n \right| \leq \sum_{n=0}^{\infty} |y_n|/|z|^n$$

$$\leq M \sum_{n=0}^{\infty} 1/|z|^n = M \frac{|z|}{|z|-1}, \quad |z| > 1.$$

In particular, the function $Y(z)$ is analytic on the region $\{z : |z| > 1\}$ because its power series in $1/z$ is absolutely convergent there. Since

$$Y(z) = F(z)X(z),$$

we see immediately that F has no poles in the region $\{z : |z| > 1\}$. Moreover, if F had a pole at a point $\lambda, |\lambda| = 1$, then

$$F(z) = \frac{G(z)}{(z-\lambda)^m},$$

where G is analytic near λ, $G(\lambda) \neq 0$, and $m \geq 1$ (see Section 5, Chapter 2). Let $x_k = \lambda^k$, $k = 0, 1, 2, \ldots$; then

$$X(z) = 1 + \frac{\lambda}{z} + \frac{\lambda^2}{z^2} + \cdots$$

$$= \frac{z}{z-\lambda}.$$

The rule for the Z-transform of the output $\{y_n\}$ corresponding to this bounded input yields

$$Y(z) = F(z)X(z), \qquad |z| > 1$$

$$= G(z)z \frac{1}{(z-\lambda)^{m+1}}, \quad |z| > 1, \ z \text{ near } \lambda.$$

Hence,

$$|Y(z)| \geq \frac{\frac{1}{2}|G(\lambda)|}{|z-\lambda|^{m+1}}, \quad z \text{ near } \lambda, \ |z| > 1.$$

By setting $z = \lambda t, t > 1$, we get

$$|Y(\lambda t)| \geq \frac{\frac{1}{2}|G(\lambda)|}{(t-1)^{m+1}}.$$

This contradicts the estimate

$$|Y(\lambda t)| \leq \frac{Mt}{t-1}$$

obtained above, if t is near 1. We conclude that F has no poles on $|z| = 1$ and so all the poles of F lie within the disc $\{z: |z| < 1\}$. $\square$

Exercises for Section 5.5.1

1.   Let p be a polynomial of degree N. Show that all the zeros of p lie within the disc $|z| < 1$ if and only if all the zeros of rational function

$$R(w) = p\left(\frac{w+1}{w-1}\right)$$

lie in the left half-plane; that is, satisfy Re $w < 0$.

2.   Let p be a polynomial of degree N; show that

$$q(w) = (w-1)^N p\left(\frac{w+1}{w-1}\right)$$

is a polynomial of degree N or less in w.

3.   Find $q(w)$ if $p(z)$ is
(a) $p(z) = Az + B$;   (b) $p(z) = 27z^4 - 24z^3 + 18z^2 - 8z + 3$;
(c) $p(z) = z^2 - z$;   (d) $p(z) = z^3 - 3z^2 + 3z - 2$.

4.   Let q be given as in Exercise 2 and suppose that p and q have the same degree. Show that all the zeros of q lie in the left half-plane if and only if all the zeros of p lie in the disc $\{z: |z| < 1\}$. (**Hint:** Combine Exercises 1 and 2.) What can happen to the degree of q? Show that degree q is less than degree p if and only if p has a zero at $z = 1$.

5.   Use Exercise 4 and, if necessary, the Routh–Hurwitz criterion (Theorem 4, Section 1.1, Chapter 3) to determine whether all the zeros of the following polynomials lie within the disc $\{z: |z| < 1\}$
(a) $p(z) = 5z^3 + 3z^2 - z + 1$;   (b) $p(z) = 10z^3 - 4z^2 + 2z$;
(c) $p(z) = 27z^4 - 24z^3 + 18z^2 - 8z + 3$;
(d) $p(z) = 59z^3 - 89z^2 + 37z - 7$;
(e) $p(z) = 10z^3 + 6z^2 - 20z + 6$.

6.   Test the following linear systems for stability:
(a) $y_n - y_{n+1} - 4y_{n+2} + 4y_{n+3} = x_n - x_{n+1}$;
(b) $y_n - y_{n+1} + 3y_{n+2} + 5y_{n+3} = x_n + 2x_{n+1} - x_{n+3}$;
(c) $y_n - y_{n+1} - 4y_{n+2} - 4y_{n+3} = x_n$;
(d) $3y_n - 8y_{n+1} + 18y_{n+2} - 24y_{n+3} + 27y_{n+4} = x_n + x_{n+1} + x_{n+2}$;
(e) $y_n - 2y_{n+1} = x_n + x_{n+1} - x_{n+2}$;
(f) $2y_n + y_{n+1} = x_n + 2x_{n+1}$;
(g) $y_n - 4y_{n+2} = 3x_n + 2x_{n+1}$;
(h) $6y_n - 20y_{n+1} + 6y_{n+2} + 10y_{n+3} = x_n + x_{n+1}$.

5.6   The Fast Fourier Transform⋆

This section is devoted to brief presentation of the salient points of the Fast Fourier Transform, an algorithm developed in the mid-1960s for efficient machine computation of the Fourier transform.*

To begin, we note that in typical applications the function u is not known; what is known is a collection of sampled values of u, where the samples were taken in some time interval $[a, b]$. It is reasonable, and convenient to assume that we know the value of u at N equally spaced points in this interval; that is, we know the numbers $x_k = u(a + k\delta)$, $k = 0, \ldots, N - 1$, $\delta = (b - a)/N$. From this data we are to approximate $\hat{u}$. More precisely, we shall set up a rule that will associate with the data $\{x_k\}_{k=0}^{N-1}$ another set of data $\{y_j\}_{j=0}^{N-1}$ in such a way that $\{x_k\}$ and $\{y_j\}$ are related to each other in much the same way that u and $\hat{u}$ are related to each other. Furthermore, $\{x_k\}$ approximates u while $\{y_j\}$ approximates $\hat{u}$.

Given the numbers $x_0, \ldots, x_{N-1}$, define

$$y_j = \sum_{k=0}^{N-1} x_k \, e^{-2\pi ijk/N}, \; j = 0, 1, \ldots, N - 1.$$

Then for $r = 0, 1, \ldots, N - 1$, we have

$$\frac{1}{N} \sum_{j=0}^{N-1} y_j \, e^{2\pi irj/N} = \frac{1}{N} \sum_{j=0}^{N-1} \left\{ \sum_{k=0}^{N-1} x_k \, e^{-2\pi ijk/N} \right\} e^{2\pi irj/N}$$

$$= \sum_{k=0}^{N-1} x_k \left\{ \frac{1}{N} \sum_{j=0}^{N-1} e^{2\pi ij(r-k)/N} \right\}$$

$$= x_r,$$

since

$$\frac{1}{N} \sum_{j=0}^{N-1} e^{2\pi ijl/N} = \begin{cases} 1 & \text{if } l = 0 \\ 0 & \text{if } l = \pm 1, \ldots, \pm(N-1). \end{cases}$$

This yields the desired pairing and gives the inversion formula for the "discrete Fourier transform":

$$y_j = \sum_{k=0}^{N-1} x_k \, e^{-2\pi ijk/N}, \; j = 0, 1, \ldots, N - 1 \tag{1}$$

$$x_k = \frac{1}{N} \sum_{j=0}^{N-1} y_j \, e^{2\pi ikj/N}, \; k = 0, \ldots, N - 1. \tag{2}$$

---

* This book does not purport to be a programming text at any level. For the practical implementation of the FFT, the reader is referred to the text by Brigham cited in the references at the end of this chapter and to the papers listed there.

A justification for (1) and (2)

Suppose that f is a piecewise smooth function on the real line. Fix L; then $g(t) = f(t(L/\pi))$ is defined for $-\pi \leqslant t \leqslant \pi$. The theory of Fourier series implies that for all x that are not breakpoints for f, we have

$$f(x) = g\left(\frac{\pi x}{L}\right) = \sum_{-\infty}^{\infty} \alpha_n \, e^{in\pi x/L},$$

where

$$\alpha_n = \frac{1}{2\pi} \int_{-\pi}^{\pi} g(t) \, e^{-int} \, dt$$

$$= \frac{1}{2L} \int_{-L}^{L} f(s) \, e^{-in\pi s/L} \, ds.$$

Let $s_n = n(\pi/L)$ and $\beta(s) = \int_{-L}^{L} f(x) \, e^{-ixs} \, dx$. Then

$$f(x) = \sum_{-\infty}^{\infty} e^{ixs_n} \alpha_n = \sum_{-\infty}^{\infty} e^{ixs_n} \left(\frac{1}{2L} \int_{-L}^{L} f(s) \, e^{-in\pi s/L} \, ds \right)$$

$$= \sum_{-\infty}^{\infty} \frac{1}{2\pi} \beta(s_n) \, e^{ixs_n} (s_n - s_{n-1}).$$

This last sum appears to be an approximating Riemann sum for the integral

$$\frac{1}{2\pi} \int_{-\infty}^{\infty} \beta(s) \, e^{ixs} \, ds.$$

We blithely ignore all the mathematical difficulties and let $L \to \infty$. This yields

$$f(x) = \frac{1}{2\pi} \int_{-\infty}^{\infty} \beta(s) \, e^{ixs} \, ds$$

and

$$\beta(s) = \int_{-\infty}^{\infty} f(x) \, e^{-ixs} \, dx.$$

These are the familiar formulas from Section 2 relating a function f to its Fourier transform, here denoted by β. Although the passage to the limit as $L \to \infty$ can be substantiated with proper hypotheses on f, that is not our concern here. Instead we take this for granted and note that we have seen that for large integers $m, f(t)$ is approximated by

$$\frac{1}{2\pi} \sum_{n=-m}^{m} \beta(s_n) \, e^{its_n} \left(\frac{\pi}{L} \right), \quad s_n = \frac{n\pi}{L}.$$

That is,

$$f(t_k) \sim \frac{1}{2L} \sum_{n=-m}^{m} \beta(s_n) \, e^{is_n t_k}$$

while

$$\beta(s_n) = \int_{-L}^{L} f(t)\, e^{-its_n}\, dt.$$

Therefore,

$$\beta(s_n) \sim \sum_{k=-m}^{m} f(t_k)\, e^{-is_n t_k}\, \Delta t_k.$$

Substituting

$$s_n = n\pi/L \quad \text{and} \quad t_k = -L + k\,\frac{2L}{2m+1}, \quad k = 0, 1, \ldots, 2m,$$

we have

$$f(t_k) \sim \frac{1}{2L} \sum_{n=-m}^{m} \beta\!\left(\frac{n\pi}{L}\right)(-1)^n\, e^{2\pi i n k/(2m+1)} \tag{3}$$

and

$$(-1)^n \beta\!\left(n\,\frac{\pi}{L}\right) \sim \frac{L}{2m+1} \sum_{k=0}^{2m} f(t_k)\, e^{-2\pi i n k/(2m+1)}. \tag{4}$$

All that remains to finalize the connection between (1), (2) and (3), (4) is to set $N = 2m + 1$, $x_k = f(t_k)$ for $k = 0, \ldots, N - 1$ and $y_n = (N/2L)(-1)^n \beta(n(\pi/L))$, $n = 0, \pm 1, \ldots, \pm m$.

The efficacy of the Fast Fourier Transform

Set $W = e^{-2\pi i/N}$ and look at the N equations in (1) that define y_j in terms of x_k:

$$y_j = \sum_{k=0}^{N-1} x_k W^{kj}, \quad j = 0, \ldots, N - 1.$$

This is a linear system of N equations, involving the N (known) numbers x_0, $\ldots$, x_{N-1} and $N \times N$ matrix (W^{jk}). Even though this matrix has only N distinct entries (because $W^N = 1$) it would still appear that to compute $y_0, \ldots, y_{N-1}$, at least N^2 arithmetic operations (multiplications and additions) would be necessary: N operations for each of the N numbers y_j. The FFT exploits the form of the matrix (W^{jk}) and the periodicity of W to reduce the number of operations to the order of $N \log N$. For N relatively large, this is an enormous savings in computation time and it is this savings that has prompted the great interest in the FFT.

Suppose that $N = 2^{r+1}$ for some nonnegative integer r. We express an integer k, $0 \leqslant k \leqslant N - 1$, in its binary representation

$$k = k_0 + 2k_1 + 2^2 k_2 + \cdots + 2^r k_r,$$

where $k_0, k_1, \ldots, k_r$ are either 0 or 1. We also write $x(k_0, \ldots, k_r)$ in place of x_k.

Hence, with $W = e^{-2\pi i/N}$ as before, we have

$$y_j = \sum_{k=0}^{N-1} x_k W^{kj}$$

$$= \sum_{k_0=0}^{1} \sum_{k_1=0}^{1} \cdots \sum_{k_r=0}^{1} x(k_0, \ldots, k_r) W^{(k_0 + 2k_1 + \cdots + 2^r k_r)j}.$$

Now write j in its binary representation:

$$j = j_0 + 2j_1 + 2^2 j_2 + \cdots + 2^r j_r.$$

Thus,

$$\begin{aligned} jk &= (j_0 + 2j_1 + \cdots + 2^r j_r)(k_0 + 2k_1 + \cdots + 2^r k_r) \\ &= k_r(2^r j_0 + 2^{r+1} j_1 + \cdots + 2^{2r} j_r) \\ &\quad + k_{r-1}(2^{r-1} j_0 + 2^r j_1 + \cdots + 2^{2r-1} j_r) \\ &\quad + \cdots + k_0(j_0 + 2j_1 + \cdots + 2^r j_r). \end{aligned}$$

When we recall that $W^{2^{r+1}} = W^N = 1$, we see that

$$W^{k_r(2^r j_0 + 2^{r+1} j_1 + \cdots + 2^{2r} j_r)} = W^{2^r k_r j_0}.$$

Likewise, the term involving k_{r-1} also simplifies:

$$W^{k_{r-1}(2^{r-1} j_0 + 2^r j_1 + \cdots + 2^{2r-1} j_0)} = W^{2^{r-1} k_{r-1}(j_0 + 2j_1)}.$$

Similar computations can be carried out for the other powers of W that appear.

Set

$$\begin{aligned} z(j_0, k_0, \ldots, k_{r-1}) &= \sum_{k_r=0}^{1} x(k_0, \ldots, k_r) W^{2^r k_r j_0} \\ &= x(k_0, \ldots, k_{r-1}, 0) + x(k_0, \ldots, k_{r-1}, 1) W^{2^r j_0}. \end{aligned}$$

Since each of $j_0, k_0, \ldots, k_{r-1}$, can be either 0 or 1, this represents $2^{r+1} = N$ equations, each carried out by one multiplication and one addition. Next, set

$$\begin{aligned} z(j_0, j_1, k_0, \ldots, k_{r-2}) &= \sum_{k_{r-1}=0}^{1} z(j_0, k_0, \ldots, k_{r-1}) W^{2^{r-1} k_{r-1}(j_0 + 2j_1)} \\ &= z(j_0, k_0, \ldots, k_{r-2}, 0) \\ &\quad + z(j_0, k_0, \ldots, k_{r-2}, 1) W^{2^{r-1}(j_0 + 2j_1)}. \end{aligned}$$

Once again this represents N equations, each carried out by two arithmetic operations. Continuing in this way we reach the $(r+1)$st (and last) step, which yields $y_j = z(j_0, j_1, \ldots, j_r)$:

$$\begin{aligned} z(j_0, j_1, \ldots, j_r) &= \sum_{k_0=0}^{1} z(j_0, \ldots, j_{r-1}, k_0) W^{k_0(j_0 + 2j_1 + \cdots + 2^r j_r)} \\ &= z(j_0, \ldots, j_{r-1}, 0) + z(j_0, \ldots, j_{r-1}, 1) W^{j_0 + \cdots + 2^r j_r}. \end{aligned}$$

In this recursive fashion we produce all of $y_0, \ldots, y_{N-1}$ in a total of $2(r+1)N$ arithmetic operations. Since $r + 1 = \log_2 N$, we see that the FFT produces y_0,

$\ldots, y_{N-1}$ in $2(\log_2 N)N$ operations.

Remarks: (1) A further savings of a factor of 2 can be effected by using the relationship $W^{a+N/2} = -W^a$ for $0 \leqslant a \leqslant N/2$.

(2) A bit of thought and some pencil work for r small shows that the FFT is really a clever factorization of the $N \times N$ matrix (W^{jk}) into the product of $r+1$ $N \times N$ matrices, each of which minimizes the number of arithmetic operations that need to be carried out.

(3) The algorithm given above is from the book by E. O. Brigham and is credited there with being the one originally developed by Cooley and Tukey. There are other versions and variations of this algorithm which may be more effective in certain specific situations.

Further reading

More on the Fourier transform can be found in Titchmarsh, E. C. *The Fourier transform.* New York: Oxford University Press, 1937; and Carslaw, H. S. *Theory of Fourier series and integrals.* New York: Dover, 1950.

The Laplace (and Fourier) transform is covered in Sneddon, I. N. *Fourier transforms.* New York: McGraw-Hill, 1951; and LePage, W. R. *Complex variables and the Laplace transform for engineers.* New York: Dover, 1980.

Several of the applications in the text come from Churchill, R. V. *Operational mathematics.* New York: McGraw-Hill, 1958; and Powers, D. L. *Boundary-value problems.* New York: Academic Press, 1972.

A good reference for the Z-transform, including many applications, is Jury, E. J. *Theory and applications of the Z-transform.* New York: John Wiley, 1964.

The Fast Fourier Transform is the subject of *The Fast Fourier Transform* (Brigham, E. O. Englewood Cliffs, N.J.: Prentice-Hall, 1974). The original paper on the subject is Cooley, J. W., Tukey, J. W. An algorithm for machine computation of the complex Fourier series. *Mathematics of Computation* 1965; 19: 297–301.

APPENDIX 1

Locating the zeros of
a polynomial

A polynomial of degree d has precisely d zeros, counting multiplicities, in the complex plane. This appendix contains several theorems that help to locate these zeros. The first result is the "grandfather" of the others.

> **Theorem 1:** If every zero of the polynomial p lies in the half-plane $\mathrm{Re}(Az + B) > 0$, then so does every zero of its derivative p'.

Proof. This result is called the **Gauss*–Lucas†** Theorem; the proof is not at all difficult. Write

$$p(z) = a \prod_1^N (z - z_j)^{m_j}, \ a \neq 0,$$

where $z_1, \ldots, z_N$ are distinct points and $m_1, \ldots, m_N$ are positive integers. A simple computation yields

$$\frac{p'(z)}{p(z)} = \sum_1^N \frac{m_j}{z - z_j}$$

$$= \sum_1^N \frac{m_j(\bar{z} - \bar{z}_j)}{|z - z_j|^2}.$$

Clearly, we need only look at those points b that are zeros of p' but not zeros of p. If b is such a point, then

$$0 = \sum_1^N \frac{m_j}{|b - z_j|^2} (\bar{b} - \bar{z}_j) = \bar{b}c_0 + c_1,$$

where $c_0 = \sum_1^N m_j |b - z_j|^{-2}$ is positive and $c_1 = -\sum_1^N m_j \bar{z}_j |b - z_j|^{-2}$. Hence,

$$\mathrm{Re}\,(Ab + B) = \mathrm{Re}\left(-A\frac{\bar{c}_1}{c_0} + B \right) = \frac{1}{c_0} \mathrm{Re}(-A\bar{c}_1 + Bc_0)$$

$$= \frac{1}{c_0} \mathrm{Re} \sum_1^N m_j\left(\frac{Az_j + B}{|b - z_j|^2} \right)$$

---

* Karl Friedrich Gauss, 1777–1855.
† Francois Édouard Anatole Lucas, 1842–1891.

$$= \frac{1}{c_0} \sum_1^N \frac{m_j}{|b - z_j|^2} \operatorname{Re}(Az_j + B) > 0. \qquad \square$$

There is another classical result on the location of the zeros of a polynomial, this time ensuring that they lie in a disc. This theorem is known as the **Eneström-Kakeya** Theorem.

> **Theorem 2:** Suppose that $0 < a_0 < a_1 < \cdots < a_n$. Then all the zeros of the polynomial $p(z) = a_0 + a_1 z + \cdots + a_n z^n$ lie in the disc $\{z : |z| < 1\}$.

Proof. Set $q(z) = z^n p(1/z) = a_n + a_{n-1} z + \cdots + a_0 z^n$; we shall show that all the zeros of q lie in the set $|z| > 1$. We have

$$(1 - z)q(z) = a_n + z(a_{n-1} - a_n) + z^2(a_{n-2} - a_{n-1})$$
$$+ \cdots + z^n(a_0 - a_1) - a_0 z^{n+1}.$$

Thus, for $|z| \leqslant 1$,

$$|1 - z| \, |q(z)| \geqslant a_n - |(a_n - a_{n-1})z + \cdots + (a_1 - a_0)z^n + a_0 z^{n+1}|$$
$$\geqslant a_n - (a_n - a_{n-1}) - \cdots - (a_1 - a_0) - a_0 = 0.$$

Furthermore, equality holds in the second inequality only if $|z| = 1$ and *all* of the terms $(a_j - a_{j-1})z^{n-j}$ have the same argument, $j = 0, 1, \ldots, n$; this can only happen when $z = 1$. But at $z = 1$, $p(1) = a_0 + a_1 + \cdots + a_n > 0$. $\qquad \square$

The zeros of cubic and quartic polynomials

The familiar quadratic formula for finding the zeros of a second-degree polynomial has an extension, albeit not so simple, which yields the zeros of a third- or fourth-degree polynomial. These formulas (or better, techniques) date back to the 16th century.

Cubic Polynomials

We wish to find the roots of the equation

$$z^3 + Az^2 + Bz + C = 0. \qquad (1)$$

We begin by making the substitution $z = w + d$ and then choosing d so that the resulting quadratic term vanishes. Specifically, if $z = w + d$, we obtain

$$w^3 + (A + 3d)w^2 + (3d^2 + 2Ad + B)w + (d^3 + Ad^2 + Bd + C) = 0.$$

The choice $d = -A/3$ yields

$$w^3 + aw + b = 0 \tag{2}$$

with

$$a = B - \tfrac{1}{3}A^2 \quad \text{and} \quad b = \tfrac{2}{27}A^3 - \tfrac{1}{3}AB + C.$$

If it happens that $a = 0$, then the equation becomes

$$w^3 + b = 0,$$

and the roots are obtained easily. If $a \neq 0$, then let λ be a square root of $-a/3$ and set $w = \lambda q$. This substitution changes (2) to the equation

$$q^3 - 3q + \beta = 0, \tag{3}$$

where $\beta = b\lambda^{-3}$. Long ago someone noticed that the substitution $q = p + 1/p$ changes equation (3) into the very simple equation

$$p^6 + \beta p^3 + 1 = 0. \tag{4}$$

Indeed, this was the whole point of putting (2) into the form of (3). Equation (4) is nothing but a quadratic in the variable p^3. Its zeros produce the q's that solve (3) and so in turn produce the zeros of (2) and finally the zeros of (1). This three-step process yields the zeros of the general cubic polynomial (1).

Example 2   The cubic $z^3 + 6z^2 + 12z + 35 = 0$ becomes

$$w^3 + 27 = 0 \tag{5}$$

after the substitution $z = w - 2$. The roots of (5) are

$$w = -3 \quad \text{and} \quad \tfrac{1}{2}(3 \pm i\sqrt{27})$$

and so the roots of the original equation are

$$z = -5 \quad \text{and} \quad \tfrac{1}{2}(-1 \pm i\sqrt{27}).$$

Example 3   The cubic $z^3 + 6z^2 + 4 = 0$ is changed to the cubic

$$w^3 - 12w + 20 = 0$$

by the substitution $z = w - 2$. With $a = -12$ and $b = 20$ we find that $\lambda = 2$ and $\beta = 5/2$, and so we must solve

$$p^6 - (5/2)p^3 + 1 = 0.$$

This gives $p^3 = 2$ or $p^3 = 1/2$; we work with the cube roots of 2, which are

$$\gamma_1 = (2)^{1/3}; \quad \gamma_2 = (2)^{-2/3}(-1 + i\sqrt{3}); \quad \gamma_3 = (2)^{-2/3}(-1 - i\sqrt{3}).$$

Hence, the roots of $w^3 - 12w + 20 = 0$ are

$$w_1 = \gamma_1 + 1/\gamma_1, \quad w_2 = \gamma_2 + 1/\gamma_2, \quad w_3 = \gamma_3 + 1/\gamma_3$$

and so the roots of the original equation are $z_j = w_j - 2, \quad j = 1, 2, 3.$

Quartic Polynomials

We wish to find the roots of $z^4 + Az^3 + Bz^2 + Cz + D = 0$. We rewrite this as

$$\left(z^2 + \frac{A}{2}z\right)^2 = \left(\frac{A^2}{4} - B\right)z^2 - Cz - D. \tag{6}$$

If the right-hand side of (6) is a perfect square, say $(Ez + F)^2$, then we obtain two quadratic equations:

$$z^2 + \frac{A}{2}z = Ez + F$$

and

$$z^2 + \frac{A}{2}z = -(Ez + F).$$

Each of these equations has two roots and these four numbers are the roots we seek. If the right-hand side of (6) is not a perfect square, we add $w^2 + 2(z^2 + (A/2)z)w$ to both sides of (6). This yields

$$(z^2 + (A/2)z + w)^2 = ((A^2/4) - B + 2w)z^2 + (Aw - C)z + (w^2 - D). \tag{7}$$

We now choose w so the right-hand side of (7) is a perfect square. This will be so if the discriminant of the right-hand side vanishes, that is, if

$$(Aw - C)^2 - 4\left(\frac{A^2}{4} - B + 2w\right)(w^2 - D) = 0. \tag{8}$$

This is a cubic in w and can be solved as outlined above. We can then take any root w^* of this cubic, substitute it into (7) and then solve the resulting equation:

$$\left(z^2 + \frac{A}{2}z + w^*\right)^2 = \left(\frac{A^2}{4} - B + 2w^*\right)\left(z - \frac{Aw^* - C}{2((A^2/4) - B + 2w^*)}\right)^2. \tag{9}$$

This will yield the four zeros of $z^4 + Az^3 + Bz^2 + Cz + D$.

Example 4   Find the zeros of $z^4 + z^2 - 4iz - 4$.

Solution   Here $A = 0$, so we obtain

$$(z^2)^2 = z^4 = -z^2 + 4iz + 4$$
$$= (iz + 2)^2.$$

Hence, we must solve the two quadratic equations

$$z^2 = iz + 2 \qquad \text{and} \qquad z^2 = -iz - 2.$$

The solutions of these are $(i \pm \sqrt{7})/2$ and $(-i \pm 3i)/2$, respectively. Hence, i, $-2i$, $(i + \sqrt{7})/2$, and $(i - \sqrt{7})/2$ are the four zeros of $z^4 + z^2 - 4iz - 4$.   □

Example 5   Find the zeros of $z^4 + 2z^3 + 4z^2 + 2z + 3$.

Solution   Here $((A^2/4) - B)z^2 - Cz - D = -3z^2 - 2z - 3$ is not a perfect square so we must find a zero of the cubic $(2w - 2)^2 - 4(-3 + 2w)(w^2 - 3)$. One root of this is $w^* = 2$. Putting this into (9) we obtain

$$(z^2 + z + 2)^2 = (1)(z + \tfrac{2}{2})^2.$$

This yields the two equations

$$z^2 + z + 2 = z + 1 \quad \text{and} \quad z^2 + z + 2 = -z - 1.$$

The first equation has roots $\pm i$ and the second has roots $-1 \pm i\sqrt{2}$. These four numbers are the zeros of the original polynomial.   □

A Final Remark: The reader may well draw the conclusion that the zeros of any polynomial may eventually be found by a succession of the steps that work for the quadratic, cubic, and quartic cases—that is, by a sequence of substitutions and extractions of roots. Quite surprisingly this is not the case if the degree is 5 or more. In fact, one of the more profound mathematical results of the early 19th century was that there is no formula of this type that will solve every 5th (or higher) degree polynomial.

Sturm sequences

Let p be a polynomial with real coefficients and let $a < b$ be two points on the real line with $p(a) \neq 0$ and $p(b) \neq 0$. We shall describe a simple technique that counts the number of distinct zeros of p in the interval (a, b); the zeros are counted *without* multiplicity.

We begin in a somewhat more general context. Let $p_1, p_2, \ldots, p_m$ be polynomials with real coefficients satisfying these three conditions:

(1) if $p_k(x_0) = 0$ for some $x_0 \in (a, b)$ and some k, $2 \leqslant k \leqslant m - 1$, then
   $p_{k-1}(x_0)p_{k+1}(x_0) < 0$.
(2) $p_m(x) \neq 0$ for $a < x < b$.
(3) $p_1(a) \neq 0$, $p_1(b) \neq 0$; if $p_1(x_0) = 0$ for some $x_0 \in (a, b)$ then the sign of $p_1(x)p_2(x)$ changes from minus to plus as x passes through x_0 from left to right.

Let $S(x)$ be the number of changes of sign in the sequence $\{p_1(x), p_2(x), \ldots, p_m(x)\}$ for $a \leqslant x \leqslant b$. A zero of some $p_k(x)$, $1 \leqslant k \leqslant m - 1$, can be counted either as a plus or a minus without affecting $S(x)$, because of conditions (1) and (3).

Theorem 3: $S(a) - S(b)$ is the number of zeros of p_1 on the interval (a, b). The zeros are counted without multiplicity.

Proof. Suppose that x is a point at which no p_k is zero. Then by the continuity of p_j, there is an interval $(x - \delta, x + \delta)$ in which no p_j is zero, and so $S(x)$ does not change in $(x - \delta, x + \delta)$. Hence, any change in $S(x)$ must occur at a zero of some p_k.

If $p_k(x_0) = 0$ for $k = 2, \ldots, m - 1$, then condition (1) shows that the number of sign changes in the triple $p_{k-1}(x)$, $p_k(x)$, $p_{k+1}(x)$ remains exactly one for all x in some small interval about x_0. Consequently, the only changes in $S(x)$ can occur, if at all, at points x_0, where $p_1(x_0) = 0$. By condition (3), however, we see that $S(x)$ decreases by exactly 1 as x passes through a zero of p_1 from left to right. Thus, the theorem is proved.   □

To fit the theorem to our context we take $p_1 = p$ and $p_2 = p'$, the derivative of p. Simple calculus shows that condition (3) holds; that is, that the sign of pp' changes from minus to plus as x passes through a zero of p from left to right. The remainder of the sequence $p_3, \ldots, p_m$ are found by long division. Write $p_1 = q_1 p_2 - p_3$, where q_1 is a polynomial and degree $(p_3) <$ degree (p_2). Continuing $p_2 = q_2 p_3 - p_4$, where degree $(p_4) <$ degree (p_3), and so on until the remainder is a polynomial of degree zero, that is, a constant. If the remainder is nonzero, then p_m is this remainder, and (2) holds. If the remainder is zero, then p_m is the previous remainder, which, therefore, divides p_{m-1} exactly. There is just one point that deserves mention: it may happen that p_m does not have degree 0 and, indeed, p_m may have zeros in $[a, b]$. If this is the case, then it is relatively easy to check that p_m actually divides all the preceding polynomials $p_{m-1}, p_{m-2}, \ldots, p_2, p_1$. Thus, p_m appears as a factor in all the terms of the sequence $\{p_1, p_2, \ldots, p_m\}$ and then the value of $S(a) - S(b)$ is just the same as it is for the sequence $\{p_1/p_m, p_2/p_m, \ldots, p_{m-1}/p_m, 1\}$.

Further Reading

An enormous amount of material on the location of zeros of polynomials is in Marden, M. *The geometry of zeros of polynomials*, Providence, R.I.: American Mathematical Society, 1966.

APPENDIX 2

A table of conformal mappings

$z = x + iy$, $w = u + iv$. Lower-case letters are mapped to upper-case letters, cross-hatched regions onto shaded regions.

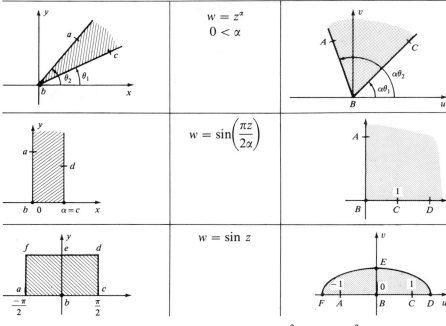

z-plane	Mapping	w-plane

$w = e^z$

The vertical segment $x = x_0$, $0 \leqslant y \leqslant \pi$ is mapped to the semicircle $u^2 + v^2 = e^{2x_0}$, $v \geqslant 0$.

$w = z^\alpha$
$0 < \alpha$

$w = \sin\left(\dfrac{\pi z}{2\alpha}\right)$

$w = \sin z$

The horizontal line $y = y_0$ goes to the ellipse $\dfrac{u^2}{\cosh^2 y_0} + \dfrac{v^2}{\sinh^2 y_0} = 1$.

Table of conformal mappings (*continued*)

z-plane	Mapping	w-plane
	$w = \log z$	
	$w = z^2$	

The vertical line $x = x_0$ is
mapped to the parabola $u = x_0^2 - \dfrac{v^2}{4x_0^2}$, $x_0 \neq 0$.

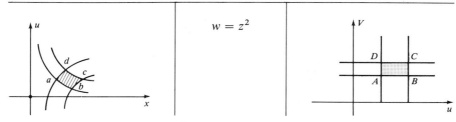

| | $w = z^2$ | |

The hyperbola $xy = \alpha$ is
mapped to the horizontal line $v = \alpha/2$.
The hyperbola $x^2 - y^2 = \beta$, $\beta > 0$, is
mapped to the vertical line $u = \beta$.

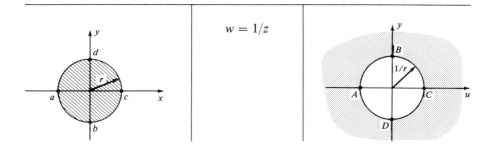

| | $w = 1/z$ | |

Table of conformal mappings (*continued*)

z-plane	Mappings	w-plane
	$w = 1/z$	
	$w = i\left(\dfrac{1+z}{1-z}\right)$ $z = \dfrac{i-w}{i+w}$ is the inverse mapping.	
	$w = i\left(\dfrac{1-z}{1+z}\right)$	
	$w = -i\,\dfrac{z^2 + 2iz + 1}{z^2 - 2iz + 1}$	

Table of conformal mappings   *(continued)*

z-plane	Mapping	w-plane
	$$w = \frac{a - z}{1 - \bar{a}z}$$ $$z = \frac{a - w}{1 - \bar{a}w}$$ is the inverse mapping. 0 and a are interchanged.	
	$$w = \cosh\left(\frac{\pi z}{h}\right)$$	
	$$w = \frac{\alpha}{2}\left(z + \frac{1}{z}\right)$$	
	$$w = z + \frac{1}{z}$$ $\alpha = 1 + r$	

The circle $|z| = \alpha$ is mapped onto the ellipse $\dfrac{u^2}{(\alpha + 1/\alpha)^2} + \dfrac{v^2}{(\alpha - 1/\alpha)^2} = 1.$

Table of conformal mappings (*continued*)

z-plane	Mapping	w-plane
	$w = z + e^z$	

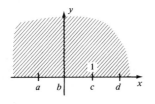

$$w = \frac{\alpha}{\pi}\left[(z^2 - 1)^{1/2} + \mathrm{Log}(z + (z^2 - 1)^{1/2})\right]$$

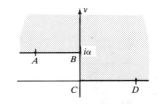

$$w = \frac{2\alpha i}{\pi}\left[z(1 - z^2)^{1/2} + \arcsin z\right]$$

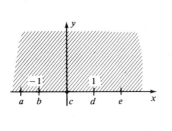

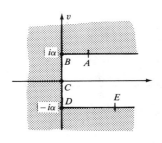

$$w = \frac{2\alpha}{\pi}\left[(z^2 - 1)^{1/2} + \arcsin(1/z)\right]$$

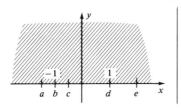

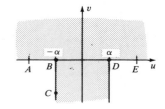

Table of conformal mappings (continued)

z-plane	Mapping	w-plane
	$w = \alpha(z^2 - 1)^{1/2}$	

$$w = \frac{i\sigma_0}{\pi}\left\{\text{Log}\left(\frac{1+\zeta}{1-\zeta}\right) - i\alpha\,\text{Log}\left(\frac{1+i\alpha\zeta}{1-i\alpha\zeta}\right)\right\}$$

$$\alpha = \frac{\tau_0}{\sigma_0}\,;\; \zeta^2 = \frac{z+1}{z-\alpha^2}$$

$$w = 2\sqrt{z+1} + \text{Log}\left(\frac{\sqrt{z+1}-1}{\sqrt{z+1}+1}\right)$$

$$w = \frac{z-\lambda}{1-\lambda z},$$

where

$$\lambda = \frac{1 + ab - \sqrt{(1-a^2)(1-b^2)}}{a+b}$$

and

$$R = \frac{1 - ab - \sqrt{(1-a^2)(1-b^2)}}{b-a}.$$

$-1 < a < b < 1;$
$a + b \neq 0$

APPENDIX 3

A table of Laplace transforms

$u(x); u(x) = 0$ if $x < 0$	$(\mathscr{L}u)(s)$
1. e^{ax}	$1/(s - a)$
2. x^m	$m!/s^{m+1}, \quad m = 0, 1, 2, \ldots$
3. $\sin bx$	$b/(b^2 + s^2)$
4. $\cos bx$	$s/(b^2 + s^2)$
5. $\sinh bx$	$b/(s^2 - b^2)$
6. $\cosh bx$	$s/(s^2 - b^2)$
7. $e^{ax} \sin bx$	$b/[(s - a)^2 + b^2]$
8. $e^{ax} \cos bx$	$(s - a)/[(s - a)^2 + b^2]$
9. $x \sin bx$	$2sb/(s^2 + b^2)^2$
10. $x \cos bx$	$(s^2 - b^2)/(s^2 + b^2)^2$
11. $\begin{cases} 1, 0 \leqslant x \leqslant \sigma \\ 0, \sigma < x \end{cases}$	$(1 - e^{-s\sigma})/s$
12. impulse function for σ	$e^{-s\sigma}$
13. x^p	$\Gamma(p + 1)/s^{p+1}, p > -1$
14. $\begin{cases} 1, 0 < x < \sigma \\ -1, \sigma < x < 2\sigma \\ 0, 2\sigma < x \end{cases}$	$(1 - e^{-s\sigma})^2/s$
15. $\dfrac{\cos(b\sqrt{x})}{\sqrt{x}}$	$\sqrt{\dfrac{\pi}{s}} \exp(-b^2/4s)$
16. $\dfrac{e^{-b^2/x}}{\sqrt{x}}$	$\sqrt{\dfrac{\pi}{s}} e^{-2b\sqrt{s}}, b \geqslant 0$
17. $\sin(b\sqrt{x})$	$\frac{1}{2}b\sqrt{\pi} s^{-3/2} e^{-b^2/4s}$

Solutions to odd-numbered exercises

Chapter 1

Section 1.1

1. (a) $7 - i$ (b) $-i$ (c) $-3 + 4i$ (d) $4 - 12i$
   (e) $4/25$ (f) $-i$ (g) $18 + 18i$

3. (a) circle of radius 3 centered at 0 (b) circle of radius 1 centered at 2 (c) circle of radius 2 centered at -2 (d) perpendicular bisector of the line segment joining 2 to -2 (e) the two points $1 \pm \sqrt{2}$ (f) the line $y = x$ (g) the line $y = -x$

5. (a) $\sqrt{2}\left(\cos\dfrac{3\pi}{4} + i \sin\dfrac{3\pi}{4}\right)$ (b) $2\left(\cos\dfrac{\pi}{3} + i \sin\dfrac{\pi}{3}\right)$

   (c) $1\left(\cos\left(-\dfrac{\pi}{2}\right) + i \sin\left(-\dfrac{\pi}{2}\right)\right)$ (d) $5\left(\cos\left(-\dfrac{\pi}{3}\right) + i \sin\left(-\dfrac{\pi}{3}\right)\right)$

   (e) 5 (f) $\sqrt{6}\,(\cos\theta_0 + i \sin\theta_0)$, where $\theta_0 = \arctan(-1\sqrt{5})$

   (g) $2\sqrt{2}\left(\cos\dfrac{5\pi}{4} + i \sin\dfrac{5\pi}{4}\right)$ (h) $1\left(\cos\left(\dfrac{-\pi}{4}\right) + i \sin\left(\dfrac{-\pi}{4}\right)\right)$

   (i) $1(\cos \pi + i \sin \pi)$

7. The roots are $z_1 = (-b + \sqrt{b^2 - 4ac})/2a$ and $z_2 = (-b - \sqrt{b^2 - 4ac})/2a$. These are complex conjugates because $b^2 < 4ac$.

9. $|z| = 1$ iff $|z|^2 = 1$ iff $z\bar{z} = 1$ iff $\bar{z} = 1/z$

11. $|z + w|^2 - |z - w|^2 = |z|^2 + 2 \operatorname{Re}(z\bar{w}) + |w|^2 - (|z|^2 - 2 \operatorname{Re}(z\bar{w}) + |w|^2)$
    $= 4 \operatorname{Re}(z\bar{w})$

13. (a) Yes (b) No (c) Yes

15. The lengths squared of the legs are $|z|^2$, $|w|^2$, $|z - w|^2$. These are equal iff $\{|z|^2 = |w|^2 = |z - w|^2\}$ iff $\{|z|^2 = |w|^2 = |z|^2 - 2 \operatorname{Re}(z\bar{w}) + |w|^2\}$ iff $\{|z|^2 = |w|^2 = 2 \operatorname{Re}(z\bar{w})\}$.

17. $\left|\dfrac{z}{w}\right|^2 = \left|\dfrac{(xs + yt) + i(ys - xt)}{s^2 + t^2}\right|^2 = \dfrac{(xs + yt)^2 + (ys - xt)^2}{(s^2 + t^2)^2}$

   $= \dfrac{x^2 s^2 + y^2 t^2 + y^2 s^2 + x^2 t^2}{(s^2 + t^2)^2} = \dfrac{(x^2 + y^2)(s^2 + t^2)}{(s^2 + t^2)^2} = \dfrac{x^2 + y^2}{s^2 + t^2}$

   $= \dfrac{|z|^2}{|w|^2}.$

19. $\cos n\theta + i \sin n\theta = (\cos \theta + i \sin \theta)^n = \sum_{k=0}^{n} \binom{n}{k}(\cos \theta)^{n-k}(i \sin \theta)^k$

For k even, there is no factor of i; for k odd there is a factor of i.

21. For all complex λ,

$$0 \le \sum_{j=1}^{n} |a_j - \lambda b_j|^2 = \sum_{1}^{n} |a_j|^2 - 2 \text{ Re } \bar{\lambda} \sum_{1}^{n} a_j \bar{b}_j + |\lambda|^2 \sum_{1}^{n} |b_j|^2$$
$$= B - 2 \text{ Re } \bar{\lambda}A + |\lambda|^2 C.$$

Apply the results of Exercise 20.

23. Equality iff $|a_j - \lambda b_j|^2 = 0$ for some λ and for all j iff $a_j = \lambda b_j$, $j = 1, \ldots, n$.

Section 1.1.1

1. $(-z)^2 = ((-x)^2 - (-y)^2, 2(-x)(-y)) = (x^2 - y^2, 2xy) = z^2$.

3. $(0, 1) = z^2 = (x^2 - y^2, 2xy)$ so $x^2 = y^2$ and $2xy = 1$. If $x = -y$, then $-2x^2 = 1$, which is impossible. If $x = y$, then $2x^2 = 1$. Hence, $x = y = \pm 1/\sqrt{2}$.

7. $z_1 z_2 = (x_1 x_2 - y_1 y_2, x_1 y_2 + x_2 y_1)$. Hence, $|z_1 z_2|^2 = (x_1 x_2 - y_1 y_2)^2 + (x_1 y_2 + x_2 y_1)^2$
$= x_1^2 x_2^2 + y_1^2 y_2^2 + x_1^2 y_2^2 + x_2^2 y_1^2 = (x_1^2 + y_1^2)(x_2^2 + y_2^2) = |z_1|^2 |z_2|^2$.

9. If $z_1 z_2 = 0$, then $0 = |z_1 z_2| = |z_1| \, |z_2|$. Hence, $|z_1| = 0$ or $|z_2| = 0$.

Section 1.2

1. The y-axis

3. The line $y = 4x + 6$

5. The ellipse $1 = \dfrac{4x^2}{25} + \dfrac{4y^2}{9}$

7. The hyperbola $4 = x^2 - y^2$

9. The points $z = \pm 1$

11. $|z - (4 + i)| = 2$

13. $\text{Re}(iz - 3) = 0$

15. $\left| z - \left(\dfrac{1 + i}{2}\right) \right| = 1/\sqrt{2}$

17. $\text{Re}((2 - i)z - 1) = 0$

19. $|z - p| = cx$ iff $x^2 + y^2 - 2px + p^2 = c^2 x^2$

(a) $0 < c < 1$ gives $\left(x - \dfrac{p}{1 - c^2}\right)^2 + \dfrac{y^2}{1 - c^2} = \dfrac{p^2 c^2}{(1 - c^2)^2}$, an ellipse

(b) $c = 1$ gives $2px = y^2 + p^2$, a parabola

(c) $1 < c < \infty$ gives $\left(x + \dfrac{p}{c^2 - 1}\right)^2 = p^2\left[1 + \dfrac{1}{(c^2 - 1)^2}\right] + \dfrac{1}{c^2 - 1} y^2$, a hyperbola

21. (a) $|z - \alpha| < |1 - \bar{\alpha}z|$ iff $|z - \alpha|^2 < |1 - \bar{\alpha}z|^2$ iff $|z|^2 + |\alpha|^2 < 1 + |\alpha|^2|z|^2$ iff

$|\alpha|^2(1-|z|^2) < 1-|z|^2$ iff $1-|z|^2 > 0$ since $0 < |\alpha| < 1$. The others are similar.

23. $z_k = \exp\left(i\left[\dfrac{\pi + 4k\pi}{10}\right]\right)$, $k = 0, 1, 2, 3, 4$

25. $z_k = -1 + (2)^{1/8}\exp\left(i\left[\dfrac{-\pi + 8k\pi}{16}\right]\right)$, $k = 0, 1, 2, 3$

27. $z_k = \exp\left(i\left[\dfrac{\pi + 2k\pi}{8}\right]\right)$, $k = 0, 1, \ldots, 7$

29. $z_k = 2\exp\left(i\left[\dfrac{2k\pi}{3}\right]\right)$, $k = 0, 1, 2$

31.

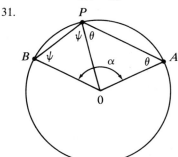

$\alpha = \pi - 2\theta + \pi - 2\psi = 2\pi - 2(\theta + \psi)$.
Hence, $\theta + \psi = \pi - \alpha/2$, a quantity independent of P.

33. The translation by β of C is the circle centered at $z_0 + \beta$ with the same radius. If the line $\mathscr{L}$ is $\{z: \text{Re}(Az + B) = 0\}$, then the translation of $\mathscr{L}$ by β is the line $\{w: \text{Re}(Aw + B - A\beta) = 0\}$.

35. $L = \{x + ia: -\infty < x < \infty\}$; $\left|\dfrac{1}{x + ia} + \dfrac{i}{2a}\right| = \left|\dfrac{2a + ix - a}{(2a)(x + ia)}\right| = \dfrac{1}{2a}$

37. $C = \{z: |z - c| = r\} = \{c + re^{i\theta}: 0 \le \theta < 2\pi\}$.

$$\left|\dfrac{1}{z} - \dfrac{c}{c^2 - r^2}\right| = \left|\dfrac{1}{c + re^{i\theta}} - \dfrac{c}{c^2 - r^2}\right| = \left|\dfrac{c^2 - r^2 - c(c + re^{i\theta})}{(c^2 - r^2)(c + re^{i\theta})}\right| = \dfrac{r}{c^2 - r^2}$$

Section 1.3

1. (a) interior $= \{x > 2 \text{ and } y < 4\}$;
    boundary $= \{x + 4i: 2 \le x < \infty\} \cup \{2 + iy: -\infty < y \le 4\}$
   (b) closed   (c) interior is connected

3. (a) interior $= \{x^2 < y\}$; boundary $= \{x^2 = y\}$
   (b) open   (c) interior is connected

5. (a) interior $= \{x^2 + y^2 > 2\}$; boundary $= \{x^2 + y^2 = 2\}$
   (b) closed   (c) interior is connected

7. (a) interior $= \{x < 0 \text{ and } |z + 1| > 1\}$;
    boundary $= \{iy: -\infty < y < \infty\} \cup \{|z + 1| = 1\}$
   (b) neither open nor closed   (c) interior is connected

9. Let $\alpha = |\alpha|e^{i\theta_0}$

   (a) $\{\alpha z + \beta: z \text{ in 1st quadrant}\} = \left\{\beta + se^{i\psi}: 0 < s < \infty,\ \theta_0 < \psi < \theta_0 + \dfrac{\pi}{2}\right\}$

   (b) $\{\beta + se^{i\psi}: 0 < s < \infty,\ \theta_0 < \psi < \theta_0 + \pi\}$

   (c) $\{\beta + se^{i\psi}: 0 \leqslant s < R|\alpha|;\ 0 \leqslant \psi \leqslant 2\pi\}$

11. A, C, D, E, G, and H are unbounded

13. Suppose that D_1, D_2 are open. Let $p \in D_1 \cap D_2$. There is an $r_1 > 0$ such that $\{|z - p| < r_1\} \subset D_1$; likewise, there is an $r_2 > 0$ such that $\{|z - p| < r_2\} \subset D_2$. Let $r = $ minimum of r_1, r_2. Then $\{|z - p| < r\}$ lies inside both D_1 and D_2. Hence, $D_1 \cap D_2$ is open. Next, let C_1, C_2 be closed. Then $\mathbb{C}\backslash(C_1 \cup C_2) = (\mathbb{C}\backslash C_1) \cap (\mathbb{C}\backslash C_2)$ and this is open, being the intersection of two open sets. Hence, $C_1 \cup C_2$ is closed. Likewise, $\mathbb{C}\backslash(C_1 \cap C_2) = (\mathbb{C}\backslash C_1) \cup (\mathbb{C}\backslash C_2)$ and this is open being the union of two open sets. Hence, $C_1 \cap C_2$ is closed.

17. The ray $\{tz_0: 0 \leqslant t < \infty\}$ is closed so D is open. Any two points p and q in D can be joined by the union of one, two, or three line segments, in turn parallel or perpendicular to the ray.

19. (a) Star-shaped with respect to 0   (b) not star-shaped   (c) not star-shaped
    (d) star-shaped with respect to any x in $(1, \infty)$

21. If $p' \in C$ and $|z_0 - p'| \leqslant r$, then the segment $\overrightarrow{pp'}$ lies in C and also in the set $\{|z_0 - z| < r\}$. Thus there is a point of C closer to z_0 than p, a contradiction.

25. Use the definition of the boundary B of D to show that the complement of B is open.

27. Use 25 and the definition of the boundary.

29. Use the definition of "interior point" of a set.

Section 1.4

1. 0   3. 0   5. No limit exists.

7. 0   9. 2   11. 1 at $z_0 = 8$; no limit exists at $8 + i$.

13. 0   15. Continuous everywhere.

17. Discontinuous only at all points of the line $y = x$.

19. Discontinuous only at all points of the circle $|z| = 1$, except $z = 1$.

21. 0   23. 4

25. f, g continuous at z_0 means $\lim_{z \to z_0} f(z) = f(z_0)$, $\lim_{z \to z_0} g(z) = g(z_0)$. Hence, $\lim_{z \to z_0} (f(z) + g(z)) = f(z_0) + g(z_0)$ since the limit of the sum is the sum of the limits. Likewise, fg and $1/g$ are continuous at z_0.

27. Let $w = f(z)$; then $\lim_{z \to z_0} f(z) = f(z_0) = w_0$ so that $\lim_{z \to z_0} g(f(z)) = \lim_{w \to w_0} g(w)$ $= g(w_0) = g(f(z_0))$.

29. $\lim_{z \to \infty} f(z) = L$ iff $|f(z) - L| < \varepsilon$ if $|z| \geqslant M$

    iff $|\tilde{f}(z) - L| < \varepsilon$ if $|z| \leqslant 1/M$

    iff $\lim_{z \to 0} \tilde{f}(z) = L$

31. Converges since $\left|\dfrac{1+2i}{6}\right| = \dfrac{\sqrt{5}}{6} < 1.$

33. Diverges since $\left|\dfrac{2+i}{\sqrt{5}}\right| = 1$ so terms do not go to 0.

35. Converges (ratio test).

37. All converge by ratio test.

39. Follow the hints.

Section 1.5

1. $\dfrac{1+i}{\sqrt{2}}$ 3. $\log 2 + i\left(\dfrac{\pi}{3} + 2k\pi\right)$, $k = 0, \pm 1, \ldots$

5. $\exp\left(-\dfrac{\pi}{4} + 2k\pi\right)[\cos(\log\sqrt{2}) + i\sin(\log\sqrt{2})]$, $k = 0, \pm 1, \pm 2, \ldots$

7. $\dfrac{1}{2} - i\dfrac{\sqrt{3}}{2}$ 9. $\log 4\sqrt{2} - i\pi/4$

11. $\cos(\pi\sqrt{3}[\tfrac{1}{2} + 2k] + i\sin(\pi\sqrt{3}[\tfrac{1}{2} + 2k])$, $k = 0, \pm 1, \ldots$

13. $4\left\{\log\sqrt{2} + i\left(-\dfrac{\pi}{4} + 2k\pi\right)\right\}$, $k = 0, \pm 1, \ldots$

15. $\log(e^z) = w$ or $e^z = e^w$. So all solutions are $z = w + 2\pi ik$, $k = 0, \pm 1, \ldots$

17. $0 = \cos z = \cos x \cosh y - i \sin x \sinh y$ iff $[0 = \cos x \cosh y$ and $0 = \sin x \sinh y]$ iff $[0 = \cos x$ and $0 = \sin x \sinh y]$ iff $[x = \pi/2 + \pi n, n = 0, \pm 1, \ldots,$ and $\sinh y = 0]$ iff $[x = \pi/2 + \pi n, n = 0, \pm 1, \ldots,$ and $y = 0]$. Likewise, $\sin z = 0$ only if $z = n\pi, n = 0, \pm 1, \ldots$

19. $\cos(iy) = \tfrac{1}{2}(e^{i(iy)} + e^{-i(iy)}) = \tfrac{1}{2}(e^{-y} + e^{y}) \to \infty$ as $y \to \infty$. Furthermore, $|\cos(x + iy)| = |\tfrac{1}{2}(e^{i(x+iy)} + e^{-i(x+iy)})| \leqslant \tfrac{1}{2}(e^{-y} + e^{y}) \leqslant e^{y}$ if $y \geqslant 0$. Likewise, $|\sin(x + iy)| \leqslant e^{y}$ if $y \geqslant 0$.

23. On the line Im $z = \pi/2$, the function e^z has the values $e^{x+i\pi/2} = e^x e^{i\pi/2} = ie^x$ and e^x assumes all values in $(0, \infty)$. Likewise, e^z maps to line Im $z = -\pi/2$ onto the negative imaginary axis.

25. Follow the hint.

27. Use the principal branch of $\log z$.

29. Let $\zeta = -i\log(iz + \sqrt{1 - z^2})$; then

$$\sin \zeta = \dfrac{1}{2i}(e^{i\zeta} - e^{-i\zeta}) = \dfrac{1}{2i}(\exp(\log[iz + \sqrt{1-z^2}]) - \exp(-\log[iz + \sqrt{1-z^2}])$$

$$= \dfrac{1}{2i}\left(iz + \sqrt{1-z^2} - \dfrac{1}{iz + \sqrt{1-z^2}}\right) = \dfrac{1}{2i}\left[\dfrac{-z^2 + 2iz\sqrt{1-z^2} + 1 - z^2 - 1}{iz + \sqrt{1-z^2}}\right]$$

$$= \dfrac{1}{2i}\left[\dfrac{2iz\sqrt{1-z^2} - 2z^2}{iz + \sqrt{1-z^2}}\right] = \dfrac{z(\sqrt{1-z^2} + iz)}{iz + \sqrt{1-z^2}} = z.$$

Section 1.6

1. 0   3. $\frac{20}{3}(1 + i)$ 5. $\frac{1}{2}(i - 1)$

7. $\int_{\gamma_1} z^2 \, dz = -2/3; \int_{\gamma_2} z^2 \, dz = -2/3$. They are equal because $\gamma = \gamma_1 - \gamma_2$ is a closed curve and $\int_\gamma z^2 \, dz = 0$ by Example 10.

9. (a) For $k \neq 0$, $\displaystyle\int_0^{2\pi} e^{ik\theta} \, d\theta = \frac{1}{ik} e^{ik\theta}\Big|_{\theta=0}^{\theta=2\pi} = \frac{1}{ik}[1 - 1] = 0.$

11. $\displaystyle\int_\gamma f(x) \, dz = \int_\gamma (u + iv)(dx + i \, dy) = \int_\gamma u \, dx - v \, dy + i \int_\gamma v \, dx + u \, dy.$

    Also

    $$\iint_\Omega \left(\frac{\partial f}{\partial x} + i \frac{\partial f}{\partial y}\right) dx \, dy = \iint_\Omega \left(\frac{\partial u}{\partial x} + i \frac{\partial v}{\partial x} + i\left[\frac{\partial u}{\partial y} + i \frac{\partial v}{\partial y}\right]\right) dx \, dy$$
    $$= \iint_\Omega \left(\frac{\partial u}{\partial x} - \frac{\partial v}{\partial y}\right) dx \, dy + i \iint_\Omega \left(\frac{\partial v}{\partial x} + \frac{\partial u}{\partial y}\right) dx \, dy.$$

13. In (8) replace u by $-v \dfrac{\partial u}{\partial y}$ and replace v by $v \dfrac{\partial u}{\partial x}$.

15. Establish the estimate: $\left| \displaystyle\int_{\gamma_R} \frac{u(z)}{(z - z_0)^2} \, dz \right| \leqslant \dfrac{C}{(R - |z_0|)^2} \, 2\pi R$ by using (3).

17. $\displaystyle\int_\gamma P \, dx + Q \, dy = \int_\gamma \frac{\partial g}{\partial x} \, dx + \frac{\partial g}{\partial y} \, dy = \int_a^b \frac{d}{dt}[g(\gamma(t))] \, dt = g(\gamma(b)) - g(\gamma(a)) = 0.$

21. $F(z) = \dfrac{1}{m + 1} z^{m+1}$, $m = 0, 1, 2, \ldots$.

Chapter 2

Section 2.1

1. (a)–(d) follow from (6) and the definition of these functions; (e) follows from (a), (b), and (4); (f) and (g) follow from the chain rule, (5).

3. $(3z^2 - 1)\exp(z^3 - z)$ 5. $-12z^2(z^3 + 100)^{-5}$

7. $e^z \cosh(e^z)$ 9. $-\dfrac{1}{z} + \dfrac{1}{3} z^3.$

11. $\frac{1}{2} \sinh(2z).$

13. $f(z_0 + h) = f(z_0) + hf'(z_0) + hE_1$, $E_1 \to 0$ as $|h| \to 0$
    $g(w_0 + p) = g(w_0) + pg'(w_0) + pE_2$, $E_2 \to 0$ as $|p| \to 0$; $w_0 = f(z_0)$

    Hence,
    $$g(f(z_0 + h)) = g(f(z_0) + h(f'(z_0) + E_1)) = g(f(z_0)) + h(f'(z_0) + E_1)[g'(w_0) + E_2]$$
    $$= g(f(z_0)) + hf'(z_0)g'(w_0) + hE_1 g'(w_0) + hf'(z_0)E_2 + hE_1 E_2$$
    $$= g(f(z_0)) + hf'(z_0)g'(w_0) + hE_3, \quad E_3 \to 0 \text{ as } h \to 0.$$

15. $0 = f'$ on D so $\dfrac{\partial u}{\partial x} = \dfrac{\partial u}{\partial y} = \dfrac{\partial v}{\partial x} = \dfrac{\partial v}{\partial y}$ at all points in D.

As in Theorem 2, f is constant on D.

17. Follow the hint.

21. $\displaystyle\int_\gamma F'(z)\,dz = \int_a^b F'(\gamma(t))\gamma'(t)\,dt = \int_a^b \dfrac{d}{dt} F(\gamma(t))\,dt = F(\gamma(b)) - F(\gamma(a)) = 0.$

23. The chain rule.

Section 2.1.1

1. Globally sourceless and irrotational; $G(z) = \frac{1}{4}z^4$.

3. Globally sourceless and irrotational; $G(z) = \sin z$.

5. Globally sourceless and irrotational; $G(z) = \frac{1}{3}z^3$.

7. Neither sourceless nor irrotational.

9. Use Exercise 21 in Section 1.

Section 2.2

1. $R = 1$ 3. $R = 2^{1/3}$ 5. $R = 1$

7. $\displaystyle\sum_0^\infty \frac{(-1)^n}{n!} z^n$ 9. $\displaystyle\sum_{n=1}^\infty z^{3n}$ 11. $\displaystyle\sum_{n=1}^\infty \frac{n}{4^{n+1}} z^{n+1}$

13. $z^2 \displaystyle\sum_{n=0}^\infty z^{2n}\left(\sum_{j=0}^n (-1)^j \frac{1}{(2j+1)!} \frac{1}{(2n+1-2j)!}\right)$

15. $\dfrac{z^3}{1-z^3}$ 17. e^{-z}

19. Let $|z - z_0| = s < R$; then $\displaystyle\sum_0^\infty \left|\frac{a_n}{n+1}\right| |z - z_0|^{n+1} \leqslant s \sum_0^\infty |a_n| s^n$,

which is finite by hypothesis.

21. (a) Follow the hint   (b) $(n+1)\rho^{n+1}/n\rho = ((n+1)/n)\rho < \frac{1}{2}(1+\rho)$ when $n > 2\rho/(1-\rho)$. Hence, $\{n\rho^n\}$ decreases geometrically fast once $n > 2\rho/(1-\rho)$.   (c) If $1 \leqslant M$, then $1 \leqslant M^{1/n} \leqslant n^{1/n}$ once $n \geqslant M$. But $n^{1/n} \to 1$ by (a). If $0 < M < 1$, then $1 \leqslant (1/M)^{1/n} \leqslant n^{1/n}$, n large.

23. Use the ratio or root test on the given series.

25. Both $\sum |a_n| s^n$ and $\sum |b_k| s^k$ converge, so their terms are bounded.

29. Follow the hint.

31. By Exercise 30, $\left|\displaystyle\sum_{n=N+1}^\infty z^n\left(\sum_{k=0}^N a_k b_{n-k}\right)\right| < \varepsilon$ if $N \geqslant N_0$. Hence,

$$\left| f(z)g(z) - \left(\sum_0^N a_k z^k\right)\left(\sum_0^\infty b_k z^k\right)\right| < \varepsilon \text{ if } N \geqslant N_1,$$

and

$$\left| \left(\sum_0^N a_k z^k \right) \left(\sum_0^\infty b_k z^k \right) - \sum_0^N c_n z^n \right| < \varepsilon \text{ if } N \geqslant N_0 .$$

33. The steps are outlined; just recall the definition of the radius of convergence.

35. By the results of Exercise 34 $\displaystyle\lim_{k \to \infty} \left| \dfrac{\binom{\alpha}{k+1}}{\binom{\alpha}{k}} \right| = 1$. Hence, $R = 1$ by Theorem 2.

Section 2.3

1. 0

3. $-2\pi i$

5. $2\pi/\sqrt{3}$

7. $2\pi/\sqrt{7}$

9. i

11. $e - 1/e$

13. By Cauchy's Theorem,

$$0 = \int_\gamma e^{iz^2} \, dz = \int_0^R e^{ix^2} \, dx + \int_0^{\pi/4} e^{i(Re^{i\theta})^2} i R e^{i\theta} \, d\theta + \int_R^0 e^{i(t(1 + i/\sqrt{2}))^2} \frac{1 + i}{\sqrt{2}} \, dt .$$

The middle integral is

$$\int_0^{\pi/4} \exp[iR^2(\cos 2\theta + i \sin 2\theta)]ie^{i\theta} R \, d\theta.$$

This is estimated above by

$$R \int_0^{\pi/4} \exp[-R^2 \sin 2\theta] \, d\theta \leqslant R \int_0^{\pi/4} \exp[-2R^2\theta/\pi] \, d\theta < \frac{\pi}{2R} .$$

Let $R \to \infty$; we obtain

$$0 = \int_0^\infty [\cos(x^2) + i \sin(x^2)] \, dx - \frac{1 + i}{\sqrt{2}} \int_0^\infty e^{-t^2} \, dt .$$

Use Exercise 20 and separate real and imaginary parts.

19. Suppose that G is analytic in $r < |z| < R$ with $G'(z) = 1/z$. Fix s, $r < s < R$, and compute

$$\int_{|z|=s} G'(z) \, dz = \int_0^{2\pi} G'(se^{i\theta})sie^{i\theta} \, d\theta = G(se^{2\pi i}) - G(se^0) = 0.$$

But

$$\int_{|z|=s} \frac{1}{z} \, dz = \int_0^{2\pi} \frac{sie^{i\theta}}{se^{i\theta}} \, d\theta = 2\pi i.$$

21. (c), (d) Cauchy's Theorem and formula.

Section 2.4

1. Zeros of order 1 at $n\pi$, $n = \pm 1, \pm 2, \dots$.

3. Zeros of order 2 at -2 and 1.

5. A zero of order 4 at 0; zeros of order 2 at $2\pi n$, $n = \pm 1, \pm 2, \dots$.

7. Zeros of order 1 at $\pi i(1 + 2\pi n)$, $n = 0, \pm 1, \ldots$, and at $\log 4 + 2\pi i m$, $m = 0, \pm 1, \ldots$.

9. $\sum_1^\infty \dfrac{z^{k+1}}{k!}$; valid everywhere.

11. $11(z - 1) + 9(z - 1)^2 + (z - 1)^3$.

13. $\sum_0^\infty (-1)^k 2^{-k-1}(z + 1)^k$.

15. $\sum_1^\infty (-1)^j \pi^{2j}[(2j)!]^{-1}(z - \tfrac{1}{2})^{2j}$.

17. $f(z) = (z - z_0)^m g(z)$; $g(z_0) \neq 0$. Thus, $f'(z) = (z - z_0)^{m-1}[mg(z) + (z - z_0)g'(z)] = (z - z_0)^{m-1}h(z)$; $h(z_0) = mg(z_0) \neq 0$. Likewise, $f^2(z) = (z - z_0)^{2m}g^2(z)$; $g^2(z_0) \neq 0$.

19. $|f'(z_0)| \leqslant 1! \, \dfrac{M}{s}$ if $|f(z)| \leqslant M$ for all z; let $s \to \infty$.

21. For $n > m$, $|f^{(n)}(z_0)| \leqslant \dfrac{n!}{s^n} A s^m$ if $|z| = s \geqslant R_0$; let $s \to \infty$.

23. Write $G(z) = \sum_0^\infty c_k z^k$ for $|z| < 1/R$. Set $w = 1/z$; then

$$F(w) = G(z) = \sum_0^\infty c_k z^k = \sum_0^\infty c_k \frac{1}{w^k}, \ |w| > R.$$

Furthermore, if $0 < r < 1/R$, then

$$c_k = \frac{1}{2\pi i} \int_{|\zeta| = r} \frac{G(\zeta)}{\zeta^{k+1}} \, d\zeta. \ \text{Let } \zeta = 1/\xi, \ |\xi| = s = 1/r > R; \text{ therefore,}$$

$$= \frac{1}{2\pi i} \int_{|\xi| = s} F(\xi)\xi^{k-1} \, d\xi, \ k = 0, 1, 2, \ldots.$$

25. Write $f(z) = \sum_0^\infty a_k z^k$; substitute into the differential equation and collect equal powers of z. This gives equations for $a_2, a_3, \ldots$, in terms of a_0 and a_1:

$$a_{2k} = \frac{(-1)^k}{(2k)!} a_0 \beta^{2k} \quad \text{and} \quad a_{2k+1} = \frac{(-1)^k}{(2k+1)!} a_1 \beta^{2k}.$$

27. Set $g(z) = \exp[-\int_{z_0}^z A(w) \, dw]$. Then

$$(fg)'(z) = f'(z)g(z) + f(z)g'(z) = A(z)f(z)g(z) - A(z)g(z)f(z) = 0.$$

Hence, $f = C/g = C \exp[\int_{z_0}^z A(w) \, dw]$.

29. Compute J'_ν, J''_ν, substitute into the given equation, and compute.

Section 2.5

1. A removable singularity at 0; the value at 0 is 1.

3. A removable singularity at 1; the value at 1 is 4.

5. A pole of order 1 at -2.

7. $\dfrac{1}{z} + \dfrac{1}{2} + \dfrac{1}{6} z + \sum_2^\infty \dfrac{1}{(k + 2)!} z^k$; $\operatorname{Res}(f; 0) = 1$.

9. $\dfrac{-1}{z - \pi} + \sum_0^\infty (-1)^{k+1} \dfrac{(z - \pi)^{2k+1}}{(2k + 1)!}$; $\operatorname{Res}(f; \pi) = -1$.

11. $\dfrac{bc - ad}{c^2} \dfrac{1}{z + d/c} + \dfrac{a}{c}$; $\text{Res}(f; -d/c) = \dfrac{bc - ad}{c^2}$ (all if $c \neq 0$).

13. $\dfrac{2}{z^2} + \dfrac{1}{6} + \dfrac{1}{120} z^2 + \cdots$; $\text{Res}(f; 0) = 0$.

15. $f(z) = (z - z_0)^{-l} g(z)$, $l \geqslant 1$. Thus, $f'(z) = (z - z_0)^{-l} [g'(z) - l(z - z_0)^{-1} g(z)]$

   and $\dfrac{f'(z)}{f(z)} = \dfrac{g'(z)}{g(z)} + \dfrac{-l}{z - z_0}$.

17. Follow the hint: if g is bounded near z_0, then z_0 is a removable singularity for g; let $A = g(z_0)$. Next, $f(z) = w + 1/g(z)$. If g has a zero at z_0 ($A = 0$), then f has a pole at z_0. If $A \neq 0$, then f is bounded near z_0.

19. Let γ be the circle $|z - z_0| = s$, $0 < s < r$. Then for any integer m

$$0 = \frac{1}{2\pi i} \int_\gamma (z - z_0)^{-m} \left(\sum_{-\infty}^{\infty} a_n (z - z_0)^n \right) dz$$

$$= \frac{1}{2\pi} \int_0^{2\pi} s^{-m} e^{-im\theta} \left(\sum_{-\infty}^{\infty} a_n s^n e^{in\theta} \right) s e^{i\theta} \, d\theta$$

$$= \sum_{-\infty}^{\infty} a_n s^{-m} s^{n+1} \frac{1}{2\pi} \int_0^{2\pi} e^{i(n+1-m)\theta} \, d\theta = a_{m-1}.$$

21. By (5)

$$\text{Res}(F; z_0) = \frac{1}{2\pi i} \int_{|z - z_0| = s} F(z) \, dz = \frac{1}{2\pi i} \int_{|z - z_0| = s} G'(z) \, dz$$

$$= \frac{1}{2\pi} \int_0^{2\pi} G'(z_0 + s e^{i\theta}) s e^{i\theta} \, d\theta$$

$$= \frac{1}{2\pi} [G(z_0 + s e^{2\pi i}) - G(z_0 + s e^{i0})] = 0.$$

23. In $1 < |z| < 2$

(a) $\displaystyle\sum_{-\infty}^{\infty} a_n z^n$, $a_n = \begin{cases} (-1)^n/3, & n = -1, -2, \ldots \\ -4/3^{n+2}, & n = 0, 1, 2, \ldots \end{cases}$

   In $2 < |z| < \infty$, $\displaystyle\sum_1^\infty \dfrac{a_k}{z^k}$, $a_k = \dfrac{(-1)^k}{3} + \dfrac{4}{3^{k+2}}$, $k = 1, 2, \ldots$.

(b) In $0 \leqslant |z| < 1$, $\displaystyle\sum_0^\infty a_k z^k$, $a_k = \dfrac{1}{24}[15(-1)^k - 3^{-k+1}]$

   In $1 < |z| < 3$, $\displaystyle\sum_{-\infty}^{\infty} a_k z^k$, $a_k = \begin{cases} (5/8)(-1)^{k+1}, & k = -1, -2, \ldots \\ (3/8)(3^{-k-1})((-1)^k - 1), & k = 0, 1, 2, \ldots \end{cases}$

(c) In $2 < |z| < 3$, $\displaystyle\sum_{-\infty}^{\infty} a_k z^k$, $a_k = \begin{cases} 2^{-k-1}, & k = -1, -2, \ldots \\ \dfrac{4}{3^{k+1}} - \dfrac{1}{5^{k+1}}, & k = 0, 1, 2, \ldots \end{cases}$

   In $5 < |z| < \infty$, $\displaystyle\sum_1^\infty \dfrac{a_n}{z^n}$, $a_n = 2^{n-1} - 4(3)^{n-1} + 5^{n-1}$, $n = 1, 2, 3, \ldots$.

25. ∞ is removable for f iff 0 is removable for $g(z) = f(1/z)$ iff g is bounded near 0 iff f is bounded near ∞.

27. (a) pole of order 2 at ∞; $3z^2 + 4 - (1/z)$.

(b) pole of order 2 at ∞; $-z^2 + 5z - 4$.
(c) pole of order 1 at ∞; $z + 4 + (4^2/z) + (4^3/z^2) + \cdots$.
(d) zero of order 2 at ∞; $-(1/z^2) + (2/z^3) - (7/z^4) + \cdots$.
(e) removable singularity at ∞; value is 1; $1 + z^{-1} + z^{-2}/2! + z^{-3}/3! + \cdots$.
(f) essential singularity at ∞; $1 - z^2 + (z^4/2!) - (z^6/3!) + \cdots$.
(g) pole of order 3 at ∞; $z^3 + 3z + (3/z) + (1/z^3)$.
(h) removable singularity at ∞; value is 0; $z^{-1} - z^{-3}/3! + z^{-5}/5! - \cdots$.
(i) removable singularity at ∞; value is 1; $1 - z^{-2}/2! - z^{-4}/4! - \cdots = \cos(1/z)$.

29. First, $\overline{G(\bar{z}; u)} = \exp[u/2; \bar{z} - 1/\bar{z}] = \exp[u/2; z - 1/z] = G(z; u)$ if u is real. Hence, $J_n(u)$ is real if u is real. Apply (13) with $s = 1$. This yields

$$J_n(u) = \text{Re}(J_n(u)) = \text{Re}\left(\frac{1}{2\pi} \int_0^{2\pi} e^{i(u \sin \theta - n\theta)} \, d\theta\right)$$

$$= \frac{1}{2\pi} \int_0^{2\pi} \cos(u \sin \theta - n\theta) \, d\theta.$$

31. $\sum_{-\infty}^{\infty} J_n(u)z^n = G(z; u) = \exp\left[\frac{u}{2}(z - 1/z)\right] = G(-1/z; u) = \sum_{-\infty}^{\infty} J_n(u)\frac{(-1)^n}{z^n}$

$$= \sum_{-\infty}^{\infty} J_{-n}(u)(-1)^n z^n. \text{ So } J_{-n}(u) = (-1)^n J_n(u).$$

Section 2.6

1. $\frac{\pi}{4}\left[\sin\left(\frac{3\pi}{8}\right)\right]^{-1}$

3. $\pi/ab(a + b)$

5. $\pi \exp(-\sqrt{2}/2)\sin(\sqrt{2}/2)$

7. $\frac{\pi}{\beta} e^{-\beta} \cos \alpha$

9. $4\pi/3\sqrt{3}$

11. $2\pi a^2/(1 - a^2)$

13. $\pi[2^\alpha - 1]/\sin \pi\alpha$

15. $\frac{\pi}{\sin \omega} \frac{\sin \omega\lambda}{\sin \pi\lambda}$

17. 0

19. $(5/32)\pi^5$

21. Follow the hint.

23. Use the technique of Examples 1 and 2; the poles in U have order 2 (for (a)) or 3 (for (b)).

25. Use the Laurent series to write $g = h + \sum_1^N P_j$, where P_j is analytic on $\{z: |z - z_j| > 0\}$ and h is analytic in D. Now proceed as in the proof of (1).

27. $C(z) = \pi \cot \pi z = \pi \frac{\cos \pi z}{\sin \pi z}$ has a pole of order 1 at each integer. Furthermore,

$$\text{Res}(C; k) = \lim_{z \to k} (z - k)\pi \frac{\cos \pi z}{\sin \pi z} = \lim_{z \to k}(\cos \pi z)\frac{\pi}{\frac{\sin \pi z}{z - k}} = \pi \frac{\cos \pi k}{(\sin \pi z)'(k)} = 1.$$

29. $\left|\int_{\gamma_N} C(z)f(z) \, dz\right| \leqslant (A/N^2) \text{ length } (\gamma_N) \leqslant \frac{B}{N} \to 0 \text{ as } N \to \infty.$

Now apply the residue theorem.

31. (a) $\dfrac{\pi \cosh \pi a}{a \sinh \pi a}$ (b) $\left(\dfrac{\pi}{\sin \pi a}\right)^2$

(c) $\pi^2/8$ (d) $\pi^4/2160$

33. In 32, the left-hand side goes to zero as $N \to \infty$; write the sum over $k = 1, \ldots, N$.

Chapter 3

Section 3.1

1. One zero in first quadrant.

3. One zero in first quadrant.

5. Two zeros in first quadrant.

7. Two zeros in the upper half-plane.

9. On the imaginary axis $f(z) = z^4 - 5z^2 + 3 - e^{-z}$ has no solution since $|e^{-iy}| = 1$ while $y^4 + 5y^2 + 3 \geqslant 3$. Let $g(z) = z^4 - 5z^2 + 3$. Then $|f - g| < |g|$ on the contour in Figure 3.5. g has two zeros in the right half-plane.

11. Set $G(z) = 1 - \exp(2\pi i/z)$; G is analytic for all z, $z \neq 0$. However, $\lim_{z\to 0} G(z)$ does not exist.

13. $z^4 - 2z - 2$ has four zeros in $1/2 < |z| < 3/2$.

15. $4z^3 - 12z^2 + 2z + 10$ has two zeros in $1/2 < |z - 1| < 2$.

17. (a) Three solutions   (b) no solutions   (c) one solution

19. The polar form of the Cauchy–Riemann equations:

$$r\frac{\partial u}{\partial r} = \frac{\partial v}{\partial \theta} \quad \text{and} \quad r\frac{\partial v}{\partial r} = -\frac{\partial u}{\partial \theta}.$$

Use this near the circle $|z| = 1$ with $u = \log|f|$, $v = \arg f$.

21. $p - q$ is a polynomial of degree n with $n + 1$ zeros.

23. The hint shows how.

25. g has no poles so it is entire.

27. The hint has it.

29. $\dfrac{P(z)}{Q(z)} z^{M-N} = \dfrac{a_0/z^N + \cdots + a_N}{b_0/z^M + \cdots + b_M} \to \dfrac{a_N}{b_M} \neq 0$ as $|z| \to \infty$.

31. $R(z) - \alpha = \dfrac{P(z) - \alpha Q(z)}{Q(z)}$.

Now $\deg(P - \alpha Q) = \max(N, M)$ except if $M = N$ and $\alpha b_N = a_N$. In this case $\deg(P - \alpha Q) < N = \deg Q = \max(N, N)$.

33. Use Cauchy's formula: let γ be a contour surrounding t. Then

$$\frac{1}{2^n}\frac{1}{2\pi i}\int_\gamma \frac{(z^2 - 1)^n}{(z - t)^{n+1}}\,dz = \frac{1}{2^n}\frac{1}{n!}\frac{d^n}{dz^n}(z^2 - 1)^n\Big|_{z=t} = P_n(t).$$

35. Let r be the largest integer that is $\leqslant n/2$. Then

$$P_n(t) = \frac{1}{2^n} \sum_{k=r}^{n} (-1)^k \frac{(2k)!}{(2k-n)!k!(n-k)!} t^{2k-n}$$

is an explicit formula for P_n. Substitute this in and compute.

37. (a) Follow the hint.
    (b) Use the definition, (8), and substitute.

Section 3.1.1

1. (a) Not stable   (b) Stable   (c) Not stable
   (d) Not stable   (e) Not stable   (f) Stable
   (g) Not stable   (h) Not stable   (i) Stable
   (j) Not stable   (k) Not stable   (l) Stable

3. (a) Stable   (b) Stable   (c) Not stable
   (d) Not stable   (e) Stable   (f) Not stable

7. Check Theorem 2, the argument principle.

Section 3.2

1. 1   3. 1/3

5. S is not continuous at $z = 1$.

7. $0 = \text{Re } F(z_0) = \dfrac{1}{2\pi} \displaystyle\int_0^{2\pi} \text{Re } F(z_0 + re^{it}) \, dt.$

   Hence, Re F must be both positive and negative somewhere on the circle $|z - z_0| = r$.

9. Use minimum modulus.

11. Follow the hint.

13. The function $g(z) = f(z)/M$ is bounded by 1 in $|z| < 1$. Let $\alpha_0 = g(z_0)$ and $\psi(z) = (z - \alpha_0)(1 - \bar{\alpha}_0 z)^{-1}$. Then $h = \psi \circ g$ is also bounded by 1 and is zero at z_0. Use 12 with z_0 in place of α.

15. Follow the hint.

17. Let $h = g/f$ and use the maximum-modulus principle for h on γ. An example of failure: $f(z) = z$, $g(z) = 1$, and $\gamma = \{w : |w| = 1\}$.

19. $\tilde{p}(z) = \overline{(p(-\bar{z}))}^{\sim} = p(z)$. Let $Q(z) = \sum_0^n a_k z^k$; then $Q = \tilde{Q}$ if and only if a_k is real for k even and pure imaginary if k is add.

21. p stable implies that $\tilde{p}(z)/p(z)$ is analytic on Re $z > 0$ and continuous on Re $z \geqslant 0$. Apply maximum modulus to $\tilde{p}/p$ on the contour $\{Re^{it} : -\pi/2 \leqslant t \leqslant \pi/2\} \cup \{iy : R \geqslant y \geqslant -R\}$; let $R \to \infty$. Conversely, if $|\tilde{p}(z)| \leqslant |p(z)|$ for Re $z \geqslant 0$ and $\tilde{p}$ and p have no common zero, then p has no zero in $\{$Re $z \geqslant 0\}$.

23. p stable implies $r = \tilde{p}/p$ maps $\{$Re $z > 0\}$ into $\{w : |w| < 1\}$. This implies $R = (1 - r)/(1 + r)$ maps $\{$Re $z > 0\}$ into $\{$Re $\zeta > 0\}$, so that R is positive. The converse is similar; use the results of Exercise 21.

Section 3.3

1. Compute

3. Compute

5. (a) $\lambda(1-i)\dfrac{z+1}{z-1}$, λ any real number

   (b) $z+1$

   (c) $iz+\frac{1}{2}$

   (d) $(z-z_0)/r$

   (e) $4i-4+4/z$

7. (a) $(1-i)z/(z-i)$

   (b) $1/(z+1)$

   (c) $(z+4)/(2z+2)$

   (d) $(z-1)/(z+1)$

9. (a) Fixed points at 0, -1; sink at 0; source at -1.
   (b) Fixed points at ± 2; sink at 2; source at -2.
   (c) Fixed points at 1, ∞; source at 1; sink at ∞; straight-line motion outward from $z=1$.
   (d) Fixed points at 0, ∞; source at ∞; sink at 0; inward clockwise spiral to origin.
   (e) Translation to the right; dipole at ∞.

11. Solve $T(1)=\dfrac{a+b}{c+d}=1$ and $T(-1)=\dfrac{-a+b}{-c+d}=-1$.

13. Use the results of Exercise 11.

15. Write $S(z)=\dfrac{az+b}{cz+d}$ and solve $z=S(S(z))$ for a, b, c, and d:

$$S(z)=\frac{az+b}{cz-a}, \text{ where } a^2+bc\neq 0.$$

17. Follow the [extensive] hints.

19. (a) and (c) are clear from the definition;   (b) if L is the real axis, then $d=|\operatorname{Im} z|$.

23. A reflection over a line: $z^*=\lambda\bar z+b$, $|\lambda|=1$; a reflection over a circle: $w^*=a+R^2/(\bar w-\bar a)$. Putting two such transforms together produces $Az+B$ or $(Az+B)/(Cz+D)$.

Section 3.4

1. $z^2=w^2$ iff $0=z^2-w^2=(z+w)(z-w)$. If $z+w=0$, then $\operatorname{Re} z=-\operatorname{Re} w$, which cannot happen. If U is open and strictly bigger than $\{\operatorname{Re} z>0\}$ then there is a $z_0\in U$ with $\operatorname{Re} z_0<0$. Hence, $-z_0\in\{\operatorname{Re} z>0\}$ and $(-z_0)^2=z_0^2$.

3. Let $\alpha_0=f(0)$; by Theorem 1, Section 2, there is a $\delta>0$ such that the disc $\{w:|w-\alpha_0|<\delta\}$ is a subset of $\{f(z):|z|<1\}$. Hence, $|f(z)-\alpha_0|\geqslant\delta$ if $|z|\geqslant 1$. (Why?) Show that $(f-\alpha_0)^{-1}$ has a removable singularity at ∞. Conclude that f has a pole at ∞ of order m, $m\geqslant 1$, and so is a polynomial of degree m. Then deduce that $m=1$.

5. (a) $K(z)=K(w)$ iff $z(1-w)^2=w(1-z)^2$ iff $z+zw^2=w+wz^2$ iff
   $z-w=zw(z-w)$ iff $z=w$ (when both $|z|<1$ and $|w|<1$.)
   (b) For $z=e^{i\theta}$, $K(e^{i\theta})=-[4\sin^2(\theta/2)]^{-1}$.
   Conclude that the range of K on $|z|<1$ is all w with w not in $(-\infty,-1/4]$.

7. (a) See Figure 3.16 with $1/2$, 2 in place of -1, 1, respectively.

(b) See Figure 3.19 translated over one unit.

(c) This is a dipole at 1 with circles centered on the real axis and on the line Re $z = 1$.

(d) See Figure 3.15 moved over and up one unit: $z^2 - 2z = (z - 1)^2 - 1$.

9. Write $p(z) = \Pi_1'(z - z_k)^{m_k}$; $\gamma_\varepsilon = \{z: \Pi_1' |z - z_k|^{m_k} = \varepsilon\}$. For the product to be small, one factor must be small so z is near some z_j (and so the other factors are bounded). For the product to be large, one factor is large (and so all the factors are large.) Let $n = $ degree p. Then $|p(z)|/|z|^n \to 1$ as $R = |z| \to \infty$ so γ_ε is "almost" the circle $|z| = R$ if R is big.

11. Follow the suggested route.

13. $g(z) = z + e^z$ so $g'(z) = 1 + e^z$ and $\text{Im}[g'(z)] = e^x \sin y > 0$ if $0 < y < \pi$. Now apply the results of Exercise 12 with $f = -ig$ to see that g is $1 - 1$.

Section 3.4.1

1. $G(iy) = -\frac{1}{2}(y^2 + (1/y^2)) \in (-\infty, -1]$ if $1 \leqslant y < \infty$; $G(e^{i\theta}) = \cos 2\theta \in [-1, 1]$ if $\pi/2 \geqslant \theta \geqslant 0$; $G(x) = \frac{1}{2}(x^2 + (1/x^2)) \in [1, \infty)$ if $1 \leqslant x < \infty$. Hence, G maps the boundary of D onto the real axis. G is also $1 - 1$ since $G(z) = G(w)$ iff $z^2 - w^2 = z^{-2}w^{-2}(z^2 - w^2)$ iff $z^2 = w^2$ (since $|z| > 1$, $|w| > 1$) iff $z = w$ (since Re $z > 0$, Re $w > 0$). The streamlines are $\Gamma_c = \{z: z^2 = t + ic + i\sqrt{1 - (t + ic)^2}$, $-\infty < t < \infty\}$, $c > 0$.

3. Refer to Exercise 5, Section 4, for the properties of K. The streamlines are

$$\Gamma_c = \left\{z: z = \left(-\frac{t^2}{4} - i\frac{tc}{4}\right) + \frac{c^2 - 1}{4}, \quad -\infty < t < \infty\right\}, c > 0.$$

5. $G(z) = G(w)$ iff $(1 - z^2)^2 = (1 - w^2)^2$ iff $(z^2 - w^2)(z^2 + w^2) = 2(z^2 - w^2)$ iff either $z^2 = w^2$ or $z^2 + w^2 = 2$. But z and w are in the first quadrant so that $0 < \text{Im}(z^2 + w^2)$. Hence, $z^2 = w^2$, so $z = w$ since Re $z > 0$, Re $w > 0$. Next, $G(iy) = -(1 + y^2)^2 \in (-\infty, -1]$ if $0 \leqslant y < \infty$; $G(x) = -(1 - x^2)^2 \in [-1, 0]$ if $0 \leqslant x \leqslant 1$; $G(x + i\sqrt{x^2 - 1}) = 4x\sqrt{x^2 - 1} \in [0, \infty)$ if $1 \leqslant x < \infty$. Hence, G maps the boundary of D onto the real axis.

7. $G'(z) = \dfrac{2}{z^2 - 1} = \dfrac{1}{z - 1} - \dfrac{1}{z + 1}$;

see Figure 3.16. The streamlines are the circles $\arg[(z - 1)/(z + 1)] = c$.

9. See Figure 3.16 with p, q in place of -1, 1, respectively.

11. Set $\gamma(t) = H(u(t))$; then $G(\gamma(t)) = u(t) \in L$ so $\text{Im } G(\gamma(t)) = c_0$. Moreover, $\gamma'(t) = H'(u(t))u'(t) = f'(\gamma(t))$ so that γ parametrizes the curve.

Section 3.5

1. $\phi(z) = z^2$

3. $\phi(z) = i\dfrac{2 - i + z}{2 + i - z}$

5. $\phi(z) = \dfrac{1}{2R}\left(z + \dfrac{R^2}{z}\right)$

7. $\phi(z) = \dfrac{\pi}{2} + \arcsin(e^z)$

9. $\phi(z) = \dfrac{4}{\pi}(\text{Log } z) - 1$

11. $f(z) = \arcsin z$

13. $f(z) = -iS(z^2 - 1)^{3/4}$

15. $f(z) = \dfrac{-a}{\pi}\left[2\sqrt{z+1} + \text{Log}\left(\dfrac{\sqrt{z+1}-1}{\sqrt{z+1}+1}\right)\right]$

17. Write $f = u + iv$; the change-of-variables formula gives

$$\text{area }(D) = \int\int_D 1 \, du \, dv = \int\int_{|z|<1} \det\begin{pmatrix} u_x & u_y \\ v_x & v_y \end{pmatrix} dx \, dy \quad \text{(apply Cauchy–Riemann)}$$

$$= \int\int_{|z|<1} (u_x^2 + u_y^2) \, dx \, dy = \int\int_{|z|<1} |f'(z)|^2 \, dx \, dy.$$

19. If $\{|\phi(z_k)|\}$ does not converge to 1, then there is a subsequence $\{z_{k_i}\}$ with $\phi(z_{k_i}) \to q, |q| < 1$. This contradicts Exercise 14, Section 4.

Chapter 4

Section 4.1

1. (a) u is harmonic; $v = 4x^3y - 4xy^3$.
   (b) u is harmonic; $v = [\sin(2xy)][e^{x^2-y^2}]$.
   (c) u is not harmonic.
   (d) u is harmonic; $v = [\cos(x^2 - y^2)][\sinh(2xy)]$.
   (e) u is not harmonic.

3. Take $f(z) = z$ on $0 < |z| < R$ and apply the conclusion of Exercise 2.

5. (a) A function u of θ alone is harmonic iff $u = A + B\theta$.
   (b) A function u of r alone is harmonic iff $u = C + D \log r, 0 < r < \infty$.

7. $u(x, y)$ has a maximum at $(0, 0)$, contradicting the maximum-modulus theorem.

9. If $z = re^{i\theta}$, then $\dfrac{1 - r^2}{1 - 2r \cos \theta + r^2} = \text{Re}\left(\dfrac{1 + z}{1 - z}\right)$.

11. Compute Δu and set it equal to 0.

13. Given $\delta > 0$, choose $r > 0$ so that $|u(z)| < \delta$ if $|z| < r$ and $\text{Im } z \geq 0$; this is possible by continuity. Let $V_\varepsilon(x, y) = u(x, y) - \varepsilon \log((x^2 + y^2)/r^2)$; consider V_ε on the contour shown. For large R, V_ε is negative on the large semicircle; V_ε is also negative on the segments $(-R, -r)$ and (r, R). On the small semicircle, V_ε equals u and so is less than δ. Hence, $V_\varepsilon < \delta$ on D by the maximum principle. Let $R \to \infty$ and then $\varepsilon \downarrow 0$. This gives $u \leq \delta$ on $\{z: \text{Im } z > 0, |z| < r\}$. Let $r \downarrow 0$ and $\delta \downarrow 0$. Then repeat the argument with $-u$ in place of u.

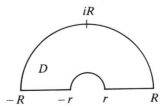

15. Let f be analytic on $|z - z_0| < \delta$ with $u = \text{Re } f$. Then $g(z) = \overline{f(\bar{z})}$ is also analytic on

$|z - z_0| < \delta$ (why?) and $f(x) = g(x)$, $x_0 - \delta < x < x_0 + \delta$. Hence, $g = f$ and so Re $g =$ Re $f = u$.

17. Follow the numerous hints.

Section 4.2

1. $\Gamma_0 = \{z: |z - i\alpha_0| = \sqrt{R^2 + \alpha_0^2}\}$. Thus, if $\sim$ is any of $<$, $=$, $>$, then $|z - i\alpha_0|^2 \sim R^2 + \alpha_0^2$ iff $|z|^2 - 2$ Re$(i\alpha_0 z) + \alpha_0^2 \sim R^2 + \alpha_0^2$ iff $|z|^2 - 2y\alpha_0 - R^2 \sim 0$.

3. Let h be another analytic function outside Γ with $h(z) = a + (c/2\pi i z) + \cdots$, Im $H = 0$ on Γ, and $H' = h$. Then $g - h$ is analytic outside Γ; near ∞, $g(z) - h(z) = a_2/z^2 + \cdots$, and $g - h$ has a single-valued indefinite integral J. Therefore, Im $J = 0$ on Γ and so J is a real constant. This gives $0 = J' = g - h$.

5. The flow has a source at -1 and a sink at λ, $|\lambda| = 1$, $\lambda \neq -1$. Thus,

$$F(z) = \frac{Q}{\pi} \log\left(\frac{1 + z}{1 - \bar{\lambda}z}\right).$$

7. Let the sources be at $\pm\pi + ip$; then $F(z) = \frac{Q}{\pi}\left[\text{Log}\left\{\sin^2\left(\frac{z}{2}\right) - \sin^2\left(\frac{p}{2}\right)\right\}\right]$.

9. Let the source be at x_0, $x_0 \in (-\pi, \pi)$; then

$$F(z) = \frac{Q}{\pi} \text{Log}\left\{\sin\left(\frac{z}{2}\right) - \sin\left(\frac{x_0}{2}\right)\right\}.$$

11. $F(z) = \frac{Q_1}{\pi} \text{Log}(z - x_1) - \frac{Q_2}{\pi} \text{Log}(z - x_2)$.

13. potential $= \phi(z) = c_0 \int_{-1}^{1} \log\left|\frac{1}{z - t}\right| dt$

$$= (2 + y)c_0 - c_0 \log|1 - z^2| - c_0 \text{ Re}\left[z \text{ Log}\left(\frac{1 + z}{1 - z}\right)\right].$$

15. $\phi(z) = (a^2 - b^2)^{-1}[az - b\sqrt{z^2 - (a^2 - b^2)}]$ maps the region in the upper half-plane and outside the ellipse onto the upper half-plane.

17. Use the conclusion of Exercise 16 for each term. Write

$$g(z) = \frac{K_2}{2\pi i} \frac{\bar{z}_1 - \bar{z}_2}{(\bar{z} - \bar{z}_1)(\bar{z} - \bar{z}_2)} + \frac{K_1 - K_2}{2\pi i} \frac{1}{\bar{z} - \bar{z}_1}.$$

For $|z|$ large, the first term is very small in comparison to the second and so the direction of the flow given by g is determined by the sign of $K_1 - K_2$: clockwise if $K_1 > K_2$, counterclockwise if $K_1 < K_2$.

19. Follow the hints; the desired flow is $F(z) = A + iB \arcsin(z/\sigma)$; A and B are chosen appropriately.

Section 4.3

1. Let the arc for temperature T_1 be $\{e^{i\theta}: 0 \leqslant \theta < \theta_0\}$. Then

$$T(z) = T_2 + \frac{T_1 - T_2}{\pi} \text{Arg}\left\{i\frac{1 + z}{1 - z} - i\frac{1 + e^{i\theta_0}}{1 - e^{i\theta_0}}\right\}.$$

3. (a) Choose M so that $|w(t)| < \varepsilon/4$ if $|t| > M$. Then

$$\frac{1}{\pi} \int_{-\infty}^{-M} \frac{\tau|w(t)|}{(\sigma - t)^2 + \tau^2}\, dt + \frac{1}{\pi} \int_{M}^{\infty} \frac{\tau|w(t)|}{(\sigma - t)^2 + \tau^2}\, dt < \frac{\varepsilon}{4}\frac{1}{\pi}\int_{-\infty}^{\infty}\frac{\tau}{(\sigma - t)^2 + \tau^2}\, dt = \frac{\varepsilon}{4}.$$

Moreover,

$$\frac{1}{\pi} \int_{-M}^{M} \frac{\tau|w(t)|}{(\sigma - t)^2 + \tau^2}\, dt \leqslant \left(\frac{1}{\pi}\int_{-M}^{M}|w(t)|\, dt\right)\left(\max_{|t| \leqslant M}\frac{\tau}{(\sigma - t)^2 + \tau^2}\right)$$

$$\leqslant \left(\frac{1}{\pi}\int_{-\infty}^{\infty}|w(t)|\, dt\right)\frac{|\zeta|}{(|\zeta| - M)^2}$$

$$< \frac{\varepsilon}{2} \text{ if } |\zeta| \text{ is very large.}$$

(b) The difference, $h = W - W_1$, is harmonic on $\{x + iy: y > 0\}$, is 0 on $\{y = 0\}$, and is bounded since both W and W_1 go to 0 at ∞; now apply Exercise 13, Section 1.

(c) $$\int_{-\infty}^{\infty}|W(\sigma + i\tau)|\, d\sigma \leqslant \int_{-\infty}^{\infty}\left(\frac{1}{\pi}\int_{-\infty}^{\infty}|w(t)|\frac{\tau}{(\sigma - t)^2 + \tau^2}\, dt\right)d\sigma$$

$$= \int_{-\infty}^{\infty}|w(t)|\left(\frac{1}{\pi}\int_{-\infty}^{\infty}\frac{\tau}{(\sigma - t)^2 + \tau^2}\, d\sigma\right)dt$$

$$= \int_{-\infty}^{\infty}|w(t)|\, dt.$$

5. (i) u is continuous at $e^{i\theta_0}$.
   (ii) Apply (ii) from Exercise 4.
   (iii) $|u(e^{it}) - u(e^{i\theta_0})| < \varepsilon/2$ since $|t - \theta_0| < \delta/4$ and (ii) from Exercise 4 holds.
   (iv) The integral of a function is less than the maximum of the function times the length of the interval.
   (v) Apply (iii) of Exercise 4 and (iv).
   (vi) Combine the two estimates on I_1 and I_2.

7. Let $u = u_1 + u_2$, where

$$u_1(x, y) = \int_{0}^{\infty} f_1(t)\left\{\frac{y}{(x - t)^2 + y^2} - \frac{y}{(x + t)^2 + y^2}\right\} dt$$

and

$$u_2(x, y) = \int_{0}^{\infty} f_2(t)\left\{\frac{x}{(y - t)^2 + x^2} - \frac{x}{(y + t)^2 + x^2}\right\} dt.$$

9. $V(z) = \dfrac{1}{2\pi} \displaystyle\int_{0}^{2\pi} f(e^{-i\theta})\dfrac{|z|^2 - 1}{|ze^{i\theta} - 1|^2}\, d\theta$, $|z| > 1$.

11. $u(re^{i\theta}) = \operatorname{Re}((1 + z)/(1 - z))$, $z = re^{i\theta}$; if $\theta \neq 0$, then $\cos\theta \neq 1$ and so $\lim_{r \uparrow 1} u(re^{i\theta}) = 0$. Follow the hint for the "hot spot."

13. $T(x + iy) = \dfrac{1}{\pi}\displaystyle\sum_{j=1}^{N} A_j\dfrac{y}{(x - x_j)^2 + y^2}$, where $A_j = \begin{cases} 1 \text{ if } x_j \text{ is "hot"} \\ -1 \text{ if } x_j \text{ is "cold"} \end{cases}$

15. Follow the hint.

17. $\displaystyle\int_{0}^{2\pi} e^{ik\theta}\, d\theta = \frac{1}{ik}e^{ik\theta}\Big|_{0}^{2\pi} = \frac{1}{ik}[e^{ik2\pi} - e^0] = 0 \text{ if } k \neq 0.$

19. $P_r(\theta) = \text{Re}\left[\dfrac{1 + re^{i\theta}}{1 - re^{i\theta}}\right] = \text{Re}[(1 + re^{i\theta})(1 + re^{i\theta} + r^2 e^{2i\theta} + \cdots)]$

$= \text{Re}(1 + 2re^{i\theta} + 2r^2 e^{2i\theta} + \cdots)$

$= 1 + r(e^{i\theta} + e^{-i\theta}) + r^2(e^{2i\theta} + e^{-2i\theta}) + \cdots$

$= \displaystyle\sum_{-\infty}^{\infty} r^{|n|} e^{in\theta}.$

21. $\displaystyle\int_0^{2\pi} U(re^{i\theta})\overline{V(re^{i\theta})}\, d\theta = \int_0^{2\pi}\left(\sum_{-\infty}^{\infty} r^{|n|}\hat{u}(n)e^{in\theta}\right)\left(\sum_{-\infty}^{\infty} r^{|m|}\overline{\hat{v}(m)}\, e^{-im\theta}\right) d\theta$

$= \displaystyle\sum_{-\infty}^{\infty}\sum_{-\infty}^{\infty} r^{|n|} r^{|m|}\hat{u}(n)\overline{\hat{v}(m)}\int_0^{2\pi} e^{i(n-m)\theta}\, d\theta$

$= 2\pi \displaystyle\sum_{-\infty}^{\infty} r^{2|n|}\hat{u}(n)\overline{\hat{v}(n)}$ by 17.

23. Do what is suggested.

25. Use the reflection principle to show that $\overline{f(\bar{z})} = f(z)$ and then that $\overline{f(-\bar{z})} = f(z)$. Hence, $f(-w) = f(w)$, as well.

27. Let $z = \phi(w) = (iw + 1)/(iw - 1)$; ϕ maps the upper half-plane onto the disc $\{|z| < 1\}$. Set $R = \phi \circ M$; then R is rational and $|R(e^{i\theta})| = 1$ for all θ. Let $a_1, \ldots,$ a_p be the zeros of R in $\{|z| < 1\}$ and $b_1, \ldots, b_q$ be the poles of R in $\{|z| < 1\}$. Show that $R(z) = \lambda\Pi_1^p((z - a_j)/(1 - \bar{a}_j z))\Pi_1^q((1 - \bar{b}_k z)/(z - b_k))$, $|\lambda| = 1$. Then recover M: $M = \phi^{-1} \circ R$.

29. One method is outlined; here is another. Let $\phi(w) = (iw + 1)/(iw - 1)$. Then $g(w) = \phi^{-1}(f(\phi(w)))$ is analytic on the upper half-plane, real on the real axis (except for those points x where $f(\phi(x)) = 1$). By the reflection principle, g extends to the lower half-plane. Hence, f extends analytically across the unit circle except possibly at a finite number of points where f might have a pole. But f is continuous at all points of the unit circle. Hence, f extends analytically over the unit circle. Now apply the argument used in the solution to Exercise 27.

Section 4.4

1. Note that $\log[(x - t)^2 + y^2]$ is a harmonic function of (x, y) for each t so that

$$\Delta u = \frac{1}{2\pi}\int_{-\infty}^{\infty} g(t)\, \Delta \log[(x - t)^2 + y^2]\, dt = 0.$$

Furthermore,

$$\frac{\partial u}{\partial y} = \frac{1}{\pi}\int_{-\infty}^{\infty} g(t)\, \frac{y}{(x - t)^2 + y^2}\, dt,$$

the Poisson integral of g and so

$$\lim_{y \downarrow 0} \frac{\partial u}{\partial y}(x, y) = g(x).$$

3. As in Exercise 1; $\Delta u = 0$ and $\dfrac{\partial u}{\partial y} = $ Poisson integral of $g - \dfrac{2Ay}{x^2 + y^2}$

Finally,

$$|u(x, y)| = \left| \frac{1}{2\pi} \int_{-M}^{M} g(t)[\log\{(x - t)^2 + y^2\} - \log(x^2 + y^2)] \, dt \right|$$

$$\leqslant \left[\frac{1}{2\pi} \int_{-M}^{M} |g(t)| \, dt \right] \max_{|t| \leqslant M} \left| \log \frac{(x - t)^2 + y^2}{x^2 + y^2} \right|$$

$$\leqslant C \max_{|t| \leqslant M} \left| \log\left(1 - \frac{2xt - t^2}{x^2 + y^2} \right) \right|; \text{ but } |\log(1 - \delta)| \leqslant 2\delta \text{ for small } \delta.$$

$$\leqslant C \frac{2|z|M + M^2}{|z|^2} \to 0 \text{ as } |z| \to \infty.$$

5. $u(x, y) = \dfrac{1}{\pi} \displaystyle\int_0^\infty f(t) \left[\dfrac{x}{(y - t)^2 + x^2} + \dfrac{x}{(y + t)^2 + x^2} \right] dt.$

7. $T(x, y) = \dfrac{T_1 + T_2}{2} + \dfrac{T_2 - T_1}{\pi} \text{Re}\left[\arcsin\left(\dfrac{z^2 - \frac{1}{2}\sigma^2}{\frac{1}{2}\sigma^2} \right) \right].$

9. Let the circles be centered on the real axis and internally tangent at $z_0 = 1$:

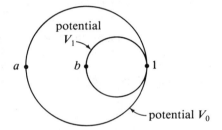

Then the potential is

$$\phi(z) = \frac{(1 - a)V_0 - (1 - b)V_1}{b - a} + \frac{(1 - a)(1 - b)}{b - a}(V_0 - V_1)\text{Re}\left(\frac{1}{z - 1} \right).$$

11. $T(z) = T_1 + \dfrac{2}{\pi}(T_2 - T_1)\text{Arg}\left(\dfrac{1 + z}{1 - z} \right).$

13. Write $z = re^{it}$ and $a = |a|e^{i\theta}$; then

$$2g(re^{it}) = \log(1 - 2r \, |a|\cos(\theta - t) + |a|^2 \, r^2) - \log(r^2 - 2|a|r \, \cos(\theta - t) + |a|^2).$$

Now compute.

15. $v(z) = \dfrac{V_0}{\pi} [\text{Arg}(\sqrt{1 + e^{-2z}} - 1) - \text{Arg}(\sqrt{1 + e^{-2z}} + 1)].$

Section 4.5

1. $G - G_1$ has no singularity, is harmonic and bounded on D, and vanishes on the boundary of D.

3. $G(z; p) = \log|(z - \bar{p})/(z - p)|.$

5. $G(z; p) = \log\left| \dfrac{z^2 - \bar{p}^2}{z^2 - p^2} \right|$

7. $G(z; p) = \log \left| \dfrac{e^z - e^{\bar{p}}}{e^z - e^p} \right|$

9. $G(z; p) = \log \left| \dfrac{\sqrt{z} + \sqrt{\bar{p}}}{\sqrt{z} - \sqrt{p}} \right|$

11. The response at q to an impulse at p equals the response at p to an impulse at q.

13. (a) 1   (b) r_0 (c) $\dfrac{b - a}{4}$ (d) $\dfrac{a + b}{2}$

15. Let h be continuous on the unit circle; then $H(re^{i\theta}) = 1/2\pi \int_0^{2\pi} h(e^{it})P_r(\theta - t) \, dt$ is harmonic in $|z| < 1$ and by (10), Section 3, $\lim_{r \uparrow 1} H(re^{i\theta}) = h(e^{i\theta})$, all θ. Take $\theta = 0$.

17. Follow the hint.

Chapter 5

Section 5.1

1. Note $\hat{u}(0) = 0$; $\hat{u}(x) = \dfrac{2i}{x}\left[b \cos bx - \dfrac{\sin xb}{x} \right]$, $x \neq 0$.

3. $\hat{u}(x) = \dfrac{4}{x^2}\left[1 - \dfrac{\sin x}{x} \right]$.

5. $\hat{u}(x) = \dfrac{\pi}{2} \exp[2(2ix - |x|)]$.

7. $\hat{u}(x) = 2\pi\left(\dfrac{\sin \sigma x}{x} \right)e^{-|x|}$.

9. $\hat{u}(x) = \sqrt{\pi}\,e^{-x^2/4} - 2\,\dfrac{\sin \sigma x}{x}\,e^{-\sigma^2}$

11. $\hat{u}(x) = \dfrac{\lambda}{v}\left[\dfrac{2 \sin(vx/2)}{x} \right]^2$.

13. (ii) Make the change of variables $a + t = s$.
    (iii) Make the change of variables $s = tb$.
    (iv) Use the conclusion of Exercise 16, Section 2, that $u(x) \to 0$ as $|x| \to \infty$. Then integrate by parts.

15. Let γ be a triangle; then

$$\int_\gamma \hat{u}(z) \, dz = \int_{-A}^{A} u(t)\left[\int_\gamma e^{-izt} \, dz \right] dt = \int_{-A}^{A} u(t) \cdot 0 \, dt = 0.$$

For

$$z = x + iy, |\hat{u}(z)| = \left| \int_{-A}^{A} u(t)\,e^{-it(x + iy)} \, dt \right|$$

$$\leq \int_{-A}^{A} |u(t)|\, e^{yt} \, dt \leq 2AM \max_{|t| \leq A} e^{yt}$$

$$= 2AM\, e^{A|y|}.$$

17. Just compute.

19.
$$\int_{-\infty}^{\infty} |u(x, y)| \, dx \leqslant \frac{1}{\pi} \int_{-\infty}^{\infty} \int_{-\infty}^{\infty} |f(t)| \, \frac{y}{(x - t)^2 + y^2} \, dt \, dx$$

$$= \int_{-\infty}^{\infty} |f(t)| \left\{ \frac{1}{\pi} \int_{-\infty}^{\infty} \frac{y}{(x - t)^2 + y^2} \, dx \right\} dt$$

$$= \int_{-\infty}^{\infty} |f(t)| \, dt = M.$$

21. Follow the hints.

23. $u(x, t) = \dfrac{1}{2} \displaystyle\int_{-t}^{t} g(x - \zeta) \, d\zeta.$

25. $u(x, y) = \dfrac{1}{2\pi} \displaystyle\int_{-\infty}^{\infty} g(t)\log((x - t)^2 + y^2) \, dt.$

27. Following the hints, the equation simplifies to
$$c^2 \frac{\partial^2 u}{\partial x^2} = \frac{\partial^2 u}{\partial t^2} + 2b \frac{\partial u}{\partial t} + au.$$

Let $U(t) = \int_{-\infty}^{\infty} u(x, t) \, e^{-ixs} \, dx$, s fixed. Then $u'' + 2bu' + (a + c^2s^2)u = 0$, so that $U(t) = A \, \exp(\gamma_1 t) + B \, \exp(\gamma_2 t)$, where $\gamma_1 = -b + \sqrt{b^2 - a - c^2s^2}$, $\gamma_2 = -b - \sqrt{b^2 - a - c^2s^2}$ are the two roots of $r^2 + 2rb + (a + c^2s^2) = 0$ and A and B are found from initial conditions to be $A = (\gamma_2 \hat{f}(s) - \hat{g}(s))/(\gamma_2 - \gamma_1)$, $B = (\hat{g}(s) - \gamma_1 \hat{f}(s))/(\gamma_2 - \gamma_1)$.
(ii) In the case that $RC = SL$, we find that $b^2 = a$; then $U = \hat{u}$ simplifies to
$$\hat{u}(s) = (\gamma_2 - \gamma_1)^{-1} \, e^{-bt}[(\gamma_2 \hat{f}(s) - \hat{g}(s)) \, e^{itcs} + (\hat{g}(s) - \gamma_1 \hat{f}(s)) \, e^{-itcs}]$$

$$= (\gamma_2 - \gamma_1)^{-1} \, e^{-bt}\{[(\gamma_2 f - g) * \delta_{-tc}]\hat{}(s) + [(g - \gamma_1 f) * \delta_{tc}]\hat{}(s)\}$$

and consequently

$$u(x; t) = \frac{e^{-bt}}{\gamma_2 - \gamma_1} \, [\gamma_2 \, f(x + tc) - \gamma_1 f(x - tc) + g(x - tc) - g(x + tc)].$$

Section 5.2

1. Let $u(t) = \begin{cases} 1, & |t| < \alpha \\ 0, & |t| > \alpha \end{cases}; \, v(t) = \begin{cases} 1, & |t| < \beta \\ 0, & |t| > \beta. \end{cases}$

Then by (3),

$$4\pi \min(\alpha, \beta) = 2\pi \int_{-\infty}^{\infty} u\bar{v} = \int_{-\infty}^{\infty} \hat{u}\bar{\hat{v}} = 4 \int_{-\infty}^{\infty} \frac{\sin \alpha x \sin \beta x}{x^2} \, dx.$$

3. $\displaystyle\int_{-\infty}^{\infty} \left[\frac{1 - \cos \sigma x}{x} \right]^2 dx = \pi\sigma.$

5. Take $u(t) = \dfrac{1}{1 + t^2}; \, \hat{u}(x) = \pi \, e^{-|x|}.$

Hence, $2\pi \displaystyle\int_{-\infty}^{\infty} \frac{dt}{(1 + t^2)^2} = \int_{-\infty}^{\infty} \pi^2 \, e^{-2|x|} \, dx = \pi^2.$
The desired value of the integral is $\pi/2$.

7. $u(t) = \dfrac{1}{2\pi} [\arctan(4 + t) + \arctan(4 - t)]$.

9. $\dfrac{\pi}{12} (3\sigma - 3\sigma^2 + \sigma^3)$

11. By Exercise 15, Section 1, $\hat{u}(z)$ is entire. Hence, $\hat{u} \equiv 0$ so $u \equiv 0$ by the inversion formula.

13. In (2) replace u by $u + v$; expand and use the known equality for u and v.

15. Follow the given steps.

17. Let $J = \lim\limits_{M \to \infty} \displaystyle\int_{-M}^{M} |u(t)| \, dt$.

Then for $A, B \geqslant M_0$, $\varepsilon > J - \displaystyle\int_{-M_0}^{M_0} |u(t)| \, dt \geqslant \int_{-\infty}^{-B} + \int_{A}^{\infty}$.

19. Since $|u(t)| \leqslant |v(t)|$ for $-\infty < t < \infty$,

$$\int_{-M}^{M} |u(t)| \, dt \leqslant \int_{-M}^{M} |v(t)| \, dt \leqslant C \text{ for all } M.$$

21. Let γ be a triangle in the strip $\tau_1 < \operatorname{Im} z < \tau_2$. Then

$$\int_{\gamma} \hat{u}(z) \, dz = \int_{\gamma} \left[\int_{-\infty}^{\infty} u(t) \, e^{-izt} \, dt \right] dz$$

$$= \int_{-\infty}^{\infty} u(t) \left[\int_{\gamma} e^{-izt} \, dz \right] dt = \int_{-\infty}^{\infty} u(t) \cdot 0 \, dt = 0.$$

23. Use the Riemann–Lebesgue lemma (Exercise 15) with

$$w(t) = \begin{cases} 0, & -\infty < t < \delta \\ \dfrac{u(x + t)}{t}, & \delta \leqslant t < \infty. \end{cases}$$

This gives

$$0 = \lim_{M \to \infty} \operatorname{Re} \hat{w}(M) = \lim_{M \to \infty} \int_{\delta}^{\infty} \frac{u(x + t)}{t} \sin Mt \, dt.$$

Similarly,

$$\lim_{M \to \infty} \int_{-\infty}^{-\delta} \frac{u(x + t)}{t} \sin Mt \, dt = 0.$$

25. Combine 23 and 24:

$$\int_{-\infty}^{\infty} u(x + t) \frac{\sin Mt}{t} \, dt = \int_{-\infty}^{-\delta} \frac{u(x + t)}{t} \sin Mt \, dt + \int_{\delta}^{\infty} \frac{u(x + t)}{t} \sin Mt$$

$$+ \int_{0}^{\delta} \frac{u(x + t) - u^+(x)}{t} \sin Mt$$

$$+ \int_{-\delta}^{0} \frac{u(x + t) - u^-(x)}{t} \sin Mt$$

$$+ [u^+(x) + u^-(x)] \int_{0}^{\delta} \frac{\sin Mt}{t} \, dt.$$

The first four terms are small if δ is small and M is large. The fifth integral converges to

$$\int_0^\infty \frac{\sin x}{x}\, dx = \pi/2$$

as $M \to \infty$, δ fixed.

Section 5.3

1. $(\mathscr{L}u)(s) = s/(A^2 + s^2)$.

3. $(\mathscr{L}u)(s) = A/[(B + s)^2 + A^2]$.

5. $(\mathscr{L}u)(s) = \dfrac{1}{s}\,[e^{-as} - e^{-bs}]$.

7. $(\mathscr{L}u)(s) = \dfrac{1}{s^2}\,[1 - s\sigma\, e^{-s\sigma} - e^{-s\sigma}]$.

9. $(\mathscr{L}u)(s) = \dfrac{1}{s^2}$.

11. $\dfrac{\sin t}{t} = \displaystyle\sum_{k=0}^\infty (-1)^k \frac{t^{2k}}{(2k+1)!}$; next, $\displaystyle\int_0^\infty t^{2k}\, e^{-st}\, dt = \frac{(2k)!}{s^{2k+1}}$.

Hence,

$$\mathscr{L}\!\left(\frac{\sin t}{t}\right)\!(s) = \sum_{k=0}^\infty (-1)^k \frac{1}{(2k+1)!}\left(\int_0^\infty t^{2k}\, e^{-st}\, dt\right) = \sum_0^\infty \frac{(-1)^k}{2k+1}\, s^{-(2k+1)}.$$

Furthermore,

$$\frac{d}{ds}\left[\mathscr{L}\!\left(\frac{\sin t}{t}\right)\!(s)\right] = \sum_0^\infty (-1)^k s^{-2k-2} = \frac{-1}{s^2+1}, \quad 1 < s < \infty$$

$$= \frac{d}{ds}\, \arctan(1/s), \quad 1 < s < \infty.$$

13. $u(t) = t(\sin At)/2A$.

15. $u(t) = \frac{1}{6}t^3\, e^t$.

17. $u(t) = i\, e^{-it}$.

19. $\mathscr{L}u$ is entire by Morera's Theorem. Furthermore,

$$(\mathscr{L}u)(s) = \int_0^A u(t)\, e^{-st}\, dt = \int_0^A u(t) \sum_0^\infty (-1)^k \frac{s^k t^k}{k!}\, dt$$

$$= \sum_0^\infty (-1)^k \frac{s^k}{k!} \int_0^A u(t)t^k\, dt.$$

21. $e^{(z-1)\log t}$ is an entire function of z; $\Gamma(z)$ is defined for Re $z > 0$ and analytic there by Morera's Theorem.

23. Integrate by parts.

25. For $n = 1$, $\Gamma(1) = \displaystyle\int_0^\infty e^{-t}\, dt = 1$. Use induction and Exercise 23 for $n \geqslant 2$.

27. Follow the suggestion.

29. Again, follow the suggestions.

Section 5.4

1. $u(t) = \frac{6}{5} e^{-t} + 2 e^{t}\{-\frac{3}{5} \cos t + \frac{17}{10} \sin t\}$.

3. $u(t) = e^{-t^2/2}$.

5. $u(t) = \begin{cases} \sin t, & 0 \le t \le \pi \\ \left(1 + \dfrac{t - \pi}{2}\right)\sin t, & \pi < t < \infty. \end{cases}$

7. $u(x; t) = \begin{cases} 0, & \text{if } x > t \\ f(t - x), & \text{if } x \le t. \end{cases}$

9. Case 1: ω is not an integer multiple of π. Then

$$u(x, t) = \frac{\sin \omega x \sin \omega t}{\sin \omega} + 2\omega \sum_{n=1}^{\infty} \frac{(-1)^{n+1}}{\omega^2 - n^2\pi^2} \sin n\pi x \sin n\pi t.$$

Case 2: $\omega = \pi$ (representative of $\omega = m\pi$).

$$u(x, t) = -2t \sin \pi x \cos \pi t - x \cos \pi x \sin \pi t - \frac{1}{2\pi} \sin \pi x \sin \pi t$$

$$+ \frac{2}{\pi} \sum_{n=2}^{\infty} \frac{(-1)^n \sin \pi n x \sin \pi n t}{n^2 - 1}.$$

11. $u(x, t) = \dfrac{\phi_0}{\sqrt{\pi k}} \displaystyle\int_0^t \frac{e^{-x^2/4ks}}{\sqrt{s}} \, ds$

Section 5.5

1. $Z(\{a_j\}) = \dfrac{z^2}{z^2 - 1}, \, |z| > 1$.

3. $Z(\{a_j\}) = e^{2/z}, \, z \ne 0$.

5. $Z(\{a_j\}) = z \operatorname{Log}\left(1 - \dfrac{1}{z}\right), \, |z| > 1$.

7. $Z(\{a_j\}) = \dfrac{z^2}{z^2 - 16} + \dfrac{3z}{9z^2 - 1}, \, |z| > 4$.

9. $a_j = \begin{cases} 0, & j \text{ even} \\ \dfrac{2^j}{j!}, & j \text{ odd}. \end{cases}$

11. $a_j = \dfrac{1}{(j + 1)!}$.

13. $a_0 = a_1 = 0; \, a_j = j - 1, j = 2, 3, 4, \ldots$.

15. $\{a_j\} * \{b_j\} = \{0\}$ so $0 = Z(\{0\}) = Z(\{a_j\} * \{b_j\}) = Z(\{a_j\})Z(\{b_j\})$. Since $Z(\{a_j\})$ and $Z(\{b_j\})$ are both analytic, one or the other must vanish identically; for that one, all its coefficients must be zero.

17. $y_0 = 1; \, y_{r+1} = -\displaystyle\sum_{k=0}^{r} (x_k + x_{k+1} + x_{k+2}), \, r = 0, 1, 2, \ldots$.

19. $y_0 = 0$; $y_1 = 1$; $y_2 = x_0 + 1$; $y_3 = -1 + x_1$; generally,

$$y_r = \begin{cases} (-1)^{m-1}(x_0 + 1) + \displaystyle\sum_{k=0}^{m-2} (-1)^k x_{r-2k-2}, & r = 2m, \ m = 2, 3, \ldots \\[2mm] 1 + \displaystyle\sum_{k=0}^{m} (-1)^k x_{r-2k-2}, & r = 2m + 1. \end{cases}$$

21. $y_0 = 1$; $y_{r+1} = -\displaystyle\sum_{k=0}^{r} \dfrac{x_k - 2x_{k+1} + x_{k+2}}{2^{r-k+1}}$, $r = 0, 1, 2, \ldots$.

Section 5.5.1.

1. $0 = R(w) = p\left(\dfrac{w+1}{w-1}\right)$ if and only if $\left|\dfrac{w+1}{w-1}\right| < 1$ if and only if Re $w < 0$.

3. (a) $q(w) = (A + B)w + (A - B)$.
   (b) $q(w) = 16[w^4 + 4w^3 + 9w^2 + 8w + 5]$.
   (c) $q(w) = 2w + 2$.
   (d) $q(w) = -w^3 + 3w^2 - 3w + 9$.

5. (a) p has a root at -1.
   (b) all roots within $\{z: |z| < 1\}$.
   (c) all roots within $\{z: |z| < 1\}$.
   (d) one root outside $\{z: |z| < 1\}$.
   (e) some root outside $\{z: |z| < 1\}$.

Index